MICROWAVE AQUAMETRY

MICROWAVE AQUAMETRY
Electromagnetic Wave Interaction
with Water-Containing Materials

Edited by

Andrzej Kraszewski

A Volume in the TAB-IEEE Press Book Series—Emerging Technologies

Willis J. Tompkins, *Editor in Chief, TAB-IEEE Press Book Series*
Ralph K. Cavin, *Editor, Emerging Technologies Series*

The Institute of Electrical and Electronics Engineers, Inc., New York

©1996 by the Institute of Electrical and Electronics Engineers, Inc.
345 East 47th Street, New York, NY 10017-2394

Printed in the United States of America

10   9   8   7   6   5   4   3   2   1

ISBN 0-7803-1146-9
IEEE Order Number: PC5617

Library of Congress Cataloging-in-Publication Data
Microwave aquametry / edited by Andrzej Kraszewski.
    p.   cm. — (TAB-IEEE Press book series.   Emerging
technologies)
    Includes bibliographical references and index.
    ISBN 0-7803-1146-9
    1. Moisture meters, Microwave.   I. Kraszewski, Andrzej.
II. Series.
TA418.64.M53   1996
681'.2 — dc20   95-43232
                                                        CIP

Contents

Preface

This book is a record of the "Workshop on Electromagnetic Wave Interaction with Water and Moist Substances" that was held in conjunction with the IEEE MTT-S International Microwave Symposium in Atlanta, Georgia, on June 14, 1993. Nearly one hundred individuals around the world involved in research, development, and application of microwave techniques for moisture content determination in solid and liquid materials were invited to participate and present results of their work at the workshop. No particular subjects were suggested and no limitations were imposed.

The response exceeded the expectations of the organizer. Twenty-five contributions were submitted before the deadline, and their summaries were printed prior to the meeting in a 100-page booklet. Several authors who were not able to come to Atlanta and attend the meeting submitted full papers for publications. The meeting itself was attended by more than eighty people from all over the world, and, in spite of the very short time allocated for each presentation, was acclaimed as a success. It provided a forum for the attendees to meet, to exchange information, and to establish contacts and acquaintances for the future. The organizer considered several avenues for publishing the contributed papers. Finally, the IEEE Press agreed to publish the material in the form of a Proceedings of the Workshop, and the book was conceived.

This book contains a collection of 29 papers that were written by actual and potential attendees of the Workshop and an Introduction and a Bibliography that were prepared by the Editor. The multinational character of the event may be best indicated by the geographical distribution of papers: five from Canada, four from both Germany and the USA, three from Japan, two each from Australia, Finland, Italy, and the United Kingdom, and one each from France, Greece, Hungary, Malaysia, and Poland. Every manuscript was carefully reviewed by at least two reviewers, and was then revised by the authors for publication. The following highly-qualified individuals helped the editor in the review process, devoting their time and expertise to improving the quality of this volume:

Dr. James Baker-Jarvis	Dr. Michael Kent
Prof. Renato G. Bosisio	Dr. Ray J. King
Dr. Robert N. Clarke	Dr. Richard G. Leffler
Prof. Allen R. Edison	Dr. Devendra K. Misra
Prof. Kenneth R. Foster	Dr. Stuart O. Nelson
Mr. David B. Funk	Dr. Ari Sihvola
Dr. Richard G. Geyer	Prof. Stan Stuchly
Prof. Ronald H. Johnston	Dr. Claude M. Weil

The value of their help cannot be overestimated, and I am extremely thankful for this assistance.

Sixty-nine persons coauthored the papers published in this book. They include physicists, chemists, and agricultural, civil, electrical, electronics, and mechanical engineers. Their ages range from nestors, who have devoted more than forty years of their professional careers to microwave aquametry, to students completing their first projects on the subject. In spite of differences in nationality, age, experience, and professional background, they share one thing in common—an enthusiasm for developing new technologies and applying them in their respective countries. I would like to express my thanks to all of these individuals for their contribution to the Workshop and for their efforts in making the Editor's job easier.

The scope of the book covers all aspects of the main subject, recently known as microwave aquametry, the broad area of metrology that includes science and technology devoted to microwave techniques applied for moisture content determination in solids and liquids. This part of metrology should be clearly distinguished from microwave hygrometry, which is devoted to humidity measurements in gases (mainly in air). First, it covers the theoretical background related to the physical properties of water, both free and in various degrees of binding in moist substances, when irradiated with electromagnetic waves of very high frequency. It also discusses the possibility of predicting the properties of water-containing materials both from theoretical and experimental viewpoints. Second, it covers all aspects of technical application of microwave techniques for on-line and laboratory moisture content measurement, including new sensors, transducers and equipment, new methods and adaptation of known methods, calibration and validation, as well as technical details related to particular materials and peculiar installations. The list of materials reported in the volume to have been successfully measured by microwave methods is, in itself, impressive. It includes gravel and sand, corn, oats, wheat and barley grains, soybean seeds and peanuts, coal, processed foods such as cheese and butter, crude oil (in reservoirs and in pipelines), textiles, cardboard, veneer and fireproof materials, snow, wood flakes, palm oil fruits, and living fish. This list provides evidence of microwave aquametry as an accepted tool in the field of moisture content monitoring and control in modern factories and laboratories.

Inclusion of the opening chapter, entitled "Microwave Aquametry—Introduction to the Workshop" may require some explanation. No such paper was presented at the Workshop, but providing a tutorial chapter as an introduction to the Proceedings seemed advisable. This chapter is provided for readers who are familiar with problems related to moisture content measurement in various situations and by various methods, but who know little about electromagnetic radiation and the concept of using microwaves for moisture determination. Also, it can be of certain value for those readers who know microwave theory and technology well, but have never devoted much thought to their use for purposes other than radar and communication applications. Because the scope of interest to be covered is broad, the picture drawn is often simplified for sake of clarity and brevity, but it is accurate and complete enough to offer the reader the background needed to study this book. Some essential references, together with the bibliography included at the end of the book, should guide readers through further study of the subject.

The purpose of publishing this collection of papers on the application of microwave theory and techniques for measurement of moisture content, which is a very important parameter in most manufacturing processes, is, first, to disseminate the information on the state of the art for potential users and, second, to generate interest in the microwave community by showing how mature technologies can be applied for purposes other than defense and communication. This second reason may be important particularly in times when "beating swords into plowshares" becomes a fact of everyday life.

A. Kraszewski
Athens, Georgia

MICROWAVE AQUAMETRY

Section *I*

Introduction

1

Andrzej W. Kraszewski
USDA, ARS, Russell Research Center, Athens, GA

Microwave Aquametry: Introduction to the Workshop

Abstract. Microwave techniques, methods, and instrumentation can be utilized to determine moisture content in various materials on-line during manufacturing processes or in laboratories. General principles and definitions are presented and future prospects are discussed.

1.1. WATER IN NATURE

Water is one of the most common components of our biosphere, and of many materials and products manufactured by mankind. The water content, to a large extent, affects physical, chemical, mechanical, and thermal properties of many nonmetallic materials in nature. For this reason, almost all branches of industry and agriculture use processes that change the amount of water by drying or moistening the materials. Thus, a quantitative determination of moisture content in solid and liquid materials is of prime importance in industry as well as in scientific research laboratories.

As an example in agriculture, grain is often dried after being harvested, since long-term storage is safe (free from microbial degradation) only when cereal grain moisture content is lower than 13–14%. However, overdrying the grain can decrease its nutritional and reproductive values and contribute to increased breakage during handling. Thus, a precise control of the drying process is needed. Before wheat is milled, its moisture content has to be increased for a short time to 15–16% to increase the elasticity of the bran and to enhance the quality of the flour. Precise control and regulation of sprinklers that add water to the grain is therefore of great importance. Also, water is a main contaminant of many liquids such as crude oil, gasoline, and jet-engine fuel. Thus, routine monitoring

3

of moisture content is necessary in various stages of product pumping, processing, storage, and trade. Similar considerations apply to most raw materials and products manufactured in today's food, building materials, and pharmaceutical industries.

Controlling moisture content in industrial, large-scale, technological processes and maintaining it in a specified range, optimum for product quality, may be beneficial for at least two reasons: (1) reducing the cost of energy used for drying the product (overdrying is the most often-used form of safe drying); (2) improving product quality through appropriate drying. On the scale of many industrial operations, these benefits could produce significant savings.

1.2. MOISTURE CONTENT

When dealing with moist materials, two terms should be clearly distinguished: *water content*, which is the amount of water in a certain amount of material, and *moisture content* (sometimes simply "moisture"), which refers to the ratio of water to the total mass of wet material. Both terms are used in practice, but moisture content is the quantity most frequently applied in manufacturing, trading, and storing of goods.

The moisture content of material may be defined on a wet basis (w.b.) as a ratio of the mass of water, m_w, to the mass of the moist material, m_m,

$$\xi = \frac{m_w}{m_m} = \frac{m_w}{m_w + m_d}, \tag{1-1}$$

or on a dry basis (d.b.), as a ratio of the mass of water in the material to the mass of dry material, m_d:

$$\eta = \frac{m_w}{m_d} = \frac{m_m - m_d}{m_d}. \tag{1-2}$$

Quite often the quantities ξ and η are expressed in percentage. In various branches of industry other quantities are used for description of water content in a material, but they are usually synonymous with those just given. For example, the percent of dry mass of the material (substance) may be expressed as

$$DM = \frac{m_d}{m_m} \times 100 = (1 - \xi) \times 100 \quad [\%], \tag{1-3}$$

which is closely correlated with eqs. (1-1) and (1-2). In manufacturing and trading sheet materials, such as paper, cardboard, textiles, and wood veneer, the moisture content is expressed as the ratio of the weight of water per unit area to the weight of dry material per unit area:

$$MC_\square = \frac{m_{w\square}}{m_{d\square}} \times 100 \quad [\%], \tag{1-4}$$

where the symbol $\square$ denotes "per square unit area."

Equation (1-1), the definition most frequently used in practice, can be rewritten as follows when the concept of moisture content is related to a certain volume of material, v:

$$\xi = \frac{m_w/v}{m_w/v + m_d/v} = \frac{k}{\rho},\qquad(1\text{-}5)$$

where k is the water concentration (mass of water in the unit volume) and ρ is the density of the moist material. Other relationships following from eqs. (1-1), (1-2), and (1-5) are

$$\xi = \frac{\eta}{1+\eta},\quad \eta = \frac{\xi}{1-\xi},\quad k = \frac{m_w}{v} = \rho\xi,\quad \frac{m_d}{v} = \rho(1-\xi) = \frac{\rho}{1+\eta}.\qquad(1\text{-}6)$$

There are many parameters of materials that can be correlated with the water concentration, k, but it is obvious in (1-5) that fluctuations in the material density, ρ, have as much influence on moisture content as the variations in k. This observation is universal, because this disturbing effect of material density does not depend on the electrical method applied for moisture content determination. Thus, when k is determined from electrical measurement, determination of moisture content from eqs. (1-1) or (1-5) requires that ρ be known. This information can be obtained by

a. Keeping the mass of wet material in the measuring space constant during the calibration procedure and during the measurement

b. Performing separate density measurements, for example by weighing a sample of given volume, or by using a γ-ray density gauge

c. Using a *density-independent* function, which will be discussed later in this chapter

The density effect is the single most important cause of uncertainties in on-line moisture content measurements and results directly from the definition of moisture content itself. It may be noticed from eq. (1-1) that the concept of moisture content, while very useful in practice, is artificial, as such a quantity does not exist by itself in nature. For this reason, measurement and monitoring of moisture content in different situations is subject to various undesirable disturbances, which will be discussed in this chapter.

1.3. MEASUREMENT OF MOISTURE CONTENT

Performing fast and accurate moisture content measurements is of great importance in the manufacturing, processing, storing, and trading of most products and raw materials. Existing methods can be divided into standard laboratory methods, which are static and usually require much time to complete, and dynamic, on-line methods, which are generally less accurate but provide results in real time. Metrological efforts are directed at speeding up standard methods and enhancing the accuracy of on-line methods. Most standard methods of moisture content determination are *direct* methods, based on the definitions of eqs. (1-1), (1-2), or (1-3) and performed in the laboratory according to rules of analytical chemistry. For most materials, these methods are accurately described in formal documents constituting professional, national, or international standards. One method involves weighing a sample of moist material, removing the water by evaporation, and reweighing the remaining dry material; another (e.g., Karl Fischer method) uses extraction. The whole procedure is precisely described, giving time and temperature for drying, exact amount of chemicals to be used, etc. These methods provide an accuracy of a few tenths of 1% moisture content, but do not provide rapid results. Drying for up to three days is required in some cases.

For rapid moisture content determination and monitoring, *indirect* methods, calibrated against the standard methods, have been used for many years. An indirect method is based on finding a property of a material that is related to its moisture content and that can be easily measured by existing methods. The quality of the correlation between the property and moisture content and the convenience of the measurement is among the main characteristics of the method determining its applicability for a given material. Indirect methods of moisture content measurement can be based on various effects, namely,

- Mechanical (sound, hardness, etc.)
- Chemical (time of reaction, amount of gas created, etc.)
- Optical (infrared scattering or absorption)
- Electrical (conductivity, capacitance, etc.)

Among these methods, those based on strong correlations between moisture content and electrical properties play an increasingly important role. Historically, there has been a trend in the development of instrumentation and equipment for moisture content measurement toward higher and higher frequencies. Moisture meters based on dc conductance measurements were developed at the beginning of the century. As high-frequency measuring techniques developed, dc meters were superseded by ac meters measuring conductance of the wet materials and, later, by instruments measuring dielectric constants at radio frequencies. At that point, the science and technology of moisture content measurement was divided into *hygrometry*, which is devoted to the humidity determination in gases (mainly in air), and *aquametry*, which is related to moisture content measurements in solid and liquid materials. Finally, parallel to the development of microwave techniques and devices during and after World War II, microwave radiation was applied to moisture content measurement. During the following years, research intensified and established some benefits and advantages of microwave methods and techniques for moisture measurement in several materials. Eventually, the science and technology related to the subject acquired the name *microwave aquametry*.

General rules of indirect methods of moisture measurement can be explained using the block diagram in Figure 1.1. The first block, often called a *sensor*, represents this part of the instrument where an interaction between the process or the material to be measured and the measuring equipment takes place. The exact meaning of the word "sense," used as a verb, is "to detect, to be aware of." In a moisture meter, this part transforms moisture content of the material into a physical property of the material which is used for the moisture measurement.

The character of $\mathcal{E}$ depends on the adopted measuring method; e.g., it may be the complex material permittivity or the intensity of transmitted or scattered infrared radiation. This transformation characterizes the relation between physical or chemical properties of the material and its water content.

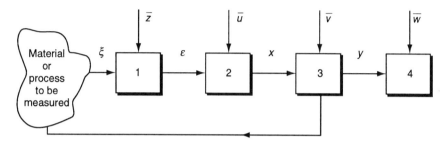

FIGURE 1.1. Block diagram of measurement process.

The second block is called a *transducer*;[1] it converts one varying physical quantity, such as material permittivity, into another, usually more convenient one, such as voltage. In many cases, a careful analysis of the measuring circuit may show more than one stage in the transducer—for example, the amplitude of the unbalance at the output of a microwave bridge, and an amplitude of dc or ac voltage at the output of a microwave detector, both being proportional to the measured quantity, the complex permittivity of the material.

Assuming that the quantity x in Figure 1.1 is a voltage, the third block, the *signal processing unit*, compares the input voltage with a standard voltage, providing the output signal y in analog or digital form, which can be displayed and/or fed back either for process control or to the transducer for improvement of overall stability or sensitivity of the measuring system. The function of the fourth block, the *data processing unit*, is to transmit data to devices where they are manipulated, processed, recorded, stored, and displayed.

During a real measuring process, every stage of the instrument is affected by external disturbances of various origins and different intensities. Disturbing factors for the sensor (z in Figure 1.1) can encompass changes in properties and composition of the material due to factors other than moisture content (e.g., its temperature, density, chemical composition, granularity). In the transducer, the disturbing factors (u) relate to all changes in measurement conditions (e.g., fluctuations of the operating frequency, changing mass of the sample, vibrations or changes in the sample position with respect to the unmatched radiating elements). Disturbances in the signal processing unit (v) contain all external effects on the measuring instrument (noise, fluctuations of the supply voltage, variations in ambient humidity and temperature, vibrations, dust). Essentially similar disturbing

---

1. There are differences in the literature concerning the definitions of "sensor" and "transducer." Sometimes, the sensor is called the first, the introductory, or the forward transducer, to distinguish it from an output or reverse transducer.

factors (w) influence the data processing unit. In the general case, there are several factors in each stage, and they can be represented as vectors, $\bar{z} = \bar{z}_1 + \bar{z}_2 + \cdots +\bar{z}_n$, etc. Sometimes the same is true for $\mathcal{E}$.

In real microwave moisture measuring instruments, blocks from Figure 1.1 are represented by microwave source and radiating elements, built of waveguide or microstrip lines, connected with electronic amplifiers, modulators, detectors, and power supplies, as well as computer boards and processing units. If the first two blocks are considered as one transducer, in reality, some means should be included for introducing the material into the measuring space and moving it out. Also, additional equipment should be provided for obtaining data on the magnitude of the external disturbing factors z and u and for stabilizing or compensating for these effects. In laboratory instruments, operations necessary for carrying out the measurements, or part of them (loading and unloading the samples, balancing a measuring bridge, etc.), are most often done by an operator, and the instruments provide discrete information. In automatic moisture meters, the operation is continuous and does not require an operator.

The metrological optimization of any moisture meter can be reduced to a requirement for the best separation of the desired signal from noise. Changes in the output signal y can be described by the expression

$$dy = \frac{\partial y}{\partial \xi} d\xi + \frac{\partial y}{\partial z} dz + \frac{\partial y}{\partial u} du + \frac{\partial y}{\partial v} dv + \frac{\partial y}{\partial w} dw, \tag{1-7}$$

where the partial derivatives are slopes of the disturbing parameters and the derivatives denote the change of the parameters. The goal of designing the moisture content meter is the transmission of the desired signal $(\partial y/\partial \xi)d\xi$, while disturbances and noise described by all other terms in eq. (1-7) are kept to a minimum. Desired limitation of the measurement uncertainties can be obtained when sensitivity of the meter to changes in the moisture content, $S_m = \partial y/\partial \xi$, is maximum and its sensitivity to the disturbances, $S_d = \partial y/\partial z + \partial y/\partial u + \partial y/\partial v + \partial y/\partial w$, is minimum. The uncertainty of measurement is defined as the range on either side of the mean value of a set of measurements between which a given fraction of observations would lie if a very large number of observations were made. In other words, it is the probability of a reading lying in a given range on either side of the mean value. The uncertainty analysis for moisture content meters is very similar to that of any other measuring instrument based on the indirect method of measurement.

One important procedure is converting the output of an electric-parameter-measuring circuit (capacitance, attenuation of electromagnetic wave, reflection of infrared radiation, etc.) into moisture content. In this procedure, called the *calibration* of the circuit, the response of a transducer to a sample containing a certain concentration of water is related to the exact value of water content, determined by a standard method for a given material. In the traditional sense, the term means the graduation of scales being spaced so as to correct for any irregularities in the instrument. Thus, the term is linked with correcting errors—that is, making the graduation intervals of unequal size. Often, when the instrument has a scale, the set of calibration measurements takes the form of a table of corrections to be applied to the instrument readings to allow for errors of one sort or another. In moisture meters, the standard method used for an instrument calibration has to be the one accepted by the profession and officially confirmed by a standard document on the national or international level. Usually, the standard method of moisture content determination is based on drying a certain amount of moist material for a specified time at a given temperature. Sometimes other requirements, such as the type of oven, drying dishes, accuracy of weighing, and temperature tolerances, are also specified. Standard procedures have to be followed exactly during the meter calibration process because it is the only way to have the meter approved for a given material.[2]

The calibration of a moisture content meter may be explained as follows. The relationship between the output quantity of block 1 or 2 in Figure 1.1 and moisture content of the material under test, determined by the standard method, is shown in Figure 1.2a. It can be, for example, the relationship between the dielectric constant and moisture content if it concerns the sensor calibration, or it can be unbalanced voltage at the output of a measuring bridge if it relates to the calibration of the first two stages. The relationship is determined experimentally by introducing samples of known moisture content into the sensor measuring space. For the whole instrument, the respective calibration curve is a function $y = f(\xi)$, which is developed experimentally by correlating the instrument response (y) in volts and the known moisture content of the material (ξ) expressed in percent. An example of such a relationship is shown in Figure 1.2b. In the calibration curves in Figure 1.2, all possible experimental errors and uncertainties are included. Thus, the relationship is known only with a certain degree of probability. To determine the unknown moisture content of the material with the calibrated meter, the inverse function, $\xi = f^{-1}(y)$, has to be found. This function is shown in Figure 1.3a, where from a given output voltage, y, the material moisture content, ξ, is determined. In most practical cases, such a relationship is linearized and y is presented in units of moisture content (percent), as shown in Figure 1.3b. Meter calibration may be much more complex than this simplified case, when disturbing factors play a more prominent role.

The figure of merit for the calibration procedure is the standard error of calibration (SEC), defined as

$$\text{SEC} = \sqrt{\frac{1}{n - p - 1} \sum_{i=1}^{n} \Delta m_i^2}, \qquad (1\text{-}8)$$

2. A typical academic question is: "How dry is the dry sample?" The proper answer should be: "It has been dried and handled according to the standard." Any changes in accepted procedure (e.g., increasing temperature and/or time of drying) may lead to removal of some additional volatile components, not necessarily water bound to the material matrix.

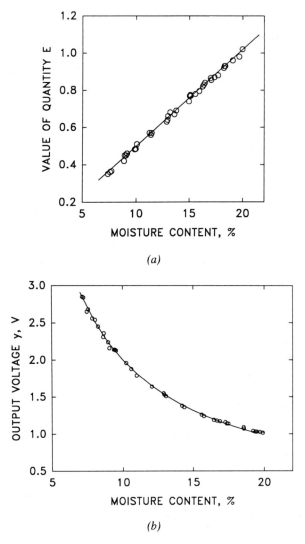

FIGURE 1.2. Examples of calibration procedure: (a) sensor output quantity $\mathcal{E}$ calibrated versus material moisture content obtained by reference method; (b) instrument output voltage y as a function of material moisture content ξ.

where n is the number of samples tested, p is the number of parameters in the equation, and Δm_i is the difference between the sample moisture content obtained by a standard method and moisture content calculated from the calibration equation for the ith sample. When a verification of the meter calibration is performed using a different set of data (obtained for samples of the same general population but not used for developing the calibration equation), the quality of the verification

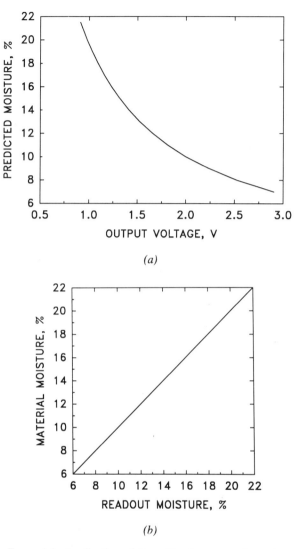

FIGURE 1.3. Application of the calibration curve shown in Figure 1.2b: (a) direct reading of the unknown value of the material moisture content from the output voltage of the instrument; (b) linearized meter output characteristic as a function of predicted material moisture content.

can be evaluated by the standard error of performance (SEP), which is a standard deviation of differences between measured and calculated values, expressed as

$$SEP = \sqrt{\frac{1}{n-1} \sum_{i=1}^{n} (\Delta m_i - \bar{m})^2}, \qquad (1\text{-}9)$$

where the mean value of the differences, sometimes called the bias, is

$$\bar{m} = \frac{1}{n} \sum \Delta m_i. \tag{1-10}$$

Practical application of these expressions is discussed later in this chapter.

1.4. PROPERTIES OF MATERIALS

The interaction of an electric field with a dielectric material has its origin in the response of charged particles to the applied field. The displacement of these particles from their equilibrium positions gives rise to induced dipoles which respond to the applied field. Such induced polarization arises from displacement of electrons around the nuclei (known as *electron polarization*) or from the relative displacement of atomic nuclei because of the unequal distribution of charge in molecule formation (*atomic polarization*). In addition to induced dipoles, some dielectric materials, known as polar dielectrics, contain permanent dipoles due to the asymmetric charge distribution of differently charged particles in a molecule, which tend to reorient under the influence of a changing electric field, thus giving rise to *dipolar* or *orientation polarization*. Finally, another source of polarization is the charge buildup at interfaces between components in heterogeneous systems, called *interfacial*, *space-charge*, or Maxwell-Wagner *polarization*. These last two mechanisms, together with dc conductivity, are the main mechanisms causing dissipation of the electromagnetic energy in dielectric materials exposed to high-frequency fields. At microwave frequencies (above 1 GHz) dipolar polarization is the dominant effect.

The parameters of materials that describe their interaction with electromagnetic fields are the permittivity ϵ^* and the permeability μ^*. The permittivity describes the material behavior in the electric field and consists of a real part ϵ', called the *dielectric constant*, and an imaginary part ϵ'', called the *loss factor*. Thus, the permittivity is expressed as

$$\epsilon^* = \epsilon' - j\epsilon'',$$

where the dielectric constant represents the ability of a material to store electric energy, and the loss factor describes the loss of electric field energy in the material. Another parameter frequently used is the dissipation factor, also called the *loss tangent*, defined as the ratio of the loss factor to the dielectric constant. The permittivity of materials is often normalized to the permittivity of a vacuum and is referred to as the *relative permittivity*:

$$\epsilon_r^* = \frac{\epsilon^*}{\epsilon_0} = \frac{\epsilon'}{\epsilon_0} - j\frac{\epsilon''}{\epsilon_0}, \tag{1-11}$$

where the permittivity of free space is $\epsilon_0 = 8.854 \times 10^{-12}$ F/m. The relative dielectric constant and loss factor are thus dimensionless quantities, and the word "relative" is often omitted and ϵ'_r is referred to as the dielectric constant ϵ'. The relative loss factor is a function of the material conductivity:

$$\epsilon''_c = \frac{\epsilon''}{\epsilon_0} = \frac{\sigma}{2\pi f \epsilon_0} \tag{1-12}$$

where σ is the conductivity (S/m) and f is the operating frequency (Hz). By analogy, any other form of loss (dipolar, Maxwell-Wagner, etc.) can be considered in any subsequent analysis to be included as part of the complex permittivity and forming its imaginary term. Therefore, if dipolar polarization was the only mechanism leading to losses, its contribution could be taken into account by writing the permittivity as

$$\epsilon^*_d = \epsilon' - j\epsilon''_d$$

where subscript d refers to the dipolar polarization loss mechanism. However, since a loss term must include all possible mechanisms, the loss factor is generally written as ϵ'' and is meant to contain losses caused by all possible mechanisms. Since all of those mechanisms show a frequency dependence, the definition of an effective loss factor is

$$\epsilon'' = \epsilon''_d(\omega) + \epsilon''_e(\omega) + \epsilon''_a(\omega) + \epsilon''_i(\omega) + \epsilon''_c(\omega)$$

where the subscripts d, e, a, and i refer to dipolar, electronic, atomic, and interfacial (Maxwell-Wagner) polarizations, respectively, and ϵ''_c is defined in eq. (1-12).

The dielectric properties of dielectric materials in nature, which are generally hygroscopic, vary predominantly with moisture content. They also depend on the frequency of the applied electromagnetic field, and the temperature, density, and structure of the materials. In granular or particulate materials, the bulk density of the air-particle mixture is another factor that influences dielectric properties. Bulk density can be altered by the shape and dimensions of particles and their surface conditions. Interest in the dielectric properties of various dielectric materials extends from soils and growing vegetation in various stages of maturity (for remote sensing), to blocks of concrete, sheets of cardboard, timber and veneer, to fruits, seeds, and nuts after harvest and during storage and trade, both as bulk materials and as individual kernels and nuts. Such a broad spectrum of material structure and quantity requires practically all existing microwave methods of permittivity measurements to be used in research, including time-domain reflectometry and broadband frequency-domain spectroscopy. The main difficulty in carrying out experiments with agricultural materials is the transient nature of their dielectric characteristics. For example, water evaporation at room temperature, accelerated evaporation at elevated temperature, and chemical processes related to plant growth and ripening all make replicating experiments with the same material virtually impossible. However, recently available broadband, computer-controlled instrumentation enables very fast measurements, and modern mathematical routines provide opportunities for using such data to develop more and more precise dielectric models.

1.4.1   Frequency Dependence

With the exception of some low-loss materials such as plastics and nonpolar liquids, the dielectric properties of most materials vary with the frequency of the applied electromagnetic field. The frequency dependence of the loss mechanisms is schematically presented in Figure 1.4 in terms of loss factors contributing to the effective loss factor of moist material. In the frequency range corresponding to radio and microwave frequencies (10^7 to 3×10^{10} Hz), the most important phenomenon contributing to the loss factor is the polarization of molecules arising from their orientation with the imposed electric field. This includes dipolar polarization in regions b (bound water relaxation) and w (free water relaxations). Loss mechanisms due to atomic and electronic polarizations, collectively termed distortion polarizations, occur at frequencies in the infrared and visible parts of the electromagnetic spectrum, and as such, play no role of interest from the microwave aquametry point of view.

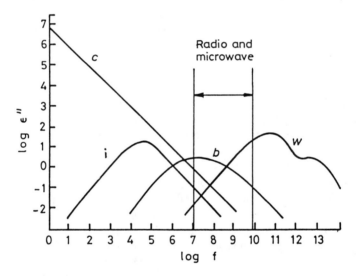

FIGURE 1.4. Mechanisms contributing to the value of the effective loss factor of moist material as a function of frequency in Hz: i, Maxwell-Wagner polarization; c, dc conductivity; b, dipolar polarization of water bound to the matrix of the material; w, dipolar polarization of free (liquid) water. *Data source [19].*

The description of the process for pure polar materials, developed by Debye (1929), is given in the form

$$\epsilon_r = \epsilon_\infty + \frac{\epsilon_s - \epsilon_\infty}{1 + j\omega\tau} \tag{1-13}$$

where ϵ_s is the static dielectric constant (i.e., the value at zero frequency— dc value), ϵ_∞ is the dielectric constant at frequencies so high that molecular orientation does not have time to contribute to the polarization, $\omega = 2\pi f$ is the angular frequency, where f is the frequency of the alternating field, and τ is the relaxation time (the period associated with the time for dipoles to revert to random orientation when the electric field is removed). Separation of eq. (1-13) into its real and imaginary parts yields

$$\epsilon_r' = \epsilon_\infty + \frac{\epsilon_s - \epsilon_\infty}{1 + \omega^2\tau^2}, \qquad \epsilon_r'' = \frac{(\epsilon_s - \epsilon_\infty)\omega\tau}{1 + \omega^2\tau^2}. \qquad (1\text{-}14)$$

The relationships defined by these equations are illustrated in Figure 1.5. At frequencies very low and high with respect to the polar molecule relaxation process, the dielectric constant has constant values ϵ_s and ϵ_∞, respectively, and the loss factor is very low. At intermediate frequencies, the dielectric constant undergoes dispersion, and dielectric losses occur with the peak loss at the *relaxation frequency* $\omega = 1/\tau$.

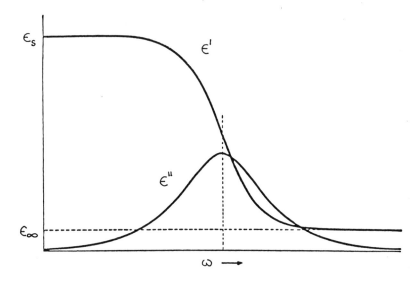

FIGURE 1.5. Dispersion and absorption curves representing the Debye model for polar substance with a single relaxation time.

The Debye equation (1-13) can be presented graphically in the complex ϵ_r''-ϵ_r' plane in the form of a semicircle with locus of points ranging from $\epsilon_r' = \epsilon_s$, $\epsilon_r'' = 0$ at the low-frequency limit to $\epsilon_r' = \epsilon_\infty$, $\epsilon_r'' = 0$ at the high-frequency limit, as shown in Figure 1.6. Such a representation of dielectric data is known as a Cole-Cole diagram. Since few materials of practical interest exhibit pure polar properties with a single relaxation time, many other expressions have been developed to better describe the frequency-dependent behavior of materials with more relaxation times or a distribution of relaxation times. One such equation is the Cole-Cole (1941) equation:

$$\epsilon_r = \frac{\epsilon_s - \epsilon_\infty}{1 + (j\omega\tau_{\text{mean}})^{1-\alpha}} \tag{1-15}$$

where τ_{mean} is the mean of different relaxation times which correspond to transitions between the different dipole positions, α denotes the relaxation time distribution parameter, and its empirical values are between 0 and 1. Any deviation from the pure polar material considered by Debye can be readily seen if the permittivity described by eq. (1-15) is plotted and compared with that given by eq. (1-14), as in Figure 1.6.

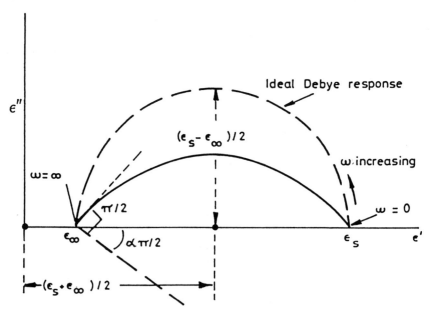

FIGURE 1.6. Cole-Cole plot for a polar substance with single relaxation time (dashed line) compared with a plot for a substance exhibiting distributed relaxation times (continuous line). *Data source [19].*

Water in its liquid state is a good example of a polar dielectric. Its properties are discussed later in the book. However, liquid water rarely appears in moist materials. Most often it is physically absorbed in material capillaries or pores or is chemically bound to other molecules of the material. The dielectric properties of bound water show significant differences from those of liquid water. Molecular relaxations of absorbed water take place at frequencies lower than the relaxation of free, liquid water. The principal relaxation occurs at 17–22 GHz for liquid water between 20 and 30°C. On the other hand, the relaxation peaks for absorbed water occur at a few megahertz to several hundred megahertz, indicating that the nature of the material absorbing water has a marked effect upon water's dielectric properties and consequently upon the interaction of the electromagnetic field with that material. Typical changes of the loss factor for a heterogeneous material containing water versus frequency are shown in Figure 1.7. The effect of dipolar polarization is apparent. The rise at the lower frequencies is attributed to the dc conductivity of the material. Possible relaxation of bound water, as shown by the dotted line, may be overshadowed by the dc conductivity.

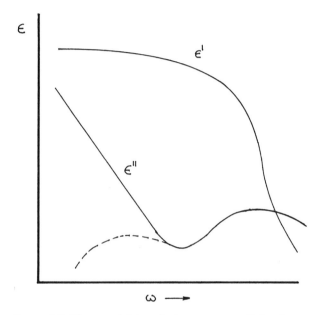

FIGURE 1.7. The permittivity of a heterogeneous dielectric material exhibiting dipolar polarization losses and conductive losses at lower frequencies.

1.4.2   Moisture Content Dependence

Water in solids is most often present in bound form, which means it is chemically combined with other molecules or physically absorbed to the surface of the material. Free water is present in paste or slurry materials of semiliquid form and as abundant water in nonhygroscopic materials. Because water molecules attached to those of the dry material exhibit much lower rotational mobility than molecules of liquid water, the dielectric properties of moist materials differ significantly from what would be expected from the fractional contribution of liquid water. A typical variation of the material permittivity with moisture content, $M = 100\xi$ (in percent), for a moist pulverized material is shown in Figure 1.8. Two different slopes of the characteristics can be related to various stages of water binding with the material. The smaller slope at low moisture content is due to strongly bound water (region 1), while the higher slope at increased moistures is due to presence of less tightly bound water molecules or even free molecules of water. The water molecules bound in the first monomolecular layer at the surface of the material are less rotationally free than the molecules in the second layer, etc. Thus, a whole spectrum of molecule binding may exist in a material, providing a smooth transition between the different stages. The change of slope occurs at certain moisture content, M_c, which is characteristic for a given material and most often may be related to an *equilibrium* moisture content reached at ambient conditions (temperature, pressure, and relative humidity). Some hygroscopic materials exhibit a gradual change of slope, making the positive identification of the two regions fairly difficult. The characteristic moisture content for highly hygroscopic materials occurs in the region between 10% and 40%, while for nonhygroscopic materials (e.g., sand) it is below 1%. When a mathematical expression is required for the material permittivity in terms of the moisture content, M (to make use of the $\epsilon = \phi(M)$ response in other expressions and optimize the design of a particular sensor or transducer), a good fit of experimental data can be obtained with (for $M < M_m$)

$$\epsilon = \epsilon_1 + \frac{AM^2}{M_m - M} \tag{1-16}$$

where the constants ϵ_1, M_m, and A are chosen to best fit the data.

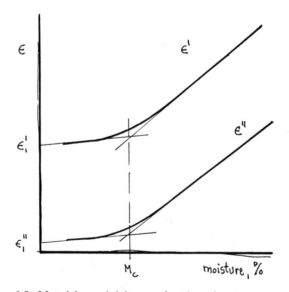

FIGURE 1.8. Material permittivity as a function of moisture content.

1.4.3 Temperature Dependence

The dielectric properties of materials are also temperature dependent, and the nature of that dependence is a function of the dielectric relaxation processes and as such is related to the operating frequency. As material temperature increases, the relaxation time decreases and the peak for the loss factor shown in Figure 1.5 will shift to a higher frequency. Thus, in a region of dispersion, the dielectric constant will increase with increasing temperature, while the loss factor may either increase or decrease, depending on whether the operating frequency is higher or lower than the relaxation frequency. The temperature dependence of ϵ_∞ is generally negligible; that of ϵ_s is larger, but its influence is minor in a region of dispersion. Below that region, the dielectric constant decreases with increasing temperature. Distribution functions can be useful in expressing the temperature dependence of dielectric properties, but their frequency and temperature-dependent behavior are in most cases complex and can best be determined by an appropriate measurement at the frequency and under the other conditions of interest.

1.4.4 Density Dependence

Since the influence of a dielectric depends upon the amount of mass interacting with the electromagnetic waves, the mass per unit volume or density, ρ, has an effect on the dielectric properties, especially in porous and particulate dielectrics such as pulverized or granular materials. In many such cases it has been observed that second-order polynomial or quadratic curves fit closely experimental data of the dielectric constant and loss factor as a function of material density. These observations are consistent with the dielectric mixture equation, proposed for two-phase mixtures in the form

$$\epsilon = (v_1\sqrt{\epsilon_1} + v_2\sqrt{\epsilon_2})^2 \tag{1-17}$$

where ϵ_1 and ϵ_2 are the permittivities of the inclusions and host medium (often air), respectively, and v_1 and v_2 are the volume-filling factors or volume fractions occupied by constituents 1 and 2, with $v_1 + v_2 = 1$. It can be shown that eq. (1-17) has the same form for both the dielectric constant and the loss factor if $\epsilon'^2 \gg \epsilon''^2$ for both components of the mixture.

For many particulate materials, air may be considered as a host medium, and therefore $\epsilon_2 = 1 - j0$ may be assumed in eq. (1-17). For the particles, $\epsilon_1 = \epsilon'_p - j\epsilon''_p$ and $v_2 = \rho/\rho_p$, where ρ is the bulk density of the material and ρ_p is the solid density. Thus, eq. (1-17) may be rewritten for the real and imaginary components of the bulk material permittivity as

$$\sqrt{\epsilon'} - 1 = \frac{\rho}{\rho_p}\left(\sqrt{\epsilon'_p} - 1\right)$$

and $\hspace{10cm}$ (1-18)

$$\sqrt{\epsilon''} = \frac{\rho}{\rho_p}\sqrt{\epsilon''_p}.$$

These simple equations allow prediction of the dielectric properties of this very simple mixture of material particles and air at different densities. For many other materials, such a solution takes a much more complex form, especially since the material density as well as its permittivity are functions of moisture content. At present no universal expressions have been developed, and the required relationships must be determined empirically.

1.5. MICROWAVE MOISTURE MEASUREMENTS

The following unique features of microwave radiation make it useful for moisture content measurement:

- Waves propagate along straight lines and reflect from metal surfaces, obeying the laws of optics.

- Microwaves can propagate through free space; thus, a physical contact between the equipment and the material under test is not required, allowing remote sensing to be accomplished.

- Many solid dielectric materials are opaque to light and infrared radiation but transparent to microwaves, which permits the probing of the whole volume of materials transported inside a dielectric tubing without the need for special windows.

- The effect of dc conductivity decreases with frequency and is much smaller at microwave frequencies than at radio frequencies, which makes moisture measurement easier and less dependent upon the material composition.

- Some materials, like alcohols and water, react specifically with selected microwave frequencies, allowing measurements of small amounts of water concentration in complex mixtures.

- Microwave radiation does not alter or contaminate the material under test as do some chemical methods, enabling fast, nondestructive, and continuous monitoring.

- In contrast to ionizing radiation, microwave methods are much safer and very fast.

- Microwave radiation is relatively insensitive to environmental conditions; thus, dust and water vapor do not affect the measurement, in contrast to infrared methods.

Some basic principles governing interaction of electromagnetic waves with matter are briefly discussed.

1.5.1 Plane Wave in Dielectric Medium

The propagation constant of a plane electromagnetic wave in a lossy dielectric medium is defined as

$$\gamma = \alpha + j\beta = \sqrt{j\omega\mu^*(\sigma + j\omega\epsilon')}, \qquad (1\text{-}19)$$

where α is the *attenuation constant* and β is the *phase constant*. For a nonmagnetic material, when $\mu^* = \mu_0$, the propagation constant may be expressed as

$$\gamma = j\frac{2\pi}{\lambda}\sqrt{\epsilon_r^*} \qquad (1\text{-}20)$$

where λ is the wavelength in free space, $f = c/\lambda$, and $c = 1/\sqrt{\epsilon_0 \mu_0}$ is the speed of light. The two components of the propagation constant can now be rewritten as

$$\alpha = \frac{2\pi}{\lambda} \sqrt{\frac{\epsilon_r'}{2} \left(\sqrt{1 + \tan^2 \delta} - 1 \right)}, \tag{1-21}$$

$$\beta = \frac{2\pi}{\lambda} \sqrt{\frac{\epsilon_r'}{2} \left(\sqrt{1 + \tan^2 \delta} + 1 \right)}. \tag{1-22}$$

The wavelength in a lossy dielectric material is then

$$\lambda_\epsilon = \lambda \sqrt{\frac{\epsilon_r'}{2} \left(\sqrt{1 + \tan^2 \delta} + 1 \right)}. \tag{1-23}$$

When a plane wave is incident normally upon a dielectric interface as shown in Figure 1.9a, part of it is reflected and part is transmitted inside the material. The reflection coefficient is

$$\Gamma = \frac{E_R}{E_0} = \frac{\eta_2 - \eta_1}{\eta_2 + \eta_1} \tag{1-24}$$

where E_0 is the incident electric field vector, E_R is the reflected electric field vector, and η_1 and η_2 are the intrinsic wave impedances of dielectric media 1 and 2, respectively;

$$\eta_1 = \sqrt{\frac{\mu_1}{\epsilon_1}} \quad \text{and} \quad \eta_2 = \sqrt{\frac{\mu_2}{\epsilon_2}}. \tag{1-25}$$

The transmission coefficient from medium 1 to 2 is

$$T = \frac{E_T}{E_0} = 1 + \Gamma = \frac{2\eta_2}{\eta_1 + \eta_2}, \tag{1-26}$$

and the transmitted power is

$$P_T = P_0(1 - |\Gamma|^2) \tag{1-27}$$

where P_0 is the incident power. If dielectric 1 is free space (or air), and dielectric 2 has $\mu^* = \mu_0$, then the reflection coefficient at the interface can be written as

$$\Gamma = \frac{1 - \sqrt{\epsilon_r^*}}{1 + \sqrt{\epsilon_r^*}}. \tag{1-28}$$

It is evident from this equation that both the modulus and phase angle of the reflection coefficient, as well as the transmission coefficient, depend upon the permittivity (i.e., both the dielectric constant and loss factor, or conductivity, of the material). Thus, the information about a test parameter that is a function of the permittivity can be obtained by measuring the modulus and/or phase of the reflection or the transmission coefficients. Choosing either of these combinations is a matter of careful consideration for each given application.

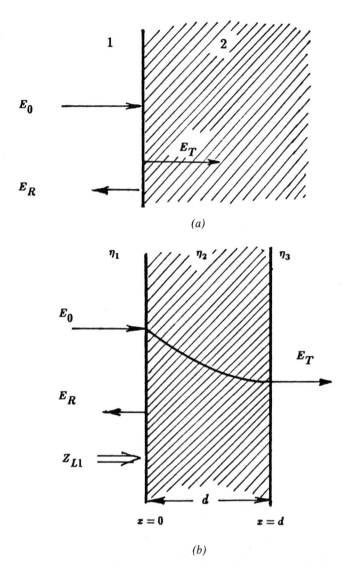

FIGURE 1.9. Plane wave normally incident on (a) semi-infinite dielectric; (b) a dielectric layer of thickness d.

For quick parameter evaluation, the following approximate expressions may be of use for those materials for which the condition $\epsilon'^2 \gg \epsilon''^2$ is valid:

$$\alpha \simeq \frac{\pi}{\lambda} \frac{\epsilon''}{\sqrt{\epsilon'}}, \quad \beta \simeq \frac{2\pi}{\lambda} \sqrt{\epsilon'}, \quad \text{and} \quad \Gamma \simeq \frac{\sqrt{\epsilon'} - 1}{\sqrt{\epsilon'} + 1} \qquad (1\text{-}29)$$

where Γ is the magnitude of the voltage reflection coefficient at the interface of air and the wet material. A careful inspection indicates that the above condition is fulfilled for most cases, and eqs. (1-29) can be used even for quite moist materials with negligible error ($<10\%$).

The reflection and transmission coefficients for a dielectric layer of thickness d and for normal wave incidence can be calculated by considering the wave and load impedances in various regions, as shown in Figure 1.9b. The reflection coefficient in region 1 is

$$\Gamma_1 = \frac{Z_{L1} - \eta_1}{Z_{L1} + \eta_1},\tag{1-30}$$

where

$$Z_{L1} = \eta_2 \left(\frac{\eta_3 \cos k_2 d + j\eta_2 \sin k_2 d}{\eta_2 \cos k_2 d + j\eta_3 \sin k_2 d} \right)\tag{1-31}$$

and $k_2 = 2\pi/\lambda_{\epsilon 2}$, and $\lambda_{\epsilon 2}$ is the wavelength in dielectric region 2. The ratio of the wave transmitted into region 3 to the incident wave in region 1 can be written as

$$T_3 = \frac{E_3}{E_0} = \frac{2\eta_3}{\eta_2 + \eta_3} \Gamma_1 e^{-jk_2 d}.\tag{1-32}$$

It may be noted from eqs. (1-30) to (1-32) that, for a constant layer thickness, information about the layer permittivity can be obtained from either reflection or transmission coefficients. In practice, radiating elements are placed relatively close to the material surface and therefore operate in the near field, and usually the surface of the material is not ideally smooth nor flat. This means that the incident waves are only approximations of plane waves and that the reflection and transmission coefficients depend on the position of the material with respect to the antenna. Moving the antenna away from the material surface, e.g., to ensure far-field conditions, requires a test sample of material with a large surface area and may cause problems because of scattering from other objects in the vicinity. In any case, however, the above equations can be used to estimate the measured quantities. Usually this approximation is good enough for practical purposes.

To compute the modulus of the transmission coefficient, the initial and final amplitude measurements are obtained for the measuring space with and without the material. From basic electromagnetic theory, it is well known that the power that propagates through a material must decrease according to the factor $e^{-2\alpha x}$. Referring to Figure 1.9, if the power at $x = 0$ is P_0, then at $x = d$ the power is given, assuming no reflection, by

$$P = P_0 e^{-2\alpha d}.$$

The total decrease in power expressed in decibels is called the layer *attenuation* and is

$$A = -20 \log |\tau| = -10 \log \left(\frac{P}{P_0} \right) = -10 \log(e^{-2\alpha d}).\tag{1-33}$$

The *phase shift* introduced by the layer of material is determined from the initial and final measurements of the wave phase angle in the reference plane, $x = d$. The initial phase measurement ϕ_0 is related to the phase constant β_0 and the thickness t of the free space through which the electromagnetic waves travel, as shown in Figure 1.9, by

$$\phi_0 = \beta_0 d, \tag{1-34}$$

assuming normal incidence. In a similar way, the final phase measurement is related to the phase constant β_1 and the thickness of material d by

$$\phi_1 = \beta_1 d. \tag{1-35}$$

The change in phase angle caused by introducing the material layer in the path of the traveling electromagnetic wave can be written as

$$\phi = \phi_1 - \phi_0 = (\beta_1 - \beta_0)d. \tag{1-36}$$

Since the phase constant is defined as 2π radians per wavelength, it can be rewritten as

$$\phi = \left(\frac{2\pi}{\lambda} - \frac{2\pi}{\lambda_0} \right) d = 2\pi d \left(\frac{1}{\lambda} - \frac{1}{\lambda_0} \right). \tag{1-37}$$

Because of the periodic nature of the wave phase through a dielectric medium, multiples of π or $2\pi n$ in phase, where n is an integer, can exist for a sample of thickness d. Thus eq. (1-37) in reality becomes

$$2\pi d \left(\frac{1}{\lambda} - \frac{1}{\lambda_0} \right) = \begin{cases} \pm 2\pi n \pm \phi, \\ \pm 2\pi n \pm (\pi - \phi) \end{cases} \tag{1-38}$$

to account for all possible combinations. Since the wavelength λ within the dielectric material is smaller than the free-space wavelength λ_0, the left side of eq. (1-38) is always positive; hence, the right side also must be positive for any n multiple of phase angle. Therefore the constraints on eq. (1-38) are

$$2\pi d \left(\frac{1}{\lambda} - \frac{1}{\lambda_0} \right) = \begin{cases} \begin{array}{ll} |\phi|, \\ |\pi - \phi|, \end{array} & n = 0, \\ \begin{array}{ll} 2\pi n \pm \phi, \\ 2\pi n \pm (\pi - \phi), \end{array} & n = 1, 2, 3, \ldots \end{cases} \tag{1-39}$$

Since there are numerous solutions to this equation, several samples of different thicknesses are used to resolve the phase-shift ambiguity.

As microwave radiation generated in the closed cavity of a microwave oven is sometime used for drying moist materials in determining their moisture content, that aspect of microwave applications will be discussed briefly. The process of microwave heating consists of dissipating the microwave energy that flows into a heated material, which, in general, is a lossy (moist) dielectric. For any heating process, the heating rate can be expressed as

$$\frac{dT}{dt} = \frac{P_d}{C\rho}, \tag{1-40}$$

where T is the temperature of the heated material (°C), t is the time (s), P_d is the power dissipated in the heated material (W/m^3), C is the specific heat of the material (J/kg °C), and ρ is the density of the material (kg/m^3). The power (W/m^3) dissipated in the material in a plane wave is

$$P_d = \frac{\omega\epsilon_0}{2} \int_V \epsilon'' |E|^2 \, dV + \frac{1}{2} \int_V \sigma |E|^2 \, dV, \tag{1-41}$$

where σ is the dc conductivity of the material (S/m) and E is the electric field intensity (peak value) (V/m). For most materials heated by microwaves, σ is small compared with $\omega\epsilon''$, and the equation for most practical applications has the form

$$P_d = 27.8 \times 10^{-6} f \int_V \epsilon'' |E|^2 \, dV, \tag{1-42}$$

where f is the operating frequency (MHz). To calculate the total power dissipated in a heated material at a given frequency, the distribution of the loss factor and the internal electric field intensity must be known. This is rarely possible in practice, and therefore some simplifying assumptions have to be made. Most frequently, it is assumed that the loss factor does not vary appreciably within the heated volume and that the heated object is large compared with the wavelength.

For a heated body in the form of a slab occupying an infinite half-space, as shown in Figure 1.9a, exposed to a plane wave at normal incidence, the incident power is partly reflected and partly transmitted and dissipated in the heated object. The distribution of the dissipated power in a homogeneous lossy material follows an exponential law; i.e.,

$$P_d = P_T e^{-2\alpha z} = P_0(1 - |\Gamma|^2)e^{-2\alpha z}, \tag{1-43}$$

where α is the attenuation constant of the material given by eq. (1-21), z is the distance along the direction of wave propagation, and Γ is the reflection coefficient determined in eq. (1-24). The *penetration depth* for given material is expressed as a distance $d = 1/2\alpha \simeq \lambda\sqrt{\epsilon'}/2\pi\epsilon''$. For objects of dimensions comparable to the wavelength, the internal electric field in eq. (1-42) becomes a complex function of the dimensions and shape of the object and the permittivity distribution inside the material.

1.5.2 Microwave Measuring Systems

A microwave moisture content measuring system may be considered as a series of information transducers connected as shown in Figure 1.1. In particular, the following blocks can be distinguished.

Microwave Sensor. In the microwave sensor block the direct interaction between the electromagnetic wave and the moist material takes place. The sensor should provide a microwave signal and convey it into the material under test. Thus, it contains a microwave source operating at a desired frequency and means for stabilizing and modulating the signal, which is next transmitted to a radiating element (antenna). Various forms of microwave radiating elements in waveguide, coaxial line, or microstrip configurations can be used in microwave sensors. They may be divided into resonant and aperiodic groups, into open and closed structures, and into reflection and transmission types. A proper choice of sensor results from a set of specific requirements, but waveguide horns and microstrip patch antennas are most often used, providing compact design and sensitive relationships between input and output quantities. Possible variations of the output signal come from fluctuations of the layer thickness, displacement of the material within the measuring space, and nonuniformity of the microwave field. As the radiating elements operate in near-field conditions, standard far-field expressions for antennas are not necessarily correct, but they often may be used as a first approximation for transmission assessments at a distance of a few wavelengths throughout the layer of wet material. The reflection of microwave energy at the air-material interface can be limited by proper matching of the antenna input impedance, and scattering and diffraction of the energy outside the measuring space can be eliminated by using properly designed lenses (dielectric or metallic). To eliminate or limit variation of the material permittivity with factors other than moisture content, the block often contains mechanical means such as vibrators, rollers, and scrapers that provide proper measuring conditions in the sensor area, i.e., the laminar flow of material, specified layer thickness of uniform density or flat material surface.

Microwave Transducer.   The microwave transducer portion of the measuring system converts the microwave signal, proportional to moisture content, into a low-frequency electrical signal. It should consist of a microwave oscillator, a transmission system, and a detector, with the sensor connected into the transmission path. Various parameters are used as a measure of moisture content in microwave meters: for example, the energy absorbed in moist material, the phase shift introduced by the material, the reflection coefficient from a wet material surface, or the resonant cavity detuning caused by the introduction of a wet sample. Each of these parameters can be precisely measured in conventional microwave circuits operating on the basis of deflection, differential, or null methods, including substitution, comparison, and compensation methods. Variation in the output electrical signal at a constant value of microwave parameter being measured may result from changes in losses and impedance mismatches in microwave components, from fluctuations of the output power and frequency of the oscillator, and from changes in conversion loss of the detector. All these variations may cause errors in the moisture content measurement.

Signal Processing Unit.   The signal processing block contains power supplies, amplifiers, rectifiers, and mixers to allow transformation of the moisture content meter output signal into analog or digital indications to be displayed, recorded, or used in automatic control systems. The performance of this unit is affected by incidental changes in sensitivity of electronic components and devices, temperature, electronic noise, and external interference. In the past, square-law detectors were most frequently used; recently, heterodyne and homodyne detectors have been used. The unit should provide high sensitivity and resolution with minimal instabilities resulting from component aging and environmental effects (short- and long-term drifts). From the user's point of view, it is usually more important to monitor the average moisture content than the instantaneous value, which may fluctuate widely because of other process variables. Therefore, the results are usually integrated over time. Some typical elements of modern data processing, transmission, and storage are more and more often included in laboratory as well as on-line microwave moisture meters and are becoming a vital part of contemporary measuring and monitoring systems.

1.5.3 Density-Independent Function

Changes in an electrical signal interacting with a moist material, regardless of operating frequency, are proportional to the water concentration (mass per unit volume), k, and are affected only to a small extent by the density of dry material, m_d/v. Thus, when k may be determined from electrical measurement, it is evident from eqs. (1-1) and (1-5) that determination of moisture content, ξ, requires knowledge of the density of the wet material, ρ. This information can be obtained from a separate density measurement, for example, by weighing a sample of given volume or by using a γ-ray density gauge. Fluctuations in the density of the test material produce effects similar to changes in water content and therefore contribute to a measurement error. This error can be limited or eliminated only if the mass of the wet material in the measuring space is held constant during the calibration procedure and during the measurement. Another way to resolve the density-variation problem is to find a *density-independent* function correlating the material moisture content with measured electrical material parameters. Efforts to eliminate effects of density fluctuations in the moisture content measurements have been made over the last 20 years; some results of those efforts are presented later in this book.

The dispersion and dissipation of electromagnetic energy interacting with dielectric material depends on the shape, dimensions, and relative permittivity of the material. When the moisture content of the material changes, a change is reflected in the wave parameters. Because the relative permittivity of water differs from that of most hygroscopic dielectric materials, its effect can be separated from the effect of the dry dielectric material. In general, this may be expressed in functional form as

$$\alpha = \Phi_1\left(\frac{m_w}{v}, \frac{m_d}{v}\right) \qquad \text{and} \qquad \beta = \Phi_2\left(\frac{m_w}{v}, \frac{m_d}{v}\right), \qquad (1-44)$$

where α and β are any two descriptive electromagnetic wave parameters. Two examples of such pairs are attenuation and phase constants of a plane wave, and resonant frequency shift and change in the Q-factor of a resonant microwave cavity. Regardless of the complexity of the analytical expressions described by eq. (1-44), it is generally possible to solve the two equations and to express the mass of water and the mass of dry material in terms of two measured parameters in the form

$$\frac{m_w}{v} = \Psi_1(\alpha, \beta) \qquad \text{and} \qquad \frac{m_d}{v} = \Psi_2(\alpha, \beta). \qquad (1\text{-}45)$$

After the analytical expressions corresponding to eq. (1-45) are substituted into eq. (1-5), the general expression for moisture content of a material can be written as

$$M = \frac{\Psi_1(\alpha, \beta)}{\Psi_1(\alpha, \beta) + \Psi_2(\alpha, \beta)} \times 100 \qquad (1\text{-}46)$$

which contains only the wave parameters determined experimentally and is totally independent of the material density. Moreover, the denominator of eq. (1-46) is the density of the material under test, so it can be determined independently of the moisture content from the expression

$$\rho = \frac{m_w}{v} + \frac{m_d}{v} = \Psi_1(\alpha, \beta) + \Psi_2(\alpha, \beta) \qquad (1\text{-}47)$$

and then used for correction or other purposes.

As an example of practical application of this concept, potential moisture content measurement in a flowing stream of wheat is discussed. Although the example relates to an agricultural product, this concept is by no means limited to such products. Similar results could be obtained in preparation for microwave monitoring of moisture content of sand in a glass factory, of crushed coal and coal dust in a coal processing plant, of artificial fertilizers, in pharmaceutical granules and powders, and in many other materials of various forms.

A bulk sample of soft red winter wheat, *Triticum aestivum* L., was selected for calibration measurements. During the microwave measurements, the sample was held in a plastic container which was located between two waveguide horn antennas that were connected to the transmitting-receiving system operating at frequency 4.8 GHz. Components of the transmission coefficient (attenuation and phase shift of the electromagnetic wave passing through the wheat sample) were measured in free space. Thirty-two wheat samples of different moisture contents were measured, ranging from 10.7% to 20.1% moisture. Each sample was measured at four to seven densities, providing 160 data points for use in developing the calibration equations. During the verification process, similar measurements provided 217 data points. The concentration of water and the dry material density were calculated from eq. (1-6). The attenuation and phase shift of microwave signals of 4.8 GHz as a function of the water concentration in grain are shown in Figure 1.10 a,b. Two linear equations fit the experimental results with high statistical significance:

$$A = -1.17 + 0.1442\frac{m_w}{v} - 0.0056\frac{m_d}{v}, \qquad r = 0.9928,$$

$$\phi = 4.0 + 2.401\frac{m_w}{v} + 0.1845\frac{m_d}{v}, \qquad r = 0.9896, \qquad (1\text{-}48)$$

where r is the correlation coefficient, and the concentration of water and density of dry material are expressed in kg m^{-3}. According to theoretical considerations, use of eqs. (1-46) and (1-47) provides the expressions for the moisture content and bulk density in the form

$$M = \frac{3.733\phi + 123.2A + 129.7}{\phi - 14.8A - 21.362}, \tag{1-49}$$

$$\rho = \frac{26.787\phi - 396.28A - 572.22}{7159.16}. \tag{1-50}$$

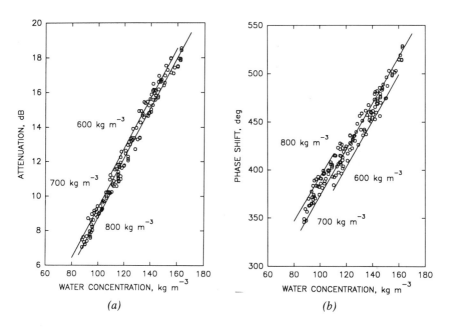

(a)   (b)

FIGURE 1.10. Attenuation (a) and phase shift (b) of soft red
winter wheat as a function of water concentration measured at
4.8 GHz and 24°C on layer of grain 87 mm thick. Lines are
for constant values of dry material density (m_d/v) indicated.

Validity of the calibration equations was checked with a hard red winter wheat, harvested the next year in a different location. Samples of 32 different moisture contents ranging from 10.3% to 19.3% were measured with the same measuring setup and the same procedures that were used for the calibration measurements. A total of 217 data points for various moisture contents and densities were obtained during these measurements. Data from those measurements were used with eqs. (1-49) and (1-50) to calculate predicted moisture and density and compared with oven moisture tests and bulk density determinations based on sample weight and volume. The histogram in Figure 1.11 shows the distribution of differences between oven moisture content and calculated moisture content for these data. The mean value of differences between oven moisture content and calculated moisture content (bias defined in eq. (1-10)) was -0.032% moisture, and the standard deviation of the differences (SEP, as defined in eq. (1-9)) was 0.27% moisture. The distribution of differences for the grain bulk density is shown in Figure 1.12.

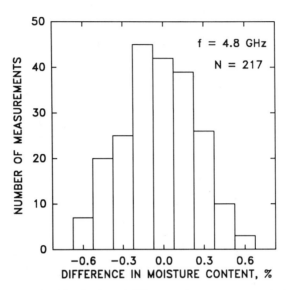

FIGURE 1.11. Distribution of differences between oven moisture content determination and moisture content calculated for hard red winter wheat from calibration equation (1-49).

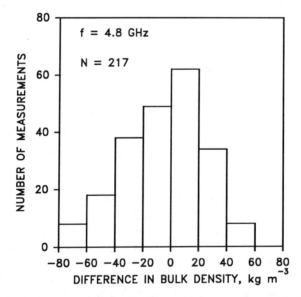

FIGURE 1.12. Distribution of differences between measured grain bulk density and the density predicted from calibration equation (1-50).

The standard error of performance for the bulk density determination was 18.6 kg m^{-3}, while the bias was -1.6 kg m^{-3}. Figures 1.11 and 1.12 indicate that the error distribution has totally random character.

It has been observed by many authors that the ratio of two measured wave parameters (attenuation and phase shift) is also a good density-independent measure of moisture content in material. Because both parameters are proportional to the grain density and both depend on the grain moisture content, it is reasonable to expect that their ratio should be independent of grain density. The dependence of the variable $X = \phi/A$ on wheat moisture content determined by the standard air-oven method for a soft red winter wheat measured at 4.8 GHz is shown in Figure 1.13. A simple empirical expression that fits the calibration set of 160 experimental data points has the form

$$\frac{1}{X} = 0.00173M + 0.00314, \qquad r = 0.9606. \tag{1-51}$$

Inspection of these expressions implies calibration equations for moisture content determination in the form

$$M = \frac{578}{X} - 1.82 \tag{1-52}$$

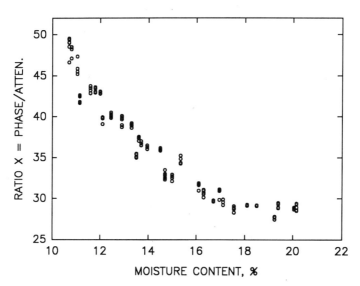

FIGURE 1.13. Ratio of two measured grain parameters (ϕ/A) as a function of moisture content for a soft red winter wheat measured at 4.8 GHz.

The validity of the calibration equation (1-52) was tested with the same data set for hard red winter wheat described above. The histograms presented in Figure 1.14 show the distribution of differences between oven moisture content and moisture content calculated from eq. (1-52). The mean value of differences (bias) is -0.028% moisture and the standard deviation of the differences (SEP) is 0.38% moisture. These results are surprisingly good, taking into account the simple form of eq. (1-52) compared to the corresponding expression (1-49). The resulting moisture content from eq. (1-52) is independent of grain density. This approach, however, does not provide information on the grain density as does the approach involving eq. (1-50).

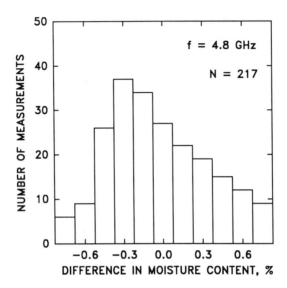

FIGURE 1.14. Distribution of differences between oven moisture content determination and moisture content calculated for a hard red winter wheat from calibration equation (1-52) for 217 measurements at 4.8 GHz.

A closer analysis of the expressions of the ratio X, using eqs. (1-29) and (1-36), yields

$$X = \frac{\phi}{A} = 13.193\frac{\epsilon' - 1}{\epsilon''}\left(\frac{\sqrt{\epsilon'}}{\sqrt{\epsilon'} + 1}\right) = 13.193\, X_1\left(\frac{\sqrt{\epsilon'}}{\sqrt{\epsilon'} + 1}\right), \qquad (1\text{-}53)$$

where $X_1 = (\epsilon' - 1)/\epsilon''$ is the density-independent function. A similar expression, consistent with the permittivity relationships with density given by eq. (1-18) could be written in the form

$$X_2 = \frac{\sqrt{\epsilon'} - 1}{\sqrt{\epsilon''}}. \qquad (1\text{-}54)$$

There are also other expressions considered as density-independent functions and used with certain success for given materials. One of them is related to a dielectric mixture formula and has the form

$$X_3 = \frac{\sqrt[3]{\epsilon'} - 1}{\sqrt{\epsilon''}}. \qquad (1\text{-}55)$$

Numerous efforts to find a more universal function correlating material moisture content with dielectric properties and eliminating the effect of the material density indicate the importance of solving of this problem for further development of aquametry and microwave aquametry in particular.

1.6. CONCLUSIONS

Microwave aquametry can be defined as a branch of metrology that investigates solid and liquid dielectric materials containing water by identifying their properties in microwave fields. Microwave aquametry has its own technical subject matter, applications, and particular research methods, as well as a characteristic instrumentation base. Microwave aquametry utilizes some well-known physical theories, for example, the theory of dielectric mixtures and the theory of bound water. Beside these cognitive purposes, microwave aquametry also has strictly defined practical objectives. These include the quantitative measurement of water content in many economically important materials. Since water appears in many raw materials, as well as in most manufactured materials, these objectives have far-reaching economic importance. Precise and real-time automatic control of moisture content is required in many technological processes, and it is impossible without equally accurate and fast moisture content measurement methods. It has been shown over the last quarter of a century that in many such cases the microwave method is the only feasible and reliable solution.

REFERENCES

Moisture Content and Its Measurement

[1] M. A. Berliner, *Measurement of Moisture Content*, (in Russian), Moscow: Izdat. Energia, 1973.

[2] A. Pande, *Handbook of Moisture Determination and Control*, vols. 1–4, New York: Marcel Dekker, 1974.

[3] D. M. Smith and J. Mitchell, Jr., *Aquametry*, Part III, *Electrical and Electronic Methods*, New York: Wiley, 1984.

Physics of Water and Moist Materials

[4] W. F. Brown, "Dielectrics," in *Encyclopedia of Physics*, vol. 17, S. Flugge, Ed., Berlin, Heidelberg: Springer-Verlag, 1956, pp. 1–154.

[5] V. V. Daniel, *Dielectric Relaxation*, London: Academic Press, 1967.

[6] F. Franks (Ed.), *Water—A Comprehensive Treatise*, New York: Plenum, 1972.

[7] J. B. Hasted, *Aqueous Dielectrics*, London: Chapman and Hall, 1973.

[8] E. H. Grant, R. J. Sheppard, and G. P. South, *Dielectric Behaviour of Biological Molecules in Solutions*, Oxford: Clarendon, 1978.

[9] A. K. Jonscher, *Dielectric Relaxation in Solids*, London: Chelsea Dielectrics Press, 1983.

[10] U. Kaatze, "Complex permittivity of water as function of frequency and temperature," *J. Chem. Eng. Data*, vol. 34, pp. 371–374, 1989.

General Metrology

[11] C. F. Dietrich, *Uncertainty, Calibration and Probability*, 2nd ed., Bristol: Adam Hilger, 1991.

[12] P. A. Payne (Ed.), *Instrumentation and analytical science*, London: Peter Peregrinus, 1989.

[13] P. H. Sydenham (Ed.), *Handbook of Measurement Science*, vols. 1, 2, New York: Wiley, 1982, 1983.

[14] M. J. Usher, *Sensors and Transducers*, Basinstoke, UK: MacMillan, 1985.

Microwave Techniques

[15] H. M. Altschuler, "Dielectric constant," in *Handbook of Microwave Measurements*, M. Sucher and J. Fox, Eds., Brooklyn: Polytechnic Press of PIB, 1963, Ch. IX.

[16] A. E. Bailey (Ed.), *Microwave Measurement*, London: Peter Peregrinus, 1985.

[17] S. W. Cheng and F. H. Levien, *Microwaves Made Simple*, Dedham, MA: Artech House, 1985.

[18] R. W. P. King and G. S. Smith, *Antennas in Matter*, Cambridge, MA: MIT Press, 1981.

[19] A. C. Metaxas and R. J. Meredith, *Industrial Microwave Heating*, London: Peter Peregrinus, 1983.

[20] E. Nyfors and P. Vainikainen, *Industrial Microwave Sensors*, Norwood, MA: Artech House, 1989.

[21] P. A. Rizzi, *Microwave Engineering*, Englewood Cliffs, NJ: Prentice Hall, 1988.

[22] J. Thuery, *Microwaves: Industrial, Scientific and Medical Applications*, Norwood, MA: Artech House, 1992.

Microwave Aquametry

[23] V. K. Benzar, *Microwave Techniques of Moisture Measurement*, (in Russian), Minsk: Izdat. Vysheyshaya Shkola (University Publishers), 1974.

[24] M. A. Berliner, "Microwave moisture meters," (in Russian), *Prib. sist. upravleniya (Control Instrum. Systems)*, (1), pp. 19–22, 1970.

[25] A. Kraszewski, "Microwave instrumentation for moisture content measurement," *J. Microwave Power*, vol. 8, nos. 3/4, pp. 323–336, 1973.

[26] A. Kraszewski, "Microwave aquametry—a review," *J. Microwave Power*, vol. 15, no. 4, pp. 209–220, 1980.

[27] A. Kraszewski, "Microwave aquametry—needs and perspective," *IEEE Trans. Microwave Theory Techn.*, vol. 39, no. 5, pp. 828–835, 1991.

[28] A. Kraszewski, "Fifteen years of literature in microwave aquametry," Sec. V, pp. 423–464, *Microwave Aquametry*, New York: IEEE Press, 1996.

Section *II*

Physical Background

2

Udo Kaatze

Universität Göttingen, Göttingen, Germany

Microwave Dielectric Properties of Water

Abstract.   A short review is given on properties of water in its gaseous, liquid, and solid state and on interactions of this unique substance with microwaves. Special emphasis is laid on the dielectric spectrum of liquid water. It is shown that in the microwave region the presently available data can be fairly well represented by a Debye-type relaxation spectral function the parameters of which are presented as a function of temperature. Also briefly discussed is the dependence of the Debye dielectric relaxation time upon hydrostatic pressure. It is indicated that a more complicated spectral function is required if complex permittivity data above about 300 GHz are taken into account. Results of recent computer simulation studies are mentioned that give significant insights into the kinetics of the fluctuating hydrogen-bonded network and the dielectric relaxation process of the associated liquid.

2.1.   THE ISOLATED WATER MOLECULE

Much interest has been directed during the past decades toward the unique properties of water, the only chemical that exists, under normal conditions, in its gaseous, liquid, and solid states on our planet. However, even now there is still a considerable gap in our understanding of the bulk behavior of this remarkable substance in terms of its molecular structure. Nevertheless, many interesting aspects of bulk water can be related, at least in a qualitative manner, to the properties of the isolated water molecule.

The water molecule is nonlinear with an H–O–H angle θ somewhat smaller than the angle 109.5° of a tetrahedron ($\Theta = 105.05°$ [1], 104.47 [2], 104.45 [3]). The mean O–H distances d_{OH} are about 0.96 Å($d_{OH} = 0.9568$ Å [1], 0.9572 Å [2], 0.9584 Å [3]). The same values are found for HDO and D_2O ($\theta = 104.45$, $d_{OD} = 0.9584$ [3]). The water molecule possesses a permanent electric dipole moment μ that is directed along the twofold axis of symmetry, the bisector of the H–O–H angle, and points from the oxygen atom to the region between the hydrogen atoms. The water molecule also possesses a polarizability α which is the sum of its atomic and electronic (displacement) polarizabilities α_a and α_e, respectively. The permanent dipole moment μ, as deduced from the variation of the static permittivity $\varepsilon(0)$ of water vapor with temperature, has a value $\mu = (1.84 \pm 0.02)$ D in reasonable agreement with values determined by other methods [2,4]. In the evaluation of the temperature-dependent $\varepsilon(0)$ data the Clausius-Mossotti equation as modified by Debye [5,6],

$$\frac{\varepsilon(0) - 1}{\varepsilon(0) + 2} \frac{M}{\rho} = \frac{4\pi N}{3} \left(\alpha + \frac{\mu^2}{3kT} \right), \tag{2-1}$$

has been used. Here M is the molar weight, ρ is the density, N is Avogadro's number, k is the Boltzmann constant, and T is the temperature in kelvins. Measurements of the Kerr constant [7] show that the anisotropy in the polarizability α of the water molecule is small. Applying the Lorentz-Lorenz form [8,9] of eq. (2-1),

$$\frac{n^2 - 1}{n^2 + 2} \frac{M}{\rho} = \frac{4\pi}{3} N\alpha, \tag{2-2}$$

one obtains the mean value $\alpha(= \alpha_e) = 1.44$ Å^3 at optical frequencies [2]. In eq. (2-2), n denotes the optical refractive index.

Quantized transitions between rotational energy levels of isolated water molecules lead to narrow resonance absorption lines with small natural linewidths. Due to nonideal conditions, such as the Doppler effect, the lines are broadened to some 10^5 Hz. As a result of effects of intermolecular collisions, the half-widths of the absorption lines additionally increase at higher pressures. The nonlinear water molecule possesses three principle moments of inertia. Since these three moments of inertia differ from each other, the water molecule as a so-called asymmetric top exhibits a most complex rotational spectrum. The rotational constants of H_2O are such that the lowest resonance has a frequency of 22.235 GHz [10] and most of the rotational spectrum lies in the submillimeter-wavelength region [11]. The rotational constants of HDO and D_2O are, of course, different [12] and so are the spectra.

The 22.235-GHz water resonance sets the upper frequency limit of the "radio window" through which we can see into the universe via radiotelescopes [13]. At higher frequencies, up to the "optical window," electromagnetic waves are absorbed by water as well as other gases like CO_2 and O_3. Other windows exist in the submillimeter range of the water spectrum, a well-known one being between 752 and 988 GHz [11,14].

2.2. LIQUID WATER

2.2.1 Static Permittivity

In its condensed phases water molecules associate to form a macroscopically percolating hydrogen-bonded network in which the free-molecule values for the intramolecular distances are not exactly maintained. Even in the low-pressure phase ice I, however, the H–O–H angle θ seems not to adopt the ideal tetrahedral value 109.5° (104.5° $\leq \theta \leq$ 109.5°). The bond length d_{OH} in ice I is increased by about 0.05 Å with respect to the vapor phase ($d_{OH} = d_{OD} = 1.01$ Å [15]). The situation of liquid water seems to be an intermediate one ($\theta \approx 104°$, $d_{OD} = 0.98$ Å [15]). Due to its permanent molecular electric dipole moment, liquid water exhibits a high low-frequency ("static") permittivity $\varepsilon(0)$ which decreases from a value of 107 $\pm$ 2 for supercooled water at −35°C [16,17] to 55.99 $\pm$ 0.06 at 99°C [18]. Some selected $\varepsilon(0)$ data from the literature are displayed as a function of temperature ϑ (in the Celsius scale) in Figure 2.1. The semilogarithmic plot there illustrates that at −20°C $\leq \vartheta \leq$ 99°C the static permittivity of liquid water can be well represented by the simple empirical relation [19]

$$\varepsilon(0) = 87.79 \exp(b\vartheta), \tag{2-3}$$

where $b = -0.00455/°C$. As realized by a careful comparative study [20] of the static permittivities $\varepsilon(0, D_2O)$ and $\varepsilon(0, H_2O)$ of heavy water and normal water, respectively, $\varepsilon(0, D_2O)$ appears to be slightly smaller than $\varepsilon(0, H_2O)$ and the relative deviation between the data for the isotopic liquids increases with T (Table 2.1).

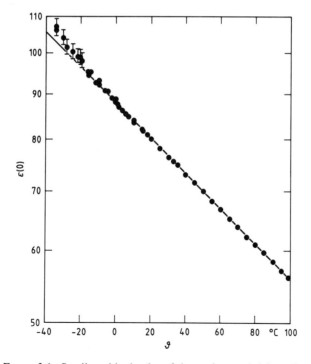

FIGURE 2.1. Semilogarithmic plot of the static permittivity $\varepsilon(0)$ of water versus temperature ϑ for some data from the literature [16–18,20]. The full line is the graph of eq. (2-3).

TABLE 2.1. Static Permittivities for H_2O and D_2O at Some Temperatures [20]

ϑ (°C)	$\varepsilon(0, H_2O)$	$\varepsilon(0, D_2O)$	$10^3(\varepsilon(0, H_2O) - \varepsilon(0, D_2O)/\varepsilon(0, H_2O)$
0	87.91	87.65	3.0
5	85.89	85.61	3.3
25	78.39	78.06	4.2
40	73.19	72.84	4.8

From the Fröhlich theory of polar liquids [21,22] the equation

$$\frac{(\varepsilon(0) - \varepsilon_\infty)(2\varepsilon(0) + \varepsilon_\infty)}{\varepsilon(0)} = \frac{4\pi N}{9kT}(\varepsilon_\infty + 2)^2 c\mu^2 g \qquad (2\text{-}4)$$

relates the static permittivity to the dipole moment μ, the concentration c of dipoles, and the Kirkwood orientation correlation factor g [23]. Eq. (2-4) could thus be used to evaluate the $\varepsilon(0)$ data in terms of g values and to extract information on the liquid structure from the measured static permittivities. Unfortunately, in eq. (2-4) relating $\varepsilon(0)$ to g, the high-frequency permittivity ε_∞ plays a significant role and the correct value to be used is not known. Kirkwood [22,23] took $\varepsilon_\infty = n^2$ and inserted for the refractive index the value $n = 1.33$ at wavelengths of visible light. He found an orientation correlation factor $g = 2.8$ at 25°C. Hill pointed out [22,24] that infrared or far-infrared values should be used instead to evaluate eq. (2-4). Presuming $g \equiv 1$, she derived ε_∞ values monotonously decreasing from 4.35 to 4.05 when the temperature of water increases from 0 to 60°C [24]. If values $\varepsilon(\nu \to \infty)$ as extrapolated from the microwave part of the spectrum (next section) are used, $g < 1$ may even result. On the basis of experimental permittivity data it is therefore difficult to comment on dipole orientation correlation effects in pure water. It is, however, possible to discuss relative changes in the g-factor if the structure of water in solutions is affected by solute particles [25].

2.2.2 Complex Dielectric Spectrum

During the past decades the microwave and near-millimeter-wavelength behavior of water has been occasionally reviewed [19, 26–30], recently in conjunction with a comprehensive compilation of individual data from the literature [19]. As an example the complex dielectric spectrum

$$\varepsilon(\nu) = \varepsilon'(\nu) - i\varepsilon''(\nu) \qquad (2\text{-}5)$$

for H_2O at 30°C is displayed in Figure 2.2. A broad dispersion $(d\varepsilon'(\nu)/d\nu < 0)$/dielectric loss $(\varepsilon''(\nu) > 0)$ region emerges at microwave frequencies, indicating relaxation characteristics as resulting from the hindrance of the water reorientational motions by hydrogen bonds. The curves are the Debye relaxation spectral function defined by

$$\varepsilon(\nu) = \varepsilon(\infty) + \frac{\varepsilon(0) - \varepsilon(\infty)}{1 + i\omega\tau_D}, \qquad (2\text{-}6)$$

where $\omega = 2\pi\nu$.

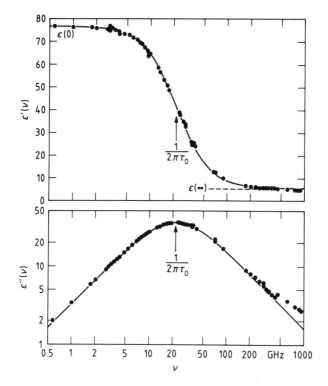

FIGURE 2.2. Real part ε' and negative imaginary part ε'' of the complex permittivity ε of water at 30°C plotted versus frequency ν. Full points indicate the data tabulated in Ref. 19. The curves represent a Debye relaxation spectral function with parameter values found by fitting eq. (2-6) to the complete set of permittivity values ($\varepsilon(0) = 76.47$, $\varepsilon(\infty) = 4.9$, $\tau_D = 7.20$ ps). No weighing factors have been used in this regression analysis.

This function with discrete relaxation time τ_D corresponds to an exponential dielectric decay function. The Debye function (eq. (2-6)), like the other relaxation spectral functions discussed below, indeed fails to describe correctly the short-time behavior of any physically meaningful relaxation process [31]. Nevertheless this function appears to appropriately represent the measured dielectric spectra of water up to remarkably high frequencies ($\nu \lesssim 200$ GHz).

Deviations from a Debye-type relaxation spectrum emerging at frequencies above about 200 GHz are accentuated by Figure 2.3, where $\varepsilon'(\nu) - \varepsilon(\infty)$ is displayed as a function of $\omega\varepsilon''(\nu)$. A Debye term is represented by a straight line with its negative slope given by the dielectric relaxation time. Two straight-line segments with significantly different slopes are found for the water spectrum, with a transition between these segments in the frequency range around 300 GHz.

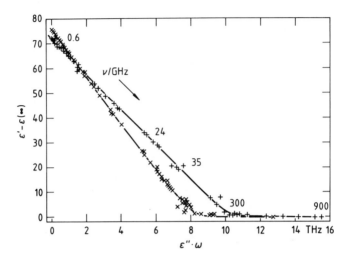

FIGURE 2.3. Plot of $\varepsilon'(\nu) - \varepsilon(\infty)$ versus $\omega\varepsilon''(\nu)$ for water at 20°C (multiplication sign) and 30°C (plus sign). The curves are graphs of the Davidson-Cole relaxation spectral function eq. (2-8) with the following parameter values from the fitting procedure: 20°C, $\varepsilon(0) = 80.29$, $\varepsilon(\infty) = 4.66$, $\tau_{DC} = 9.83$ ps, $\beta = 0.04$; 30°C, $\varepsilon(0) = 76.72$, $\varepsilon(\infty) = 4.20$, $\tau_{DC} = 7.67$ ps, $\beta = 0.06$.

It was thus obvious to attempt to describe the water spectrum at $\nu \lesssim 1000$ GHz by a double Debye function, as given by the relation

$$\varepsilon(\nu) = \varepsilon(\infty) + \frac{\Delta\varepsilon_1}{1 + i\omega\tau_{D1}} + \frac{\Delta\varepsilon_2}{1 + i\omega\tau_{D2}}. \qquad (2\text{-}7)$$

In doing so, the following values for the relaxation time τ_{D1} of the high-frequency term with low amplitude ($\Delta\varepsilon_1/(\Delta\varepsilon_1 + \Delta\varepsilon_2) \approx 0.02$) have been found at 25°C: 0.27 ps [32], 0.53 ps [33], and 1.02 ps [34]. Hence, τ_{D1} largely agrees with the lifetime of individual hydrogen bonds in water for which Rayleigh scattering studies yielded 0.54 ps [35]. The high-frequency relaxation term in eq. (2-7) may thus be taken to be due to the mechanism of hydrogen bond breaking and re-forming [34].

The microwave and near-millimeter-wavelength spectra of water can also be well represented by a continuous relaxation time distribution. The curves in Figure 2.3 illustrate that the Davidson-Cole relaxation spectral function can also be used to represent the water spectra at different temperatures. This function is given by [36]

$$\varepsilon(\nu) = \varepsilon(\infty) + \frac{\varepsilon(0) - \varepsilon(\infty)}{(1 + i\omega\tau_{DC})^{1-\beta}} \qquad (2\text{-}8)$$

where β denotes a parameter ($0 \leq \beta < 1$) which measures the width of the underlying relaxation time distribution. The Davidson-Cole spectral function corresponds to an asymmetric relaxation time distribution in which τ_{DC} is the largest relaxation time. Even at high frequencies ($\nu > 300$ GHz) $\varepsilon'(\nu)$ of water decreases steadily with ν. There are thus no indications for a negative slope

$$S(\infty) = \lim_{\nu \to \infty} \frac{d\varepsilon''(\nu)}{d(\varepsilon'(\nu) - \varepsilon(\infty))} < 0 \qquad (2\text{-}9)$$

in the permittivity data as expected for physical systems with orientational relaxation characteristics [31]. Probably, the experimental accuracy in the near-millimeter-wavelength region is too small to allow the limiting relaxation behavior as predicted by eq. (2-9) to be extracted from data that might, in addition, contain contributions from vibrational and librational motions.

Normally, studies of the microwave dielectric properties of liquids are restricted to frequencies below about 100 GHz. It is thus sufficient to represent the dispersion/dielectric loss region of water by a simple Debye relaxation spectral function with only three parameters. Values of these parameters at different temperatures are presented in Table 2.2. The relations between the dielectric relaxation time and other molecular mobility parameters are displayed in Figure 2.4 at different temperatures to show that these quantities exhibit nearly the same dependence upon ϑ.

TABLE 2.2. Parameters of the Debye Relaxation Spectral Function for Water at Various Temperatures[a]

ϑ (°C)	$\varepsilon(0) \pm \Delta\varepsilon(0)$	$\varepsilon(\infty) \pm \Delta\varepsilon(\infty)$	$\tau_D \pm \Delta\tau_D$ (ps)
− 4.1	89.3 ± 0.2	5.9 ± 0.3	21.1 ± 0.2
0	87.9 ± 0.2	5.7 ± 0.2	17.6 ± 0.2
5	85.8 ± 0.2	5.7 ± 0.2	14.9 ± 0.2
10	83.9 ± 0.2	5.5 ± 0.2	12.7 ± 0.1
15	82.1 ± 0.2	6.0 ± 0.5	10.8 ± 0.2
20	80.2 ± 0.2	5.6 ± 0.2	9.36 ± 0.05
25	78.36 ± 0.05	5.2 ± 0.1	8.27 ± 0.02
30	76.6 ± 0.2	5.2 ± 0.4	7.28 ± 0.05
35	74.9 ± 0.2	5.1 ± 0.3	6.50 ± 0.05
40	73.2 ± 0.2	3.9 ± 0.3	5.82 ± 0.05
50	69.9 ± 0.2	4.0 ± 0.3	4.75 ± 0.05
60	66.7 ± 0.2	4.2 ± 0.3	4.01 ± 0.05

[a] Values have been obtained by fitting eq. (2-6) to microwave dielectric spectra, mostly from this laboratory [30]. For a recent analysis of water spectra see Ref. 54. Relaxation times τ_D for H_2O and D_2O at temperatures up to 260°C, as derived from complex permittivity measurements at 9.3 GHz along the coexistence curve, are tabulated in Ref. 55.

In the microwave frequency range D_2O spectra and spectra for mixtures of D_2O with H_2O can be well described by a discrete relaxation time τ_D (Figure 2.5). At 25°C, τ_D linearly increases as a function of the mole fraction of D_2O (Table 2.3). That finding is in contrast with the behavior of the shear viscosity η_s, the

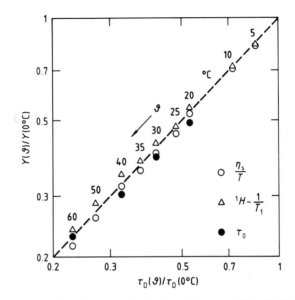

FIGURE 2.4. Double logarithmic plot of the ratio $Y(\vartheta)/Y(0°C)$ versus $\tau_D(\vartheta)/\tau_D(0°C)$ for the proton nuclear spin lattice relaxation rate $(Y = 1/T_1, \triangle)$, the shear viscosity over temperature $(Y = \eta_s/T, o)$, and the structural relaxation time $(Y = \tau_o, \bullet)$ for water [29].

TABLE 2.3. Parameters of the Debye Relaxation Spectral Function (eq. (2-6)) for H_2O/D_2O Mixtures at 25°C with Different Mole Fraction x of D_2O [37]

x	$\varepsilon(0) \pm \Delta\varepsilon(0)$	$\varepsilon(\infty) \pm \Delta\varepsilon(\infty)$	$\tau_D \pm \Delta\tau_D$ (ps)
0	78.36 ± 0.05	5.2 ± 0.1	8.27 ± 0.02
0.25	78.1 ± 0.3	5.6 ± 0.3	8.78 ± 0.09
0.5	78.1 ± 0.3	5.5 ± 0.3	9.32 ± 0.10
0.78	78.1 ± 0.3	5.5 ± 0.3	9.83 ± 0.10
1	78.3 ± 0.3	4.8 ± 0.3	10.37 ± 0.11

values of which for the mixtures increase somewhat less strongly than predicted by a linear dependence upon x. This result is noteworthy in view of Debye's diffusive theory of relaxation [22] that predicts a linear relationship between τ_D and η_s.

As shown by Figure 2.6 (see page 46) the dielectric relaxation time decreases if water is exposed to hydrostatic pressure. Around the temperature of the highest water density the relative change $(1/\tau_D)(d\tau_D/dp)$ in the dielectric relaxation time exceeds that of other parameters reflecting molecular motions. On first glance, it may be surprising that the water mobility is enhanced on application of hydrostatic pressure. This behavior, however, can be consistently explained by recent computer simulation studies of water, as discussed below.

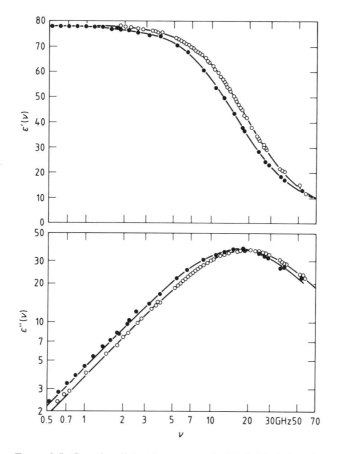

Figure 2.5. Complex dielectric spectrum for H_2O (circles) and a mixture of H_2O and D_2O with mole fraction $x = 0.75$ of heavy water (points) at 25°C [37]. The curves are graphs of the Debye relaxation spectral function (eq. (2-6)) with the parameter values given in Table 2.3.

2.2.3 Relaxation Mechanism

Recent computer simulation studies [39–46] illustrated the molecular motions in water and of the interrelation of these motions with structural properties of the liquid. At a given moment water molecules are connected by j hydrogen bonds ($j = 0, \ldots, 4$) forming a random network well above the percolation threshold. The bond order j of the individual molecules nearly follows a binomial distribution function [43]. With correlation times as small as 0.1 to 1 ps, the strength of the hydrogen bonds fluctuates rapidly [40,41]. Orientational motions through significant angles, however, are only performed if the angular distribution of rotational barriers is flattened by the presence of extra nearest neighbors and if, as a second precondition, an appropriate site for a new hydrogen bond is offered. Such sites are

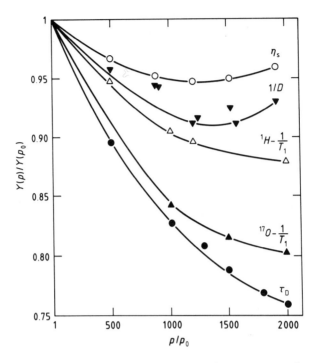

FIGURE 2.6. Ratio $Y(p)/Y(p_0)$ at hydrostatic pressure p and normal pressure p_0 displayed as a function of p/p_0 for the following quantities Y reflecting molecular mobility characteristics of water [38]: shear viscosity η_s (o, 5°C), reciprocal self-diffusion coefficient $1/D$ (▼; 4°C), proton ^{1}H (Δ; 5°C) and oxygen ^{17}O (▲, 10°C) nuclear magnetic spin lattice $1/T_1$, and dielectric relaxation time τ_D (•, 5°C).

normally provided by the presence of an additional neighbor molecule constituting a defect in the random tetrahedral network. Structural defects by "fifth neighbors" [39] lead to nonlinear H bonds. Bifurcated structures in which a proton is engaged in two bonds appear by which the activation enthalpy for orientational motions is lowered. If an additional H-bonding partner is present, the reorientation of a water molecule occurs within about 0.1 ps. Hence, the dielectric relaxation time τ_D is predominantly determined by the time for which water molecules have to wait until thermal activations result in favorable conditions for reorientation.

As shown in Figure 2.7, in pure water under normal conditions ($\rho = 1$ g/cm^3) about one-third of all molecules appears to be fivefold coordinated and another 13% even sixfold. These high coordination numbers are the reason for the comparatively small time τ_D. At reduced density almost perfect tetrahedral arrangements of molecules are adopted and the molecular dynamics slows [46]. In correspondence with these results from computer simulation studies, the mobility of water molecules increases if, due to hydrostatic pressure, the density and, thereby, the local availability of hydrogen bonding sites are enhanced.

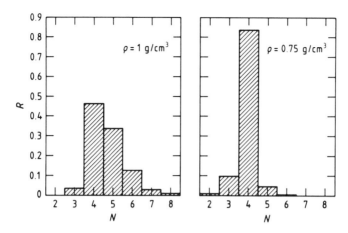

FIGURE 2.7. Fraction R of molecules with N neighbors within a distance $r = 3.3$ Å for water at $0°C$ [46].

2.3. ICE

As water freezes at normal pressure it forms a crystalline lattice in which each oxygen atom is tetrahedrally coordinated to four neighboring oxygen atoms [47–50]. In the normal low-pressure phase I_h the tetrahedrals are regular and the oxygen atoms form puckered six-membered rings. These hexagons, on their part, form puckered planes which are joined to neighboring puckered planes to establish an open tridymite-like structure. Besides this ordinary hexagonal phase I_h, a low-pressure cubic phase I_c has been observed which exists only below about $-70°C$. At high hydrostatic pressure a variety of different ice phases has been found, the structural characteristics and thermodynamic properties of which were summarized previously [47,49].

Within the ice I crystal lattice the hydrogen atoms lie nearly along an 0–0 axis [49], but are situated closer to one oxygen (1.01 Å) than to the other (1.75 Å), respectively. Also with solid water one may thus imagine H_2O molecules, each possessing a permanent dipole moment. It is likewise obvious that these molecules are connected by hydrogen bonds.

The static permittivity of polycrystalline ice I at the freezing point exceeds that of liquid water at the same temperature by about 5%, though the density is substantially smaller in the solid state than in the liquid ($\varepsilon(0) = 92$ [51], $\rho = 0.91671$ g cm^{-3} [47], polycrystalline ice I, $0°C$; $\varepsilon(0) = 87.9$ (Table 2.2), $\rho = 0.99984$ g cm^{-3} [52], water, $0°C$). For single crystals anisotropic static permittivities $\varepsilon_\perp(0)$ and $\varepsilon_\parallel(0)$ are found, with the permittivity parallel to the crystallographic c-axis slightly larger than that perpendicular to the c-axis ($(\varepsilon_\parallel(0) - \varepsilon_\perp(0))/\varepsilon_\perp(0) = 1.14$, $0°C$ [47,48]). The $\varepsilon(0)$ values of ice I increase with decreasing T. Both the magnitude of the static permittivity at $0°C$ and the temperature dependence of $\varepsilon(0)$ have been discussed in terms of dipole orientation correlation effects [47–50]. Notice that antiferroelectric structures without net electric moment of the elementary cell are assumed for some ordered high-pressure ice configurations for which $\varepsilon(0) = \varepsilon(\infty)$ [50].

The finding of $\varepsilon(0) \neq \varepsilon(\infty)$ with normal ice I and other phases of solid water indicates the ability of protons to change their configuration relative to the oxygen lattice. These changes require the existence of structural defects of which two different types are generally accepted. L defects consist of an oxygen-oxygen link without a hydrogen atom, and D defects represent situations in which the space between two oxygen positions is occupied by two hydrogen atoms. Following Bjerrum, these defects are suggested to arise from the rotation of water molecules [48,49]. The relative concentration of defects is assumed to be small (10^{-7}), and the energy of formation per mole pair of defects has been estimated at 16 kcal. The assumption of the low content of defects and the estimate of the activation energy are in conformity with the dielectric relaxation behavior of ice. As shown in Figure 2.8, dielectric relaxation in frozen water is about six orders of magnitude

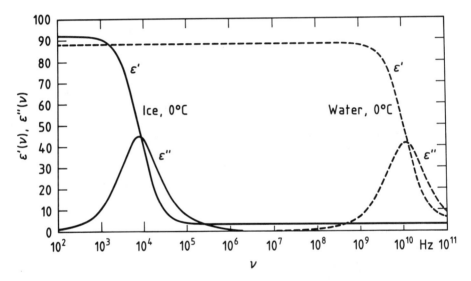

FIGURE 2.8. Semilogarithmic plot of the complex dielectric spectrum of ice (full curves) and of water (dashed curves) at 0°C. The parameters of the Debye-type spectra are $\varepsilon(0) = 92.0$ [51], $\varepsilon(\infty) = 3.17$ [53], and $\tau_D = 20\ \mu s$ [47] for ice as well as $\varepsilon(0) = 87.9$, $\varepsilon(\infty) = 5.2$, $\tau_D = 17.67$ ps (Table 2.2) for liquid water.

slower than in liquid water. The constant real part of the complex permittivity of ice at microwave frequencies ($\varepsilon'(\nu) = \varepsilon(\infty) = 3.2$, $\nu > 1$ MHz) results from librational motions and intra- as well as intermolecular vibrations. Only the low-frequency tail-end of the absorption bands related to these processes and the high-frequency tail-end of the loss due to the relaxation around 10 kHz contribute to the microwave value of $\varepsilon''(\nu)$. Hence, this value is small ($\varepsilon''(\nu) = 0.0014$,

$\tan \vartheta = \varepsilon''/\varepsilon' = 4.5 \times 10^{-4}$, 10 GHz [48]) and, consequently, ice hardly absorbs microwaves.

2.4. CONCLUSIONS, OUTLOOK

Due to the dominant role that water plays in a multitude of biological, ecological, and industrial processes, much effort has been undertaken to elucidate the outstanding features of this most important substance. However, despite a variety of ingenious experimental and theoretical studies on water and on a multitude of aqueous systems, our understanding of the unusual properties even of the pure substance is incomplete. This is particularly true for the complex liquid state, although its detailed knowledge on a molecular basis is of considerable significance for many developments in medicine, environmental engineering, and chemical technology.

The dielectric properties of water are known with considerable accuracy (globally ±1%) over a broad frequency range including the microwave regime (300 MHz $\leq \nu \leq$ 300 GHz). Up to about 300 GHz the dielectric spectrum of liquid water can be fairly well represented by the semiempirical Debye-type relaxation function reflecting a discrete relaxation time. The parameters of this relaxation function have been determined over a significant range of temperatures and, at a few temperatures, as a function of hydrostatic pressure. However, the theoretical representation of the static permittivity $\varepsilon(0)$ of the liquid by the dipole moment μ of an isolated water molecule is still insufficient. At present it is unclear which distortion polarization contributions have to be considered in the theoretical expression relating $\varepsilon(0)$ to μ. Consequently, the Kirkwood dipole orientation correlation factor and related structural information cannot be derived from the static permittivity of water. It is expected that computer simulation studies currently performed with great success for water and other associated liquids will yield valuable insights into relevant structural properties and into distortion polarization mechanisms of hydrogen-bonded networks. Present computer simulation studies predominantly aim at an explanation of the dielectric relaxation process and the existence of an almost discrete relaxation time. In conjunction with dielectric spectroscopy on relevant hydrogen-bonded liquids in progress, these computer studies appear to substantially complete our ideas of network kinetics and molecular reorientation mechanisms.

Since the theoretical description of the water dielectric spectrum by a Debye relaxation function must be incorrect for physical reasons, even more precise broadband measurements of the complex permittivity of water as a function of frequency are most desirable. It would be particularly useful to reduce the experimental error by a factor of 10 in such a study and to additionally extend with comparable accuracy the measurements up to the far-infrared region. At present there seems to be no laboratory with appropriate equipment for the complete frequency range. Hence, joint efforts of several groups would be required to drastically enhance the experimental accuracy in the microwave dielectric properties of water.

Acknowledgment

I thank Professor R. Pottel for many helpful discussions.

References

[1] O. J. Samoilow, *Die Struktur wässriger Elektrolytlösungen und die Hydration vol Ionen*, Leipzig: Teubner, 1961, ch. 2, pp. 25–56.

[2] J. B. Hasted, *Aqueous Dielectrics*, London: Chapman and Hall, 1973, ch. 1, pp. 1–31.

[3] C. W. Kern and M. Karplus, "The water molecule," in *Water, a Comprehensive Treatise*, vol. 1, F. Franks, Ed., New York: Plenum, 1972, ch. 2, pp. 21–91.

[4] A. H. Price, "The dielectric properties of gases," in *Dielectric Properties and Molecular Behaviour*, N. E. Hill, W. E. Vaughan, A. H. Price, and M. Davies, Eds., London: Van Nostrand Reinhold, 1969, ch. 3, pp. 191–231.

[5] P. Debye, "Einige Resultat einer kinetischen Theorie der Isolatoren," *Phys. Z.*, vol. 13, pp. 97–100, 1912.

[6] P. Debye, "Nachtrag zur Notiz über eine kinetische Theorie der Isolatoren," *Phys. Z.*, vol. 13, p. 295, 1912.

[7] H. Gränicher, C. Jaccard, P. Scherrer, and A. Steinemann, "Dielektrische Eigenschaften des Eises bei sehr tiefen Frequenzen und der Einfluß eines Vorfeldes," *Helv. Phys. Acta*, vol. 28, pp. 300–303, 1955.

[8] H. A. Lorentz, "Über die Beziehung zwischen der Fortpflanzungsgeschwindigkeit des Lichtes und der Körperdichte," *Ann. Phys. NF*, vol. 9, pp. 641–665, 1880.

[9] L. Lorenz, "Über die Refractionsconstante," *Ann. Phys. NF*, vol. 11, pp. 70–103, 1880.

[10] A. N. Leontakianakos, "A K-band oscillator locked to the first water resonance," *IEEE Trans. Microwave Theory Techn.*, vol. 40, pp. 191–195, 1992.

[11] G. W. Chantry, *Submillimetre Spectroscopy*, London: Academic, 1971, ch. 5, pp. 141–234.

[12] C. H. Townes and A. L. Schawlow, *Microwave Spectroscopy*, New York: McGraw-Hill, 1955, app. 6, pp. 613–642.

[13] H. H. Vogt, *Abriß der Astronomie*, Mannheim: Bibliographisches Institut, 1980.

[14] L. E. S. Mathias, A. Crocker, and M. S. Wills, "Pulsed gas laser sources," in *Millimetre and Submillimetre Waves*, F. A. Benson, Ed., London: Iliffe Books, 1969, ch. 7, pp. 125–149.

[15] J. C. Dore, "Structural studies of water by neutron diffraction," in *Water Science Reviews*, F. Franks, Ed., Cambridge: Cambridge University Press, 1985, ch. 1, pp. 3–92.

[16] J. B. Hasted and M. Shahidi, "The low frequency dielectric constant of supercooled water," *Nature*, vol. 262, pp. 777–778, 1976.

[17] I. M. Hodge and C. A. Angell, "The relative permittivity of supercooled water," *J. Chem. Phys.*, vol. 68, pp. 1363–1368, 1978.

[18] C. G. Malmberg and A. A. Maryott, "Dielectric Constant of Water from 0° to 100°C," *J. Res. Nat. Bur. Stand.*, vol. 56, pp. 1–7, 1956.

[19] K. Lamkaouchi, "L'eau: ètalon dièlectrique. Etude de lois dièlectriques appliquèes à l'eau et à des èmulsions de pètrole mesurèes en micro-ondes," Thesis, University of Bordeaux I, 1992.

[20] G. A. Vidulich, D. F. Evans, and R. L. Kay, "The dielectric constant of water and heavy water between 0 and 40°C," *J. Phys. Chem.*, vol. 71, pp. 656–662, 1967.

[21] H. Fröhlich, *Theory of Dielectrics*, Oxford: Clarendon, 1958, ch. 2, pp. 15–61.

[22] N. E. Hill, "Theoretical treatment of permittivity and loss," in Ref. 4, ch. 1, pp. 1–107.

[23] J. G. Kirkwood, "The dielectric polarization of polar liquids," *J. Chem. Phys.*, vol. 7, pp. 911–919, 1939.

[24] N. E. Hill, "The temperature dependence of the dielectric properties of water," *J. Phys. C: Solid State Phys.*, vol. 3, pp. 238–239, 1970.

[25] U. Kaatze, "Microwave dielectric properties of liquids," *Rad. Phys. Chem.*, vol. 45, pp. 549–566, 1995.

[26] P. S. Ray, "Broadband complex refractive indices of ice and water," *Appl. Opt.*, vol. 11, pp. 1836–1844, 1972.

[27] J. B. Hasted, "Liquid Water: Dielectric Properties," in Ref. 3, ch. 7, pp. 255–309.

[28] P. R. Mason, J. B. Hasted, and L. Moore, "The use of statistical theory in fitting equations to dielectric dispersion data," *Adv. Mol. Relax. Processes*, vol. 6, pp. 217–232, 1974.

[29] U. Kaatze and V. Uhlendorf, "The dielectric properties of water at microwave frequencies," *Z. Phys. Chem. N.F.*, vol. 126, pp. 151–165, 1981.

[30] U. Kaatze, "Complex permittivity of water as a function of frequency and temperature," *J. Chem. Eng. Data*, vol. 34, pp. 371–374, 1989.

[31] J. G. Powles, "Cole-Cole plots as they should be," *Adv. Mol. Relax. Processes*, vol. 56, pp. 35–47, 1993.

[32] T. Manabe, H. J. Liebe, and G. A. Hufford, "Complex permittivity of water between 0 and 30 THz," in *IEEE Con. Dig. 12th Conf. on IR Millmetre Waves*, pp. 220–230, 1987.

[33] M. S. Zafar, J. B. Hasted, and J. Chamberlain, "Submillimetre wave dielectric dispersion in water," *Nature, Phys. Sci.*, vol. 243, pp. 106–109, 1973.

[34] J. Barthel, K. Bachhuber, R. Buchner, and H. Hetzenauer, "Dielectric spectra of some common solvents in the microwave region. Water and lower alcohols," *Chem. Phys. Lett.*, vol. 165, pp. 369–373, 1990.

[35] O. Conde and J. Teixera, "Hydrogen bond dynamics in water studied by depolarized Rayleigh scattering," *J. Phys. (Paris)*, vol. 44, pp. 525–529, 1983.

[36] D. W. Davidson and R. H. Cole, "Dielectric relaxation in glycerine," *J. Chem. Phys.*, vol. 18, p. 1417, 1950.

[37] U. Kaatze, "Dielectric relaxation of H_2O/D_2O mixtures," *Chem. Phys. Lett.*, vol. 203, pp. 1–4, 1993.

[38] R. Pottel, E. Asselborn, R. Eck, and V. Tresp, "Dielectric relaxation rate and static dielectric permittivity of water and aqueous solutions at high pressures," *Ber. Bunsenges. Phys. Chem.*, vol. 93, pp. 676–681, 1989.

[39] A. Geiger, P. Mausbach, and J. Schnitker, "Computer simulation study of the hydrogen-bond network in metastable water," in *Water and Aqeuous Solutions*, G. W. Neilson and J. E. Enderby, Eds., Bristol: Hilger, 1986, pp. 15–30.

[40] H. Tanaka and I. Ohmine, "Large local energy fluctuations in water," *J. Chem. Phys.*, vol. 87, pp. 6128–6139, 1987.

[41] I. Ohmine, H. Tanaka, and P. G. Wolynes, "Local energy fluctuations in water. II: Cooperative motions and fluctuations," *J. Chem. Phys.*, vol. 89, pp. 5852–5860, 1988.

[42] D. Bertolini, M. Cassettari, M. Ferrario, P. Grigolini, G. Salvetti, and A. Tari, "Diffusion effects of hydrogen bond fluctuations. I: The long time regime of the translational and rotational diffusion of water," *J. Chem. Phys.*, vol. 91, pp. 1179–1190, 1989.

[43] F. Sciortino and S. L. Fornili, "Hydrogen bond cooperativity in simulated water: time dependence analysis of pair interactions," *J. Chem. Phys.*, vol. 90, pp. 2786–2792, 1989.

[44] F. Sciortino, A. Geiger, and H. E. Stanley, "Isochoric differential scattering functions in liquid water: the fifth neighbour as a network defect," *Phys. Rev. Lett.*, vol. 65, pp. 3452–3455, 1990.

[45] F. Sciortino, A. Geiger, and H. E. Stanley, "Effects of defects on molecular mobility in liquid water," *Nature*, vol. 354, pp. 218–221, 1991.

[46] F. Sciortino, A. Geiger, and H. E. Stanley, "Network defects and molecular mobility in liquid water," *J. Chem. Phys.*, vol. 96, pp 3857–3865, 1992.

[47] D. Eisenberg and W. Kauzmann, *The Structure and Properties of Water*, Oxford: Clarendon Press, 1969, ch. 3, pp. 71–149.

[48] Ref. 2, ch. 4, pp. 100–116.

[49] F. Franks, "The properties of ice," in Ref. 3, ch. 4, pp. 115–149.

[50] C. Jaccard, "Structural information from dielectric properties of ice," in *Structure of Water and Aqueous Solutions*, W. A. P. Luck, Ed., Weinheim: Verlag Physik–Chemie Verlag, 1974, ch. VI. 3, pp. 409–424.

[51] R. P. Auty and R. H. Cole, "Dielectric properties of ice and solid D₂O," *J. Chem. Phys.*, vol. 20, pp. 1309–1314, 1952.

[52] G. S. Kell, "Thermodynamic and transport properties of fluid water," in Ref. 3, ch. 10, pp. 363–412.

[53] M. D. Blue, "Permittivity of ice and water at millimetre wavelength," *J. Geophys. Res.*, vol. 85, pp. 1101–1106, 1980.

[54] M. G. M. Richards, "The development of techniques to measure the complex permittivity of liquids and biological tissues at 90 GHz," Thesis, University of London, 1993.

[55] O. A. Nabokov and Y. A. Lubimov, "The dielectric relaxation and the percolation model of water," *Mol. Phys.*, vol. 65, pp. 1473–1482, 1988.

O. Barajas
University of Calgary, Calgary, Alberta, Canada

H. A. Buckmaster
University of Victoria, Victoria, B.C., Canada

Calculation of the Temperature Dependence of the Debye and Relaxation Activation Parameters from Complex Permittivity Data for Light Water

Abstract. This paper discusses predictive techniques for deriving the temperature dependence of the Debye and activation parameters using complex permittivity data for light water obtained at 9.355 GHz. It is shown that the temperature variation of ϵ_∞ and the dielectric relaxation time τ can be derived from this data using the Debye model. The relaxation times were used to determine the temperature dependence of the Gibbs free energy ΔG^*, enthalpy ΔH^*, and entropy ΔS^* relaxation activation thermodynamic parameters for double-distilled deionized light water from 1°C to 90°C. The results are compared with those reported previously by the authors using a different method of analysis. It is concluded that the temperature dependence of the relaxation activation parameters is significant and is relaxation-process-model-sensitive.

3.1. INTRODUCTION

Water is an extremely important but very common liquid which plays a crucial role in biological systems [1]. It has been studied extensively using complex permittivity techniques since the pioneer work by Collie et al. [2]. Those studies prior to 1973 have been reviewed by Hasted [3,4]. Buckmaster [5,6] has reviewed continuous-wave techniques for the measurement of high-loss liquids such as water. The acquisition of high-accuracy complex permittivity data is an important tool for obtaining a detailed understanding of the physical processes and structures in molecular studies of condensed matter. During the past decade, McAvoy and

Buckmaster [7], Zaghloul and Buckmaster [8], van Kalleveen and Buckmaster [9], Barajas and Buckmaster [10], and Hu et al. [11] have reported 9.355-GHz complex permittivity measurements for double-distilled deionized light water with a 1σ precision which now approaches ∼0.01% and an accuracy of ∼0.1% using a new type of complex permittivity spectrometer developed initially by McAvoy and Buckmaster [12] with various subsequent improvements [7–11]. The latest measurements [10,11] were made over the temperature interval from ∼1°C to 90°C in ∼2°C steps and represent the most comprehensive and precise data available for light and heavy water. This precision enabled Barajas and Buckmaster [10] to demonstrate, for the first time, that the enthalpy ΔH^* and entropy ΔS^* of activation parameters characterizing the relaxation process are temperature-dependent, contrary to Conway [13], Zaghloul and Buckmaster [8], and Steinhoff [14], who assumed that they were temperature-independent over temperature intervals of at least 20°C. They also showed that the independent calculation of these parameters from the permittivity ϵ' and the dielectric loss ϵ'' data could be used to test the consistency and accuracy of complex permittivity data sets.

The temperature dependence of the dielectric relaxation time as well as the Gibbs free energy ΔG^*, enthalpy ΔH^*, and entropy ΔS^* activation parameters for the relaxation process provide information concerning the molecular processes and structures occurring within the liquid dielectric [15]. This paper describes a new technique for analyzing complex permittivity data using Debye theory [1,16,17], which permits easy calculation of the above-mentioned parameters. The results are compared with those calculated previously [10].

3.2. THEORY

The complex permittivity ϵ^* is

$$\epsilon^* = \epsilon' - i\epsilon''. \tag{3-1}$$

The first theoretical equations for the complex permittivity of dielectrics was proposed by Debye [16]. Later, Cole and Cole [18] introduced the empirical parameter α into the Debye equation, which improved the consistency with experimental data for different dielectrics. This parameter was a measure of the spread in the dielectric relaxation time τ. However, Kaatze [19] has concluded from an analysis of recent measurements of ϵ' and ϵ'' for light water by many researchers at various temperatures and frequencies that these measurements were more consistent with the original Debye equation, $\alpha = 0$. However, his conclusion may arise because of the inhomogeneity of his data set.

The complex permittivity according to Debye theory is

$$\epsilon^* = \frac{\epsilon_0 - \epsilon_\infty}{1 - i\omega\tau}. \tag{3-2}$$

Taking the real and imaginary parts of eq. (3-2) yields

$$\epsilon'(\omega, t) = \epsilon_\infty + \frac{\epsilon_0 - \epsilon_\infty}{1 + (\omega\tau)^2}, \tag{3-3a}$$

$$\epsilon''(\omega, t) = \frac{(\epsilon_0 - \epsilon_\infty)\omega\tau}{1 + (\omega\tau)^2}. \tag{3-3b}$$

Taking the magnitude of eq. (3-2) yields

$$(\epsilon' - \epsilon_\infty)^2 + (\epsilon'')^2 = \frac{(\epsilon_0 - \epsilon_\infty)^2}{1 + (\omega\tau)^2}, \tag{3-4}$$

which is the equation of the circle with center at ϵ_∞ on the ϵ'-axis and radius $(\epsilon_0 - \epsilon_\infty)^2/(1 + (\omega\tau)^2)$. Equation (3-4) leads to the Cole-Cole diagram [18]. Equation (3-3a) can now be rewritten as

$$\frac{1}{1 + (\omega\tau)^2} = \frac{\epsilon' - \epsilon_\infty}{\epsilon_0 - \epsilon_\infty}. \tag{3-5}$$

Substituting eq. (3-5) into eq. (3-4) and solving for ϵ_∞ yields

$$\epsilon_\infty = \frac{(\epsilon'')^2}{\epsilon' - \epsilon_0} + \epsilon'. \tag{3-6}$$

Substituting eq. (3-5) into eq. (3-3b) yields

$$\epsilon'' = (\epsilon' - \epsilon_\infty)\omega\tau. \tag{3-7}$$

Replacing ϵ_∞ with eq. (3-6) and solving for the relaxation time τ yields

$$\tau = \frac{\epsilon_0 - \epsilon'}{\omega\epsilon''}. \tag{3-8}$$

The values for ϵ_∞ and τ at any temperature can be calculated using eqs. (3-6) and (3-8) from the values of ϵ_0, ϵ', and ϵ'' at that temperature for fixed frequency. The most precise ϵ' and ϵ'' data for water as a function of temperature at a single frequency have been reported by Barajas and Buckmaster [10] and Hu et al. [11]. The temperature dependence of $\epsilon_0(t)$ has been fitted to an exponential function determined by Vidulich et al. [20] using data for the temperature interval from 0°C to 50°C:

$$\epsilon_0(t) = \epsilon_0(0) \exp(bt), \tag{3-9}$$

where $\epsilon_0(0) = 87.9103$, and $b = -0.00458$ (°C^{-1}), and t is in degrees Celsius. The estimated error in $\epsilon_0(t)$ was ±0.05.

Glasstone et al. [15] used the theory of absolute reaction rates to obtain an expression for the temperature dependence of the dielectric relaxation time $\tau(t)$:

$$\tau(t) = [h/k(t + 273.15)] \exp[\Delta G^*/R(t + 273.15)], \tag{3-10}$$

where h is Plank's constant, k is Boltzmann's constant, and R is the gas constant. This temperature dependence has been discussed by Conway [13], Hasted [3], Grant et al. [1], and others. It follows that eq. (3-10) can be used to calculate the Gibbs free energy of activation $\Delta G^*(t)$ for the relaxation process at each temperature if $\tau(t)$ is known. The Gibbs free energy of activation ΔG^* is related to the enthalpy of activation ΔH^* and the entropy of activation ΔS^* by

$$\Delta G^* = \Delta H^* - \Delta S^*(t + 273.15). \tag{3-11}$$

Equation (3-11) can be used to calculate ΔS^* and ΔH^* from ΔG^* if it is assumed that ΔS^* and ΔH^* change only slightly over a small temperature interval $t_1 - t_2$:

$$\Delta G_1^* = \Delta H^* - \Delta S^*(273.15 + t_1) \tag{3-12a}$$

and

$$\Delta G_2^* = \Delta H^* - \Delta S^*(273.15 + t_2), \tag{3-12b}$$

where $t_1 \approx t_2$. Equations (3-12a) and (3-12b) can be used to calculate

$$\Delta S^* = \frac{\Delta G_1^* - \Delta G_2^*}{t_2 - t_1} \tag{3-13a}$$

and

$$\Delta H^* = \frac{\Delta G_1^*(273.15 + t_2) - \Delta G_2^*(273.15 + t_1)}{t_2 - t_1}. \tag{3-13b}$$

3.3. RESULTS AND DISCUSSION

Table 3.1 lists the values for ϵ_∞ obtained by using eq. (3-6) and the ϵ' and ϵ'' values calculated from the attenuation α and phase coefficients measurements at 9.355 GHz reported by Barajas and Buckmaster [10]. Figure 3.1 is a graph of the temperature dependence of these ϵ_∞ values. It also includes $\epsilon_\infty(t)$ calculated using

$$\epsilon_\infty(t) = 5.77 - 2.74 \times 10^{-2}t, \tag{3-14}$$

due to Kaatze [19], which summarizes the results of many complex permittivity measurements at different frequencies from 1.1 GHz to 57 GHz from 0°C to 50°C, reported by many researchers, derived on the basis that the Cole-Cole relaxation time spread factor α is zero. The values in this paper are up to 25% greater than those reported by Kaatze [19] up to 50°C and up to 75°C if eq. (3-14) is assumed to remain valid. No explanation can be offered for the difference except to note that our measurements were at a fixed frequency of 9.355 GHz and should be more reliable because only one set of complex permittivity data was used. The negative values of ϵ_∞ obtained above $\sim$80°C indicate that the data measured at these temperatures are unreliable [10]. This is probably due to the creation of water vapor bubbles in the liquid as the sample approaches the boiling point ($\sim$95°C at 1-km altitude).

Table 3.1 also lists the values of the dielectric relaxation time τ calculated using eq. (3-8) and the same measured data [10]. Figure 3.2 is a graph of these relaxation time values as a function of temperature as well as the values calculated by Barajas and Buckmaster [10]. The relaxation times calculated using

TABLE 3.1. Values of ϵ_∞, τ, and ΔG^* for Double-Distilled Deionized Water at 9.355 GHz

t (°C)	ϵ_∞	τ (ps)	ΔG^* (kJ mol^{-1})	t (°C)	ϵ_∞	τ (ps)	ΔG^* (kJ mol^{-1})
1.13	7.621(42)	17.918(19)	10.5562(25)	46.27	9.008(346)	5.590(35)	9.6047(165)
3.07	7.474(47)	16.709(20)	10.4866(27)	48.26	8.552(384)	5.354(37)	9.5659(184)
4.99	7.327(54)	15.580(20)	10.4138(30)	50.29	8.160(397)	5.130(36)	9.5283(190)
6.96	7.246(61)	14.586(20)	10.3504(32)	52.31	8.259(422)	4.977(38)	9.5228(207)
8.91	7.210(67)	13.700(20)	10.2918(35)	54.26	7.944(482)	4.780(41)	9.4861(235)
10.84	7.216(75)	12.899(21)	10.2361(38)	56.26	7.702(502)	4.606(42)	9.4592(250)
12.82	7.213(86)	12.140(21)	10.1798(41)	58.24	7.300(524)	4.430(42)	9.4252(261)
14.78	7.236(95)	11.465(22)	10.1289(46)	60.25	7.074(563)	4.258(43)	9.3894(280)
16.75	7.283(106)	10.830(22)	10.0774(49)	62.25	7.188(600)	4.130(45)	9.3773(304)
18.73	7.230(115)	10.251(23)	10.0294(53)	64.24	6.086(675)	3.918(47)	9.3017(336)
20.71	7.317(129)	9.713(23)	9.9815(59)	66.22	7.416(684)	3.900(49)	9.3596(354)
22.71	7.428(146)	9.217(25)	9.9379(67)	68.23	4.726(768)	3.589(49)	9.1961(387)
24.69	7.677(154)	8.770(25)	9.8978(70)	70.21	4.614(829)	3.500(51)	9.1943(416)
26.67	8.175(168)	8.419(26)	9.8783(76)	72.25	4.066(885)	3.334(52)	9.1264(448)
27.61	8.311(171)	8.230(25)	9.8603(77)	74.30	4.597(916)	3.270(54)	9.1417(477)
29.59	8.543(188)	7.866(27)	9.8279(85)	76.20	2.491(1.067)	3.069(57)	9.0232(539)
31.57	8.274(207)	7.480(28)	9.7812(94)	78.21	0.729(1.153)	2.920(57)	8.9465(570)
33.55	8.314(232)	7.139(29)	9.7423(105)	80.25	1.356(1.195)	2.861(59)	8.9555(606)
35.57	9.328(240)	6.930(30)	9.7470(111)	82.30	0.720(1.267)	2.750(60)	8.9076(645)
36.29	8.897(242)	6.791(29)	9.7237(110)	84.30	−0.507(1.345)	2.614(60)	8.8237(682)
38.31	9.949(263)	6.550(31)	9.7104(124)	86.31	−2.108(1.469)	2.490(61)	8.7448(732)
40.31	9.167(286)	6.263(32)	9.6727(134)	88.27	−5.359(1.721)	2.320(64)	8.5964(772)
42.30	9.114(303)	6.020(33)	9.6469(142)	90.31	−8.899(2.059)	2.157(68)	8.4417(953)
44.27	8.670(331)	5.756(34)	9.6053(156)	91.33	−15.795(2.632)	1.949(71)	8.1667(1103)

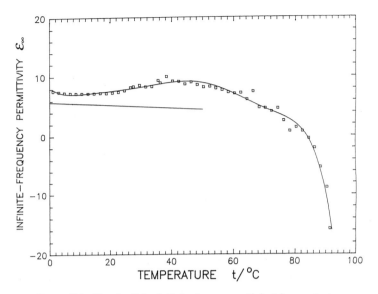

FIGURE 3.1. Graph of the infinite frequency dielectric constant ϵ_∞ as a function of the temperature as obtained using eq. (3-6) and the ϵ' and ϵ'' data reported by Barajas and Buckmaster [10] for double-distilled deionized light water at 9.355 GHz. The graph of $\epsilon_\infty = 5.77 - 2.74 \times 10^{-2}t$ (straight line) obtained by Kaatze [19] is also included for temperatures up to 50°C.

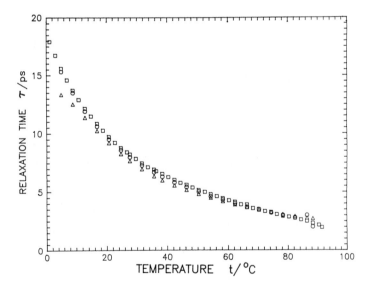

FIGURE 3.2. Graphs of the dielectric relaxation time τ as a function of the temperature for double-distilled deionized light water as obtained from eq. (3-8) (□) as well as from ϵ' data (○) and the ϵ'' data (△) reported by Barajas and Buckmaster [10].

eq. (3-8) are consistent, within 1σ, with those calculated using the ϵ' and ϵ''
data [4], particularly below $\sim 40°$C. This supports the conclusion of Barajas and
Buckmaster [10] that their ϵ' data are more reliable at lower temperatures than
their ϵ'' data. The dielectric relaxation times reported here are 2% to 12% longer
than those of Kaatze [19].

The Gibbs free energy of activation parameter for the relaxation process ΔG^*
can be calculated as a function of temperature from the dielectric relaxation time
by using eq. (3-10). Table 3.1 gives ΔG^* values determined with the values of τ
in this table. Figure 3.3 graphs the values of the free energy of activation given in
Table 3.1 as a function of temperature. It also includes the ΔG^* values derived
from ϵ' and ϵ'' values reported by Barajas and Buckmaster [10]. The new values
of ΔG^* are consistent with those values calculated from their ϵ' data [10] up to
$\sim 80°$C. The values of ΔG^* calculated in this paper deviate significantly from those
reported by Barajas and Buckmaster [10] above $\sim 80°$C.

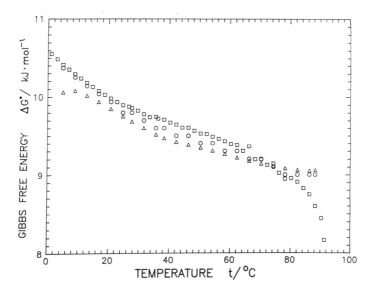

FIGURE 3.3. Graphs of the Gibbs free energy of activation ΔF
as a function of the temperature for double-distilled deionized
light water as derived from τ using eq. (3-10) ($\square$) as well as
from ϵ' data ($\circ$) and the ϵ'' data ($\triangle$) reported by Barajas and
Buckmaster [10].

The values for the Gibbs free energy of activation parameter ΔG^* for the
relaxation process given in Table 3.1 were then used with eqs. (3-13a) and (3-13b)
to calculate the entropy of activation, ΔS^*, and the enthalpy of activation, ΔH^*,
parameters respectively. Table 3.2 lists the calculated values for these parameters.
The temperatures given are the average of t_1 and t_2. The temperature dependence
of the ΔH^* and ΔS^* activation parameters in Table 3.2 as well as the values
derived from ϵ' and ϵ'' data reported by Barajas and Buckmaster [4] are graphed

TABLE 3.2 Values of ΔH^* and Entropy ΔS^* for Double-Distilled Deionized Water

t (°C)	ΔH^* (kJ mol^{-1})		ΔS^* (J mol^{-1} K^{-1})		t (°C)	ΔH^* (kJ mol^{-1})		ΔS^* (J mol^{-1} K^{-1})	
	Raw	Smoothed	Raw	Smoothed		Raw	Smoothed	Raw	Smoothed
2.10	20.396(522)	22.320	35.88(1.90)	42.78	47.27	15.833(3.978)	13.409	19.50(12.42)	11.91
4.03	20.960(582)	21.508	37.92(2.10)	39.84	49.28	15.519(4.201)	13.524	18.22(13.03)	12.56
5.98	19.365(622)	20.747	32.18(2.23)	37.11	51.30	10.409(4.512)	13.704	2.72(13.91)	12.82
7.94	18.768(683)	20.021	30.05(2.43)	34.51	53.29	15.648(5.241)	13.920	18.82(16.06)	13.49
9.88	18.432(757)	19.332	28.86(2.68)	32.07	55.26	13.890(5.633)	14.170	13.45(17.16)	14.25
11.83	18.311(804)	18.658	28.43(2.82)	29.70	57.25	15.116(6.030)	14.500	17.17(18.25)	15.25
13.80	17.606(902)	18.029	25.97(3.14)	27.50	59.25	15.328(6.329)	14.870	17.81(19.04)	16.37
15.77	17.656(985)	17.448	26.14(3.41)	25.48	61.25	11.407(6.908)	15.297	6.05(20.66)	17.65
17.74	17.105(1.060)	16.884	24.24(3.65)	23.54	63.25	22.119(7.657)	15.765	37.99(22.77)	19.05
19.72	17.091(1.172)	16.353	24.19(4.00)	21.72	65.23	−0.5644(8.340)	16.275	−29.24(24.65)	20.56
21.71	16.388(1.316)	15.878	21.80(4.46)	20.10	67.23	81.343(8.879)	16.846	81.34(26.09)	22.24
23.70	15.930(1.452)	15.430	20.25(4.89)	18.59	69.22	9.506(9.823)	17.496	0.91(28.70)	24.14
25.68	12.831(1.559)	15.039	9.84(5.22)	17.27	71.23	20.623(10.318)	18.178	33.28(29.97)	26.13
27.14	15.620(3.456)	14.740	19.15(11.51)	16.28	73.28	6.549(11.056)	18.925	−7.46(31.92)	28.29
28.60	14.782(1.747)	14.508	16.36(5.79)	15.51	75.25	30.812(13.194)	19.738	62.37(37.88)	30.63
30.58	16.968(1.943)	14.218	23.59(6.40)	14.55	77.21	22.354(13.672)	20.508	38.16(39.03)	32.94
32.56	15.768(2.175)	13.956	19.65(7.11)	13.69	79.23	7.396(14.368)	21.475	−4.41(40.78)	35.59
34.56	9.029(2.327)	13.736	−2.33(7.56)	12.97	81.28	17.213(15.298)	22.448	23.37(43.17)	38.34
35.93	19.738(6.708)	13.591	32.36(21.70)	12.50	83.30	23.819(16.727)	23.500	41.95(46.93)	41.30
37.30	11.761(2.546)	13.507	6.58(8.206)	12.23	85.31	22.855(17.839)	24.600	39.25(49.77)	44.38
39.31	15.581(2.852)	13.405	18.85(9.13)	11.90	87.29	35.961(19.561)	25.722	75.71(54.28)	47.50
41.31	13.737(3.085)	13.329	12.96(9.81)	11.66	89.29	36.004(21.777)	26.927	75.83(60.12)	50.83
43.29	16.308(3.387)	13.302	21.12(10.71)	11.57	90.82	106.433(52.004)	27.871	269.61(142.90)	53.43
45.27	9.701(3.615)	13.342	0.30(11.35)	11.70					

in Figure 3.4 and Figure 3.5 respectively. The values of these parameters at lower temperatures are consistent with those values derived from the ϵ' data but not the ϵ'' data. Above $\sim 20°$C the values for both thermodynamic parameters scatter and no trend is discernible. This scatter occurs because small irregularities in the ΔG^* data are amplified when the ΔS^* and ΔH^* values are calculated. This scatter can be eliminated by fitting the temperature variation of the ΔG^* to a seventh-order polynomial. The smoothed values for ΔG^* were used to calculate the enthalpy of activation ΔH^* and the entropy of activation ΔS^*. This procedure resulted in more reliable values and graphs for ΔH^* and ΔS^*, which are given in Table 3.2 and Figures 3.4 and 3.5 respectively. These graphs clearly illustrate that these thermodynamic parameters have a parabolic temperature dependence with minima near $\sim 43°$C. This minima is consistent with previous results found by Barajas and Buckmaster [10]. Neutron scattering and x-ray diffraction experiments by Ohtomo et al. [21] and references therein indicate that water, between 20°C and 90°C, is more crystal-like at lower temperatures and more liquid-like at higher temperatures. The former may consist of tetrahedral pentamers and the latter monomers. The minimum detected in ΔS^* and ΔH^* is consistent with a model of water consisting primarily of tetrahedral pentamers below $\sim 43°$C and primarily of nonbonded monomers above $\sim 43°$C. However, recent measurements reported by Mashimo and co-workers [22], and in references therein, using time-domain reflectometry, support a hexagonal model for light water at low temperatures which

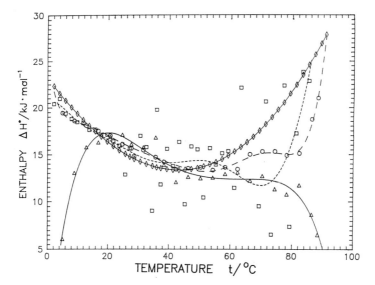

FIGURE 3.4. Graphs of the enthalpy of activation ΔH^* as a function of the temperature for double-distilled deionized light water as derived from raw ΔG^* data ($\square$) and smoothed ΔG^* data ($\diamond$) using eq. (3-13b), as well as from ϵ' data ($\circ$) and the ϵ'' data ($\triangle$) reported by Barajas and Buckmaster [10].

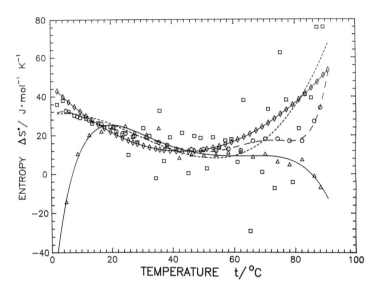

FIGURE 3.5. Graphs of the entropy of activation ΔS^* as a function of the temperature for double-distilled deionized light water as derived from raw ΔG^* data (□) and smoothed ΔG^* data (◇) using eq. (3-13a), as well as from ϵ' data (○) and the ϵ'' data (△) reported by Barajas and Buckmaster [10].

is similar to that for ice near $0°C$ as well as a weaker higher-order structure involving one water molecule with 30 neighbors similar to an ordinary ice lattice.

One advantage of this approach of analyzing complex permittivity data is that $\epsilon_\infty(t)$ and $\tau(t)$ can be calculated at each temperature from the values of $\epsilon_0(t)$, $\epsilon'(t)$, and $\epsilon''(t)$, where the latter two are determined at a single frequency. This shows that it is very important to have accurate, precise, and comprehensive data sets for these three parameters as a function of temperature. It is straightforward to derive the Gibbs free energy ΔG^*, enthalpy ΔH^*, and entropy ΔS^* of activation parameters from the dielectric relaxation time τ. The results indicate clearly that ΔH^* and ΔS^* for the relaxation process are temperature-dependent. Further investigations are required to determine the frequency dependence of these parameters. The need for comprehensive complex permittivity data sets from the point of view of temperature and frequency dependence is crucial to obtaining a clearer understanding of the molecular processes and structures of water.

3.4. CONCLUSION

It has been demonstrated that the temperature dependence of $\epsilon_\infty(t)$ and $\tau(t)$ can be calculated directly from $\epsilon'(t)$ and $\epsilon''(t)$ data using Debye theory, provided $\epsilon_0(t)$ is known. It has also been confirmed that ΔG^*, ΔH^*, and ΔS^* are temperature-dependent, with the latter two having minima at $\sim43°C$, which suggests that the transition of water from a crystal-like structure to a liquid-like one occurs near this temperature minimum. The Gibbs free energy decreases monotonically by $\sim 10\%$ over the measurement temperature interval. It should be emphasized that it has been assumed that the Debye model is valid for light water. However, it has been shown previously [10,11] that the values of $\tau(t)$, $\Delta G^*(t)$, $\Delta H^*(t)$, and $\Delta S^*(t)$ are not very sensitive to whether the analysis assumes the Debye model or the Cole-Cole modification with $\alpha = 0.013$ [23]. The advantage of using the former model is that $\epsilon_\infty(t)$ can be calculated from single-frequency data sets for $\epsilon'(t)$ and $\epsilon''(t)$. The values of $\tau(t)$ are in good agreement with those reported by other workers.

ACKNOWLEDGMENT

This research was supported by research grants awarded by The Natural Sciences and Engineering Research Council of Canada (A00000716) and The University of Calgary to one of us (H. A. B.).

REFERENCES

[1] E. H. Grant, R. J. Sheppard, and G. P. South, *Dielectric Behavior of Biological Molecules in Solution*, Oxford: Clarendon Press, 1978.

[2] G. H. Collie, J. B. Hasted, and D. M. Ritson, "The dielectric properties of water and heavy water," *Proc. Phys. Soc.*, vol. 60, pp. 145–160, Feb. 1948.

[3] J. B. Hasted, *Aqueous Dielectrics*, London: Chapman and Hall, 1973, pp. 32–64.

[4] J. B. Hasted, "Liquid water dielectric properties," in *The Physics and Physical Chemistry of Water: A Comprehensive Treatise*, vol. 1, F. Franks, Ed., New York: Plenum Press, 1972, pp. 255–309.

[5] H. A. Buckmaster, "Precision microwave complex permittivity measurements of high loss liquids," *J. Electromagn. Waves Appl.*, vol. 4, pp. 645–659, 1990.

[6] H. A. Buckmaster, "9 GHz Instrumentation System For High Precision Complex Permittivity Measurements of High Loss Liquids," *Trends Microwave Theory Tech.*, vol. 1, pp. 87–101, 1991.

[7] J. G. McAvoy and H. A. Buckmaster, "The 9 GHz complex permittivity of water at 293 K and 298 K," *J. Phys. D: Appl. Phys.*, vol. 16, pp. 2519–2523, 1983.

[8] H. Zaghloul and H. A. Buckmaster, "The complex permittivity of water at 9.356 GHz from 10 to 40°C," *J. Phys. D: Appl. Phys.*, vol. 18, pp. 2109–2118, 1985.

[9] T. H. T. van Kalleveen and H. A. Buckmaster, "The 9.355 GHz complex permittivity of water in the temperature interval 0–10°C," *Can. J. Chem.*, vol. 66, pp. 672–675, 1988.

[10] O. Barajas and H. A. Buckmaster, "9.355 GHz complex permittivity of water from 1°C to 90°C," *J. Phys.: Condens. Matter*, vol. 4, pp. 8671–8682, 1992.

[11] X. Hu, H. A. Buckmaster, and O. Barajas, "The 9.355 GHz complex permittivity of light and heavy water from 1°C to 90°C using an improved high precision instrumentation system," *J. Chem. Eng. Data*, vol. 39, pp. 625–638, 1994.

[12] J. G. McAvoy and H. A. Buckmaster, "A new, high precision 9 GHz dual sample cell complex permittivity instrumentation system for high loss liquids," *J. Phys. E: Sci. Instrum.*, vol. 18, pp. 244–249, 1985.

[13] B. E. Conway, "Some observations on the theory of dielectric relaxation in associated liquids," *Can. J. Chem.*, vol. 37, pp. 613–628, 1959.

[14] H. J. Steinhoff, A. Redhardt, K. Lieutenant, W. Chrost, G. Hess, J. Schlitter, and H. J. Neumann, "High precision measurements of the permittivity of water in the microwave range," *Z. Naturforsch.*, vol. 45a, pp. 677–686, 1990.

[15] S. Glasstone, K. J. Laidler, and H. Eyring, *The Theory of Rate Processes*, New York: McGraw-Hill, 1941.

[16] P. Debye, *Polar Molecules*, Cleveland: Chemical Catalog Company, 1929, p. 91.

[17] H. Frohlich, *Theory of Dielectrics — Dielectric Constant and Dielectric Loss*, 2nd ed., Oxford: Clarendon Press, 1958, pp. 72ff, 137ff.

[18] K. S. Cole and R. H. Cole, "Dispersion and absorption in dielectrics," *J. Chem. Phys.*, vol. 9, pp. 341–351, April 1941.

[19] U. Kaatze, "Complex permittivity of water as a function of frequency and temperature," *J. Chem. Eng. Data*, vol. 34, no. 4, pp. 371–374, 1989.

[20] G. A. Vidulich, D. F. Evans, and R. L. Kay, "The dielectric constant of water and heavy water between 0 and 40°C," *J. Phys. Chem.*, vol. 71, pp. 656–662, Feb. 1967.

[21] N. Ohtomo, K. Tokiwano, and K. Arakawa, "The structure of liquid water by neutron scattering. III: Calculation of the partial structure factors," *Bull. Chem. Soc. Jpn.*, vol. 57, no. 2, pp. 329–333, 1984.

[22] S. Mashimo and N. Miura, "Higher order and local structure of water determined by microwave dielectric study," *J. Chem. Phys.*, vol. 99, no. 12, pp. 9874–9881, 1993.

[23] P. R. Mason, J. B. Hasted, and L. Moore, "The use of statistical theory in fitting equations to dielectric dispersion data," *Adv. Mol. Relax. Process.*, vol. 6, no. 3, pp. 217–232, 1974.

P. Pissis
A. Anagnostopoulou-Konsta
L. Apekis
A. Kyritsis
*National Technical University of Athens,
Athens, Greece*

R. Pelster
A. Enders
G. Nimtz
Universitat Köln, Köln, Germany

4

Dielectric Properties of Water Dispersed and Confined in Different Systems

Abstract.   We report on dielectric relaxation spectroscopy measurements on water confined in small volumes of mesoscopic dimensions in three different systems: phospholipid dimyristoilphosphatidylcholine bilayers (hydrophilic, one-dimensional confinement in water layers), poly(hydroxyethyl acrylate) hydrogels (hydrophilic, two-dimensional confinement in pores), and butyl rubber (hydrophobic, three-dimensional confinement in droplets). We use broadband ac techniques in the 5 Hz–10 GHz frequency range and thermally stimulated depolarization currents techniques in the 77–300 K temperature range. Additional information on the structure and the properties of water and the matrix itself is obtained from differential scanning calorimetry, small-angle X-ray scattering, and dynamic and equilibrium water sorption and desorption measurements. The rotational mobility of water molecules is found to be reduced in the phospholipid bilayers and the hydrogels, compared to that of bulk water, and to be enhanced in the BR-water system.

4.1.   INTRODUCTION

This work deals with detailed investigations of changes in the structure and the properties of water induced by dispersion and confinement in small volumes [1,2] in three systems: dimyristoilphosphatidylcholine (DMPC) bilayers, poly(hydroxyethyl acrylate) (PHEA) hydrogels, and butyl rubber (BR) containing hydrophilic components. Different dielectric relaxation spectroscopy (DRS) techniques are used: broadband ac techniques in the 5-Hz–10-GHz frequency range and thermally stimulated depolarization currents (TSDC) techniques in the 77–300 K temperature range. The latter correspond to measuring dielectric losses versus temperature at fixed frequencies of 10^{-2}–10^{-4} Hz and are characterized by high sensitivity, high resolving power, and special procedures for experimentally analyzing complex relaxation processes [3,4]. Additional information on the structure and properties of water and the matrix itself is obtained from differential scanning calorimetry (DSC), small-angle X-ray scattering (SAXS), and dynamic and equilibrium water sorption and desorption measurements.

Several aspects of hydration [5] and water confinement have been taken into account in selecting the systems. DMPC bilayers are characterized by hydrophilic interactions of water molecules with the head groups and one-dimensional confinement of water in layers [6–8]. In PHEA hydrogels we have strong hydrophilic interactions of water with polar side-chain groups and two-dimensional confinement of water in pores [9,10]. In hydrophobic butyl rubber, water diffuses into the rubber to the hydrophilic components (metal oxides and salts), which act as adsorption sites, and forms three-dimensionally confined mesoscopic water droplets [11,12].

Information from DRS concerning the structure and properties of water in different systems is obtained from the following two regions of the relaxation spectra.

1. From the reorientation of water molecules themselves [13] which occurs in the MHz–GHz frequency range in the liquid phase in ac measurements and at 110–150 K in the solid-ice phase in TSDC measurements. The characteristics of this relaxation (relaxation time, activation energy, entropy factor) reflect the influence of the surrounding on the reorientating water molecules. In subzero measurements, in the solid-ice phase, aspects of freezing come into play and have to be taken into account in the interpretation of the results [4].

Since we will discuss the above-mentioned characteristics of the relaxations, but not absolute values of the complex permittivity, we do not have to apply effective medium formulas to separate the dielectric properties of the matrix from those of water. Such an analysis would yield the same peak positions for the dispersed water.

2. From the influence of water on the relaxation and conductivity mechanisms of the matrix itself (main-chain relaxation, side-chain relaxations, Maxwell-Wagner, and space-charge relaxations). It has been shown that this influence depends in a very sensitive way on the binding modes of water molecules (expressed, e.g., in terms of tightly bound, loosely bound, and free water molecules [5]) and can be used to obtain relevant information [4].

4.2. EXPERIMENTAL

Details of the preparation of the samples and the adjustment of water content, h, have been given elsewhere [8,10,12]. The quantity h is defined as grams of water per gram of dry sample (dry basis). It was varied up to about 0.5 in the phospholipid DMPC bilayers and the PHEA hydrogels, and up to 0.22 in BR.

Three ac techniques have been used to measure the complex permittivity $\epsilon = \epsilon' - j\epsilon''$ as a function of frequency and temperature:

   1. A novel broadband measurement and calibration method from 5 Hz to 2 GHz at 25°C [13]. A condenser-like measuring cell is placed in a trans-

mission line, and the transmission coefficient of the setup is measured with two network analyzers (HP 3577B for 5 Hz–200 MHz and HP 8510C for 200 MHz–2 GHz).

2. A commercial reflection method (HP 85070A dielectric probe) from 100 MHz to 10 GHz at 25°C. The sample represents the termination of an open coaxial line, the reflection coefficient of which is measured (network analyzer HP 8510C).

3. A microwave bridge at a fixed frequency of 9.1 GHz from 90 to 300 K [14].

For details of the measurements and the evaluation procedures see Refs. 12–14.

The TSDC method consists of measuring the thermally activated release of stored dielectric polarization. The sample is polarized by a dc electric field and then cooled to a sufficiently low temperature (liquid nitrogen temperature, 77 K, in our measurements) to freeze in the polarization. The field is then switched off and the sample is warmed up at a constant rate while the depolarization current, as the polarization relaxes, is measured. Thus, for each polarization mechanism, an inherent current peak is detected. The theory, apparatus, and procedures used to determine the parameters characterizing the dielectric behavior of a sample have been described elsewhere [3,4].

4.3.   RESULTS AND DISCUSSION

4.3.1   Phospholipid Bilayers

In the DMPC-water system the water content, h, was varied up to about 0.50, corresponding to approximately $n = 18$ molecules of water per molecule of DMPC. Detailed investigations of the ac dielectric response of phospholipid bilayers in the gigahertz frequency region [7] have shown that the rotational mobility of the water molecules in the bilayers is reduced compared to that of bulk water, with the degree of reduction decreasing with increasing h. Figure 4.1 shows the low-temperature (LT) TSDC peak at about 140 K measured on a sample with $h = 0.44$. The peak shifts slightly to higher temperatures with decreasing h, from 140 K for fully hydrated samples ($h = 0.49$) to 144 K for $h = 0.10$. In analogy to many other hydrated systems [4] the peak is attributed to the reorientation of water molecules in the frozen water clusters around the primary hydration sites and layers.

The LT TSDC peak in macroscopic polycrystalline pure ice due to the reorientation of water molecules is located at 119 K [4], so our results suggest a restriction of the rotational mobility of water molecules in the bilayers. This restriction increases slightly with decreasing h.

The LT TSDC peak is significantly broader (by a factor of about 2 [8]) in the bilayers than in bulk ice, suggesting a relatively broad distribution of dielectric relaxation times for water in the bilayers, in agreement with a physically realistic picture: obviously, the reorientating water molecules in the clusters have

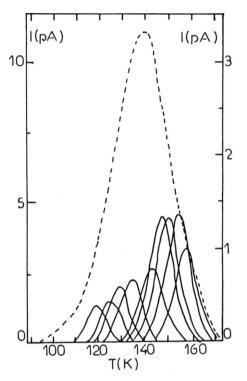

FIGURE 4.1. The LT TSDC peak (- - -, left current scale) with
water content $h = 0.44$ and thermal sampling responses isolated in this peak (__, right current scale).

different microscopic environments. Thermal sampling (TS) and partial heating
(PH) techniques were extensively used to investigate this question. The TS technique consists of sampling the relaxation processes within a narrow temperature
window by polarizing the sample, during cooling, only within this temperature
window [3]. In Figure 4.1 we show TS responses isolated in the LT peak. The polarization window was 5 K while the temperature was scanned in steps of 5 K. By
the PS technique the sample was polarized as usual and then partially depolarized
by partial heatings (separated by rapid cooling) up to a series of temperatures that
spanned the whole temperature range of the peak [3].

Figure 4.2 shows the activation energy W of the TS responses, calculated by
fitting the expression for the depolarization current density to the peaks [3,4], and
of the PH responses, calculated by the initial rise method [3]. The results suggest
broad and continuous distributions with mean values of W higher in the bilayers
(0.35 and 0.40 eV from TS and PH experiments, respectively) than in bulk ice
(0.31 eV [8], in agreement with ac measurements [16]).

The high sensitivity of the TSDC technique allows us to accurately determine
that the LT peak in the bilayer appears only for water contents higher than a
critical one, $h_c = 0.08$, and that the normalized magnitude I_n of the peak (i.e.,

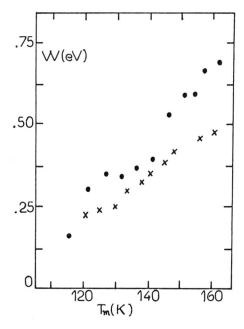

FIGURE 4.2. Activation energy W vs. peak temperature T_m of the TS responses (x) or PH cut temperature (.) for a DMPC sample with $h = 0.32$.

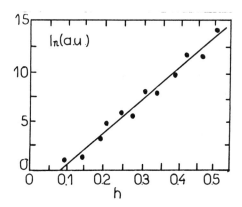

FIGURE 4.3. Normalized magnitude I_n of the LT TSDC peak in DMPC vs. water content h. The line serves as a guide.

the maximum depolarization current divided by heating rate and polarizing field), which is a measure of the number of relaxing units contributing to the peak [3,4], increases linearly with h (Figure 4.3). The first result suggests that water up to $h_c = 0.08$, corresponding to $m_c = 3.0$ molecules of water per molecule of DMPC, are irrotationally (tightly) bound in primary hydration sites. This fraction does not freeze during cooling [17]. It remains open whether it undergoes a glass transition like bound water in other systems [18]. The quantity h_c is approximately equal to the h-value where the adsorption isotherms in DMPC, of which an example at 70°C is shown in Figure 4.4, start their upward swing, indicating the beginning of the creation of clusters [5]. The linear increase of I_n with h in Figure 4.3 suggests that the polarizability of water molecules is independent of h in the range $h = 0.08–0.49$.

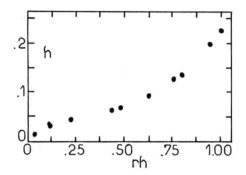

FIGURE 4.4. Water adsorption isotherms (water content h vs. relative humidity, rh) for DMPC at 70°C.

4.3.2 PHEA Hydrogels

In the PHEA hydrogels h was varied up to about 0.5. Broadband ac measurements at 25°C (Figure 4.5) show an overall increase of ϵ' and ϵ'' with h. In Figure 4.6 we show the ac response at 25°C of a sample at two water contents in the frequency range of interest for the dielectric relaxation molecules themselves (0.1– 10.0 GHz). The maxima in ϵ'' might be attributed to relaxations of loosely bound water molecules shifting to higher frequencies with increasing h toward that of bulk water (at 19.2 GHz at 25°C). However, there is some doubt about this interpretation,

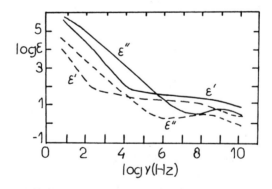

FIGURE 4.5. Logarithmic plot of the complex dielectric permittivity $\epsilon' - j\epsilon''$ for a PHEA sample at two different water contents $h = 0.42$ (___) and 0.22 (- - - -) at 25°C.

because side-chain relaxations (γ and β_{sw} [9,19]) which are influenced by water might contribute to the dispersions in Figure 4.6.

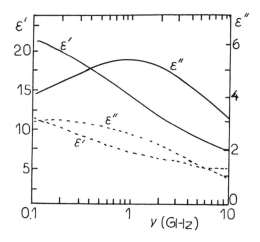

FIGURE 4.6. Semilogarithmic plot of the complex dielectric permittivity $\epsilon' - j\epsilon''$ at high frequencies for a PHEA sample at two different water contents, $h = 0.42$ (____) and 0.02 (- - - -) at 25°C.

To further clarify this point we measured the TSDC thermograms of PHEA samples by systematically varying h. The thermograms consist of two dispersions: a complex dispersion at low temperatures (dispersion I) and a dispersion at higher temperatures (dispersion II), which shifts significantly to lower temperatures with increasing h and is associated with the glass transition [10]. In Figure 4.7 we show dispersion I at four h values. It consists of two contributions, γ and β_{sw}, in order of increasing temperature [10,19]. With increasing h the magnitude of β_{sw} increases, while that of γ decreases and β_{sw}, which dominates for $h \geq 0.05$, shifts to lower temperatures. However, Figure 4.8 shows that there is a critical water content of about 0.30, where there is an increase in the slope of the I_n-h plot, suggesting that an additional relaxation mechanism contributes to dispersion I at $h \geq 0.30$. We interpret these changes by assuming that, for $h \geq 0.30$, clusters of loosely bound water molecules are formed around the primary hydration sites and the water molecules in these clusters contribute by their reorientation to the ac response at 25°C (Figure 4.6) and to the TSDC response (Figures 4.7 and 4.8). In both cases the relaxation of loosely bound water molecules is slower than the relaxation of water molecules in bulk water and bulk ice respectively.

Support for this interpretation comes from three observations: (1) A detailed investigation by DRS and DSC of the shift of the glass transition temperature T_g to lower temperatures with increasing h shows that the rate of this shift becomes significantly reduced for $h \geq 0.30$ [20], suggesting that water up to this h-value is molecularly distributed, whereas the excess above that forms clusters. (2) Detailed DSC investigations and specific heat measurements of the PHEA hydrogels

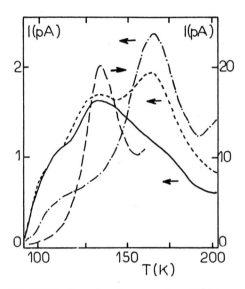

FIGURE 4.7. TSDC dispersion I measured on PHEA samples
with $h = 0.01$ (___), 0.02 (- - - -), 0.10 (_.._), and 0.36 (_ _ _).

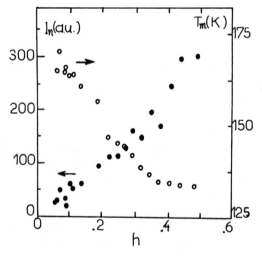

FIGURE 4.8. Normalized amplitude I_n (.) (i.e., maximum depolarization current I_m divided by heating rate b) and peak
temperature T_m (o) of dispersion I in PHEA vs. water content h.

at different h-values suggest that the hydrogels behave like homogeneous mixtures
for $h \leq 0.20$, while for $h \geq 0.20$–0.30 a pure water phase coexists with the homogeneous mixtures and crystallizes at low temperatures [20]. (3) Water sorption
isotherm measurements on PHEA suggest molecular distribution of water at low
h-values and formation of clusters at higher h-values [21].

4.3.3 Butyl Rubber (BR)

The water content in the BR samples was varied up to 0.22. The samples contained hydrophilic metal oxides which act as adsorption sites for water diffusing into the rubber matrix. DSC measurements show, upon cooling, two exothermic minima at about $-35°C$ and $-43°C$, which are attributed to freezing of mesoscopic water droplets of two sizes dispersed in the rubber matrix [12]. SAXS experiments indicate the existence of two kinds of droplets with radii about 2 nm [12].

Figure 4.9 shows ϵ' and ϵ'' for a BR sample with $h = 0.195$ at a fixed frequency of 9.1 GHz as a function of temperature during cooling and heating and for a dry sample. Comparison with measurements of the temperature dependence of ϵ' and ϵ'' in bulk water [22] suggests that (1) the rapid changes occurring at about $-40°C$ during cooling and at about $0°C$ during heating are due to freezing-melting processes in the water droplets and that (2) the frequency of maximum dielectric losses ν_{max} of water in the droplets has been shifted to higher frequencies (compared to 19.2 GHz at $25°C$ for bulk water). Measurements up to 10 GHz at $25°C$ support this latter result by showing no indications of the beginning of a relaxation. This shifting of the relaxation of water dispersed in BR to higher frequencies compared to bulk water must be attributed to a decrease of the mean number of intact hydrogen bonds resulting from the combined effects of confinement in small droplets (large surface-to-volume ratio) and hydrophobic matrix (water molecules on the surface of the droplets not bound to the matrix).

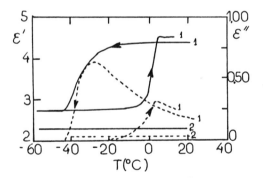

FIGURE 4.9. Real ϵ' (___) and imaginary ϵ'' (- - - -) part of the complex dielectric permittivity at 9.1 GHz vs. temperature for BR with water content $h = 0.195$ (1) and for a dry BR sample (2).

TSDC measurements on BR show an LT peak at 138 K. The shift to higher temperatures compared to bulk ice (119 K [4]) is attributed to the high concentration of physical imperfections due to supercooling breakdown, similar to ice microcrystals dispersed in oil [23]. The plot of the normalized magnitude of the LT TSDC peak in BR versus h (Figure 4.10) indicates that there is no molecularly distributed nonfreezable water. Support for this result comes from measurements of T_g by DSC and DRS, which show T_g to be independent of h.

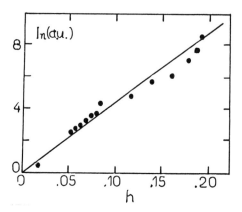

FIGURE 4.10. Normalized amplitude I_n of the LT TSDC peak in BR vs. water content h. The line serves as a guide.

TSDC measurements on BR samples with H_2O replaced by D_2O (Figure 4.11) show a shift of the LT peak to higher temperatures by about 7 K, confirming the assignment of the peak to the relaxation of water molecules in the ice nanocrystals [4].

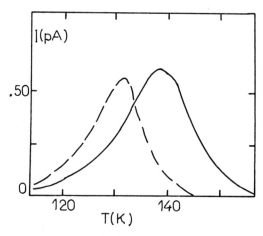

FIGURE 4.11. The LT TSDC peak measured on two BR samples with 0.04 H_2O (- - - -) and 0.04 D_2O (___).

The shape of the peak does not reflect the existence of two sizes of water droplets, indicated by DSC and SAXS. However, TSDC measurements with the samples being annealed prior to measurement at subzero temperatures, after freezing the water droplets of one size (cooling to $-38°C$) [24], show two peaks (Figure 4.12). The peak at lower temperatures is assigned to droplets frozen by cooling to $-38°C$ and subsequently annealed, so the concentration of physical imperfections is reduced and the peak shifts toward that of bulk ice [24]. The peak at higher temperatures is assigned to the other set of droplets. From the relative magnitudes of the two peaks the ratio of the amount of water in the two sets of droplets can be determined to be about 2:1, independently of h, in agreement with measurements [12].

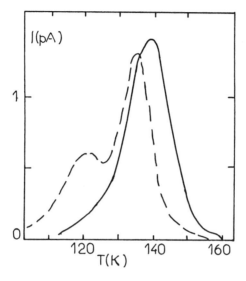

FIGURE 4.12. Evolution of the LT TSDC peak measured on a BR sample with $h = 0.065$ to a double peak (- - - -) by cooling the sample to $-38°C$ and annealing at $-7°C$ for 45 min prior to the TSDC measurement.

4.4. CONCLUSIONS

The measurements show significant differences in the dielectric properties of water in phospholipid DMPC bilayers and PHEA hydrogels on the one hand, and in BR-containing hydrophilic components, on the other. We think that these reflect differences in the organization of water in the two groups of systems, indicated also by the results of measurements by other techniques, mainly DSC. In the phospholipid bilayers and the hydrogels a significant fraction of water is molecularly distributed and tightly bound at primary hydration sites. This fraction does not crystallize upon cooling and does not contribute by reorientation to the dielectric spectra. In BR, on the other hand, no fraction of water is molecularly distributed; all water present in the samples crystallizes upon cooling and contributes by reorientation to the dielectric spectra. In the bilayers and the hydrogels the rotational mobility of water in excess of the tightly bound fraction is reduced compared to bulk water. There is no fraction of dielectrically free (bulk) water up to the highest water contents of about 0.5 (w/w). In BR, however, the rotational mobility of water is enhanced compared to bulk water. These differences are attributed to the presence of the hydrophilic head groups in the phospholipids and the hydrophilic side chains in the hydrogels in contrast to the hydrophobic BR matrix rather than to the different dimensionalities of the confinement. However, the enhancement of the rotational mobility of water in BR above that of bulk water might be a combined effect of three-dimensional mesoscopic confinement in a hydrophobic matrix. Measurements on other selected systems are needed prior to generalizing the results.

ACKNOWLEDGMENT

We gratefully acknowledge financial support by the Volkswagen-Stiftung (Project I/68 356).

REFERENCES

[1] J. M. Drake and J. Klafter, "Dynamics of confined molecular systems," *Phys. Today*, vol. 43, pp. 46–44, 1990.

[2] J. C. Li, D. K. Ross, P. L. Hall, and R. K. Heenan, "Small angle neutron scattering studies of adsorbed water in porous vycor glass," *Phys. B*, vol. 156–157, pp. 185–188, 1989.

[3] J. van Turnhout, "Thermally stimulated discharge of electrets," in *Electrets, Topics in Applied Physics*, vol. 33, G. M. Sessler, Ed., Berlin: Springer, 1980, ch. 3.

[4] P. Pissis, A. Anagnostopoulou-Konsta, L. Apekis, D. Daoukaki-Diamanti, and C. Christodoulides, "Dielectric effects of water in water-containing systems," *J. Non-Cryst. Solids*, vol. 131–133, pp. 1174–1181, 1991.

[5] I. D. Kuntz, Jr., and W. Kauzmann, "Hydration of proteins and polypeptides," *Adv. Protein Chem.*, vol. 28, pp. 239–345, 1974.

[6] D. Exerowa and D. Kashchiev, "Hole-mediated stability and permeability of bilayers," *Contemp. Phys.*, vol. 27, pp. 172–177, 1986.

[7] G. Nimtz, "Magic numbers of water molecules bound between lipid bilayers," *Phys. Scripta*, vol. T13, pp. 173–177, 1986.

[8] P. Pissis, A. Enders, and G. Nimtz, "Hydration dependence of molecular mobility in phospholipid bilayers," *Chem. Phys.*, vol. 171, pp. 285–292, 1993.

[9] K. Pathmanathan and G. P. Johari, "Dielectric and conductivity relaxations in poly(hema) and of water in its hydrogels," *J. Polym. Sci. Polym. Phys.*, vol. 28, pp. 675–689, 1990.

[10] A. Kyritsis, P. Pissis, J. L. Gomez Ribelles, and M. Monleon Pradas, "Hydration properties of PHEA/Kevlar composites studied by dielectric, DSC and sorption isotherm measurements," in *Proc. 7th Int. Symp. on Electrets*, R. Gerhard-Multhaupt, W. Kunstler, L. Brehmer, and R. Danz, Eds., New York: IEEE, 1991, pp. 215–220.

[11] E. Southern and A. G. Thomas, "Diffusion of water in rubbers," in *Water in Polymers*, S. P. Powland, Ed., Washington: American Chemical Society, 1980, pp. 375–386.

[12] R. Pelster, A. Krops, G. Nimtz, A. Enders, H. Kietzmann, P. Pissis, A. Kyritsis, and D. Woermann, "On mesoscopic water droplets dispersed in butyl rubber," *Ber. Bunsenges. Phys. Chem.*, vol. 97, pp. 666–675, 1993.

[13] R. Pelster, Ph.D. thesis, Koeln 1993 and International Patent Application PCT/EP 92/02711 (25.11.92).

[14] A. Enders, "An accurate measurement technique for line properties, junction effects, and dielectric and magnetic material parameters," *IEEE Trans.*, vol. MTT 37, pp. 598–605, 1989.

[15] U. Kaatze and R. Pottel, "Dielectric and ultrasonic spectroscopy of liquids. Comparative view for binary aqueous solutions," *J. Mol. Liq.*, vol. 49, pp. 225–248, 1991.

[16] A. Enders and G. Nimtz, "Dielectric relaxation study of dynamic properties of hydrated phospholipid bilayers," *Ber. Bunsenges. Phys. Chem.*, vol. 88, pp. 512–517, 1984.

[17] M. Grunert, L. Borngen, and G. Nimtz, "Structural phase transitions due to a release of bound water in phospholipid bilayers at temperatures below $0°C$," *Ber. Bunsenges. Phys. Chem.*, vol. 88, pp. 608–612, 1984.

[18] F. X. Quinn, V. J. McBrierty, A. C. Wilson, and G. D. Friends, "Water in hydrogels. 3. Poly(hydroxyethyl methacrylate)/saline solution systems," *Macromolecules*, vol. 23, 4576–4581, 1990.

[19] J. L. Gomez Ribelles, J. M. Meseguer Duenas, and M. Monleon Pradas, "Dielectric relaxations in poly(hydroxyethyl acrylate): Influence of the absorbed water," *Polymer*, vol. 29, 1124–1127, 1988.

[20] A. Kyritsis, P. Pissis, J. L. Gomez Ribelles, and M. Monleon Pradas, "Dielectric relaxation spectroscopy in poly(hydroxyethyl acrylate) hydrogels," *J. Non-Cryst. Solids*, vol. 172, pp. 1041–1046, 1994.

[21] A. Kyritsis, unpublished results.

[22] O. Barajas and H. A. Buckmaster, "9.355 GHz complex permittivity of water from $1°C$ to $90°C$," *J. Phys.: Condens. Matter*, vol. 4, pp. 8671–8682, 1992.

[23] P. Pissis, L. Apekis, C. Christodoulides, and G. Boudouris, "Dielectric study of dispersed ice microcrystals by the depolarization thermocurrent technique," *J. Phys. Chem.*, vol. 87, pp. 4034–4037, 1983.

[24] P. Pissis, "A thermally stimulated depolarization technique for studying the freezing of water dispersed within emulsions," *J. Phys. D*, vol. 17, pp. 787–791, 1984.

T. K. Bose
R. Chahine
Universite du Quebec à Trois-Rivières
P.Q., Canada

R. Nozaki
Hokkaido University, Sapporo, Japan

5

Dielectric Relaxation Study of Water and Water/Oil Microemulsion System

Abstract. Extensive dielectric measurements have been performed at 25°C on a water/oil microemulsion system composed of water, oil, a surfactant, and a cosurfactant (alcohol). Recent improvements in time-domain reflectometry have made it possible to make accurate measurements of complex permittivity over a wide frequency range (about 30 MHz to 20 GHz). The relaxation observed in the high-frequency range (1 GHz) is attributed to the presence of water and alcohol in the vicinity of the reverse micellar surface. The strength of the dielectric relaxation process in the very high frequency region of about 16 GHz is dependent on the water content and is related to pure water outside the interfacial layer. The dielectric relaxation study of the water/oil microemulsion system is compared with that of pure water.

5.1. INTRODUCTION

There has been much interest lately in microemulsions both from the point of view of experiment [1,2,3] as well as theory [4,5]. Experimentally, the study of microemulsions has been investigated extensively by light scattering, nuclear magnetic resonance, viscosity, and electrical conductivity. Our purpose is to study the state of microemulsions from the point of view of dielectric relaxation. Our measurements have been carried out by time-domain reflectometry (TDR).

A microemulsion is composed of water, oil, a surfactant (detergent), and a cosurfactant (alcohol). It exists as a transparent isotropic fluid. It is simply composed of "swollen micelles"—water droplets in oil or the inverse; the water/oil interface is saturated by surfactant + cosurfactant molecules. A microemulsion is formed spontaneously with the right amount of compositions. The order of mixing the components is not important. The spontaneous formation of a microemulsion simply means that no external work is necessary to form the droplets. As suggested by Schulman, the main role of the surfactant and cosurfactant is to reduce the superficial tension to zero. The sizes of microemulsion droplets are generally much smaller than those of normal emulsions. They may vary from 50 to 1500 Å. Another important difference between the microemulsion and the emulsion is the long-time stability of the former. The stable nature of microemulsions is the principal factor in the tremendous possibilities of practical applications [6].

We discuss the dielectric properties of pure water and that of water/oil microemulsion systems composed of water, toluene, 1-butanol, and SDS (sodium dodecylsulfate). This microemulsion system has been chosen because its phase diagram is well known.

Time-domain reflectometry is an important technique for the measurement of the complex permittivity of materials over a wide band. Figure 5.1 shows the basic concept of TDR. A steplike pulse produced by a pulse generator propagates through the coaxial line and is reflected from the sample section placed at the end of the line. The difference between the reflected and the incident pulses recorded in the time domain contains the information on the dielectric properties of the sample.

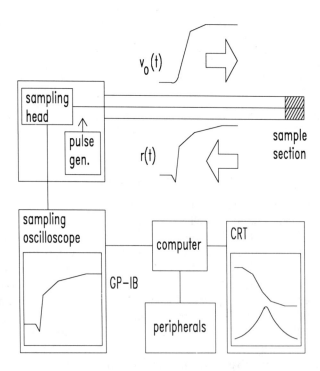

FIGURE 5.1. Block diagram of the TDR system.

The advantage of the time-domain method is that one measurement covers a wide frequency range, sometimes over two decades. The TDR can measure the complex permittivity not only in the radio frequency [7,8] range but also in the region of microwave [9].

The development in the basic theory of TDR and the analysis of the experimental results have been quite substantial in the last 10 years. Earlier, TDR theory had some limitations when measuring the complex permittivity in the frequency region higher than 1 GHz [10,11]. However, Cole and his co-workers improved the measurement of complex permittivity to 10 GHz by the introduction of the total reflection method using linear transmission theory [3,9].

Recent developments in high-speed sampling techniques and digital processing started a new generation in sampling oscilloscopes (HP 54120T) which made it possible to measure the time-domain signal with a time resolution of 0.25 ps with sufficient stability.

This article is divided into three sections. In the first section, we describe the basic theory of TDR. The second section describes the TDR system. The third section discusses experimental results and the microemulsion system as well as pure water.

5.2. BASIC TDR THEORY

A steplike voltage $v_i(t)$ produced by a pulse generator and the corresponding reflected voltage $v_r(t)$ from the sample section are propagated through the coaxial line in opposite directions (Figure 5.1). The input admittance Y_{in} of the line is given by the voltage-current relation

$$Y_{in} = G_0[V_i(s) - V_r(s)]/[V_i(s) + V_r(s)], \qquad s = j\omega, \qquad (5\text{-}1)$$

where $V_i(s)$ and $V_r(s)$ are the Laplace transforms of the incident and reflected signals, $v_i(t)$ and $v_r(t)$ respectively, and G_0 is the characteristic conductance of the line. For a linear transmission line with length d, terminated by a load admittance Y_d,

$$Y_{in} = Y_0[Y_d + Y_0 \tanh(\gamma d)]/[Y_0 + Y_d \tanh(\gamma d)], \qquad (5\text{-}2)$$

where Y_0 is the characteristic admittance of the line and $\gamma = \alpha + j\omega(\epsilon^*\mu^*)^{1/2}/c$ is the propagation constant. Here ϵ^* and μ^* are complex permittivity and permeability, respectively, α indicates the line losses, and c is the speed of light. In a nonmagnetic sample,

$$\gamma = \alpha d + jx/d,$$

where $x = (\omega d/c)(\epsilon^*)^{1/2}$. Since for the sample section with an open-ended configuration, $\gamma d = 0$, we get

$$Y_{in} = \frac{Y_0 x}{x \coth(\alpha d + jx)} = \frac{G_{0s}(\epsilon^*)^{1/2}x}{x \coth(\alpha d + jx)},$$

where $G_{0s} = Y/(\epsilon^*)^{1/2}$ is the characteristic conductance of the sample section filled with air ($\epsilon^* = 1$). After neglecting sample line losses, we get

$$Y_{in} = G_{0s} \left(\frac{j\omega d}{c} \right) \left(\frac{\epsilon^*}{x \cot x} \right), \tag{5-3}$$

where

$$\coth(jx) = -j \cot(x).$$

Finally, taking a measurement reference plane at the end of the coaxial line and combining eqs. (5-1) and (5-3) gives a useful expression for ϵ^* using measurable quantities such as

$$\epsilon^* = \rho x \cot x \tag{5-4}$$

where

$$\rho = \frac{c}{gd} \frac{V_i - V_r}{j\omega(V_i + V_r)} \tag{5-5}$$

and $g = G_{0s}/G_0 \, (= Z_0/Z_{0s})$ is the normalized characteristic conductance between the line and the sample section. The incident pulse can be replaced by reflection from an empty sample section using the relation $v_i(t) = v_r(t) \exp(-2d/c)$.

In an actual measurement, the upper limit of the time epoch has to be set at which $v_r(t)$ reaches $v_i(t)$ to attain a common value. In this case, the Laplace transform of $v_i(t) - v_r(t)$ is easy, but that of $v_i(t) + v_r(t)$ cannot be performed directly because of truncation error. However, the term $j\omega(V_i + V_r)$, in the denominator of eq. (5-5), can be replaced by the transform of $d[v_i(t) + v_r(t)]/dt$, and this makes it possible to avoid the truncation error if the plateau of $v_i(t) + v_r(t)$ is flat enough at the end of the time epoch.

Normally, for the sample section a coaxial-type cell with good symmetry is used. This is quite useful because it allows the selection of a variety of cell dimensions by controlling g and d.

The effect of gd can be discussed using the magnitude of the sample response M, which can be roughly obtained from eq. (5-4) by taking the limit as $\omega \to 0$:

$$M \equiv \lim_{\omega \to 0} [V_i - V_r]$$

$$= \int_0^\infty [v_i(t) - v_r(r)] \, dt = \frac{2\epsilon_s v_i(\infty) gd}{c}, \tag{5-6}$$

where ϵ_s is the dc permittivity of the sample and $v_i(\infty)$ represents the amplitude of both $v_i(t)$ and $v_r(t)$ at $t = \infty$. This equation shows that M depends on ϵ_s and gd. For the measurement of a sample with small permittivity, a large value of gd is recommended.

The magnitude of the response also depends on the actual time epoch used to acquire the TDR waveform. Should the magnitude M be small, it is necessary to increase gd in order to increase M to ensure precision measurement at low frequencies.

However, the term $x \cot x$ has divergent points at every $x = n\pi$ ($n = 1, 2, 3, \ldots$). It is very difficult to evaluate $x \cot x$ around such divergent points. Therefore, it would be reasonable to restrict x to values less than $\pi/2$ at which $x \cot x$ has the first zero. This restriction gives the highest frequency, f_h, to be

$$f_h = c/(4\epsilon^{1/2}d), \tag{5-7}$$

where ϵ is the permittivity of the sample at f_h.

The value of gd can be determined by using eq. (5-6) or the low-frequency behavior of the complex permittivity of a standard sample, and d can be evaluated from the high-frequency behavior because the high-frequency correction term, $x \cot x$, includes d independently.

5.3. TDR SYSTEM

The experimental setup used is quite simple (Figure 5.1). A steplike voltage pulse generated by a step recovery diode is propagated through the coaxial line and is applied to the sample cell located at the end of the coaxial line. The cell configuration is discussed in the next section. The reflected voltage pulse from the sample cell is also propagated through the same line and sampled at the sampling head. The generator and the sampling head are included in the four-channel test set (HP 54121A). The form of the voltage pulse thus measured is digitized and averaged by the digitizing oscilloscope mainframe (HP 54120A). A personal computer is used to carry out numerical computations such as Laplace transforms.

5.4. RESULTS AND DISCUSSION

In this section, we give the complex permittivity results for the water/oil microemulsion system and pure water.

For samples with unknown permittivity ϵ_x^* and known permittivity ϵ_s^*, eqs. (5-4) and (5-5) can be rewritten as

$$\epsilon_k^* = \rho_k x_k \cot x_k$$

$$\rho_k = \frac{c}{gd} \frac{V_i - V_k}{j\omega(V_i + V_k)},$$

where $k = s, x$. Combining these equations, we get

$$\epsilon_x^* = \frac{(\rho_1 + \epsilon_s/f_s)f_x}{1 - g^2(\omega d/c)^2 \rho_1 \epsilon_s/f_s}, \tag{5-8}$$

$$\rho_1 = \frac{c}{gd} \frac{V_s - V_x}{j\omega(V_s + V_x)}, \tag{5-9}$$

where $f_s = x_s \cot x_s$ and $f_x = x_x \cot x_x$, respectively. This equation does not include the Laplace transform of the incident voltage pulse. Therefore, two measurements of the time-domain pulse which are reflected from a standard liquid and an unknown sample give the complex permittivity of the sample. Normally, air ($\epsilon' = 1$) is a good standard.

If the dielectric sample in the sample section is conductive, the feature of the reflected pulse is of quite different form from that of a nonconductive sample, as shown in Figure 5.2. In a nonconductive sample, the voltage of the reflected pulse at long time, $v_s(\infty)$, is always the same as that for the nonconductive standard sample, $v_x(\infty)$. On the other hand, $v_x(\infty)$ for the conductive sample never reaches the value of $v_s(\infty)$, and the difference between them is the contribution of the dc conductivity. In this case, the Laplace transform of $v_s(t) - v_x(t)$ cannot be performed directly because of the nonzero value of $v_s(\infty) - v_x(\infty)$.

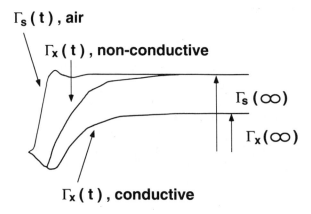

FIGURE 5.2. Reflected signal behavior for air, nonconductive sample, and conductive sample.

Introducing $h = v_x(\infty)/v_s(\infty)$ can make it possible to eliminate the problem discussed above and evaluate the dc conductivity [9,12,13]. In this case, $V_s - V_x$ in eq. (5-9) is replaced by $hV_s - V_x$, which is zero at $t = \infty$. Finally, for $\xi = (1-h)/(1+h)$, modified equations corresponding to eqs. (5-8) and (5-9) are

$$\epsilon_x^* = \frac{[\rho_2(1 + \xi) + (c/j\omega gd)\xi + \epsilon_x/f_s] f_x}{1 - g^2(\omega d/c)^2 [\rho_2(1 + \xi) + (c/j\omega gd)\xi] \epsilon_s/f_s}, \tag{5-10}$$

$$\rho_2 = \frac{c}{gd} \frac{hV_s - V_x}{j\omega(V_s + V_x)}. \tag{5-11}$$

These equations are exactly the same as eqs. (5-8) and (5-9) when $\xi = 0(h = 1)$.

The complex permittivity of the conductive dielectric sample is represented by

$$\epsilon^*(\omega) = \epsilon_d^*(\omega) - j(\sigma/\omega\epsilon_0), \tag{5-12}$$

where ϵ_d^* is the complex permittivity due only to the dielectric process and σ is the dc conductivity. On the other hand, if ω is small, eq. (5-10) can be reduced to

$$\epsilon_x^* = (1 + \xi)\rho_2 + \epsilon_s - j(c/\omega gd)\xi, \tag{5-13}$$

where ϵ_o is the permittivity of free space. Hence, the value of dc conductivity can be evaluated without a Laplace transform as

$$\sigma = (c/gd)\epsilon_0\xi. \tag{5-14}$$

Figure 5.3 shows the frequency dependence of the complex permittivity of a microemulsion system composed of water, 1-butanol, SDS, and toluene. Table 5.1 gives the mass fractions of two compositions of the microemulsion system. The lower-frequency part between 30 MHz and 1 GHz was measured using the cell in Figure 5.4, and the complex permittivity was obtained using eqs. (5-10) and (5-11). The points for ϵ'', indicated by triangles, are actual measurements, and those indicated by squares are the result after subtraction of the dc conductivity, which was determined by the imaginary part of the complex permittivity at a very low frequency. On the other hand, the higher-frequency part between 2 and 20 GHz was measured using the cell in Figure 5.5.

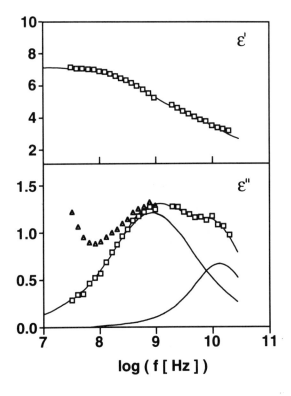

FIGURE 5.3. Frequency dependence of the complex permittivity of SDS microemulsion system. Below 1 GHz, triangles indicate experimental values of ϵ'' and squares indicate the values after subtraction of dc conductivity. Above 1 GHz, squares indicate experimental values. Curves are obtained from eq. (5-15).

TABLE 5.1. Mass Fraction Percent of Constituents of Microeumlsions

System	Water	1-Butanol	Toluene	SDS
B_1	4.000	10.698	80.003	8.307
B_2	6.207	10.615	68.871	8.307

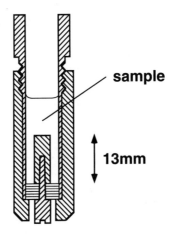

FIGURE 5.4. Cell for microemulsion measurement at lower frequencies. Bottom part is composed of APC7 connector.

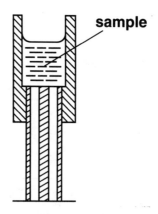

FIGURE 5.5. Flat-end capacitor cell. Bottom part is the end of 2-mm copper coaxial cable connected to the four-channel test set.

According to Figure 5.3, it can be considered that there are at least two relaxation processes, each one having a distribution of relaxation times. Therefore, to describe the data, we used two Cole-Cole relaxations [14]. The equation is

$$\epsilon^* = \frac{\Delta\epsilon_1}{1 + (j\omega\tau_1)^{1-\alpha_1}} + \frac{\Delta\epsilon_2}{1 + (j\omega\tau_2)^{1-\alpha_2}} + \epsilon_\infty, \tag{5-15}$$

where α is the Cole-Cole parameter which describes the broadness of the relaxation. Subscripts 1 and 2 indicate the lower- and higher-frequency processes, respectively. The results of the fitting are given in Table 5.2 for two mass fractions of the microemulsion system.

It can be seen from Table 5.2 that for the lower-frequency relaxation α_1 is about 0.75, and for the high-frequency α_2 is equal to 1. This clearly indicates the presence of more than one relaxation in the low-frequency region and a Debye-like relaxation for the high frequency.

TABLE 5.2. Dielectric Parameters of Microemulsions

SYSTEM	ϵ_s	f_1 (GHz)	f_2 (GHz)	$1 - \alpha_1$	$1 - \alpha_2$	$\Delta\epsilon_1$	$\Delta\epsilon_2$
B_1	5.669	0.893	15.915	0.780	1.000	1.644	0.535
B_2	6.561	1.247	15.915	0.749	1.000	2.004	0.844

The lower-frequency process has been observed in many microemulsion systems and can be attributed to more than one relaxation mechanism. We have indicated earlier [15] that this low-frequency relaxation process shows a Maxwell-Wagner relaxation mechanism, an intermediate ionic relaxation due to the presence of counterions, and finally the relaxation due to the interaction between bound water and 1-butanol inside the layer [3,16]. However, the higher-frequency relaxation process is very similar to that of free water and Debye-like. Therefore, it is quite probable that the higher-frequency relaxation found here is due to free water existing in the micelle.

We have shown that the water inside the microemulsion system could be studied up to about 20 GHz with a TDR system. This was possible because the dielectric constant of the microemulsion system is low, and, as such, the cutoff frequency of the fundamental mode is quite high. In a cylindrical waveguide [12] the cutoff frequency is f_c (GHz) $= 1.9 \times 10^2 / D\epsilon^{1/2}$, where D is the diameter of the outer conductor in millimeters and ϵ is the dielectric constant of the substance filling the waveguide.

For pure water with a static dielectric constant of 80, we get a cutoff frequency of 10 GHz with a coaxial line of 2 mm outer diameter. We are, of course, assuming that the cylindrical waveguide cutoff frequency is valid in the fringing field space. In fact, as shown in Figure 5.6, the experimental results of pure water seem to justify the validity of our assumption. Although the experimental results for pure water are not valid [17] after 10 GHz, it is possible to determine the relaxation frequency, which is much higher than 10 GHz. Since water is Debye-like, we can represent the complex dielectric constant of water by the well-known Debye equation

$$\epsilon^* = \epsilon' - j\epsilon'' = \epsilon_\infty + \frac{\epsilon_0 - \epsilon_\infty}{1 + j\omega\tau},$$

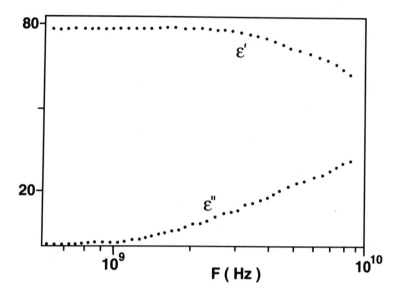

FIGURE 5.6. Frequency dependence of the real and imaginary part of the complex permittivity of water at 20°C.

where τ is the relaxation time. Equating the real parts on both sides of the equation, we get

$$\epsilon' = \epsilon_0 - \omega\epsilon''\tau.$$

Plotting ϵ' versus $\omega\epsilon''$ gives a straight line of slope $-\tau$ (Figure 5.7). We determined the relaxation time $\tau = 9.27$ ps or about 17 GHz at 20°C from the slope. This

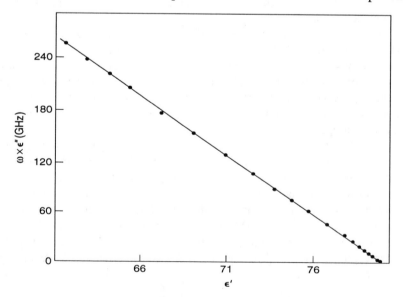

FIGURE 5.7. Plot of $\omega\epsilon''$ as a function of ϵ'.

way of plotting the results enables one to determine the relaxation time outside the measuring frequency range. Table 5.3 gives the static dielectric constant as well as the relaxation time of water for seven temperatures. The agreement with the literature values [18] is good.

TABLE 5.3. Dielectric Parameters of Water
as a Function of Temperature

	ϵ_0		τ (ps)	
T (°C)	EXPT	LIT. [18]	EXPT	LIT. [18]
0	89.02	88.3	17.71	17.9
5	85.82		14.42	
10	84.24	84.1	12.48	12.6
15	81.98		10.51	
20	80.50	80.4	9.27	9.3
25	79.20		8.33	
30	77.16	76.8	7.93	7.2

For the microemulsion system of water, 1-butanol, SDS, and toluene, we see from Table 5.2 that a proportionality exists between the dielectric increment $\Delta\epsilon_2$ and the water mass fraction [15]. More precisely,

$$\frac{(\phi_W)_{B_1}}{(\phi_W)_{B_2}} = \frac{(\Delta\epsilon_2)_{B_1}}{(\Delta\epsilon_2)_{B_2}} \approx 0.6.$$

This observation leads us to conclude that the higher-frequency process should be attributed to the relaxation of pure water. This result is important because such water in micellar phases can be looked upon as a model of the biological water in living cells.

5.5. CONCLUSION

We have shown that it is not possible to characterize pure water in the microwave region above 10 GHz by a transmission line system. Because of the high dielectric constant of pure water, the cutoff frequency for the fundamental mode is limited to 10 GHz. One way to avoid the frequency limitation is to study water in a heterogeneous form where the dielectric constant of the total system is low. We have applied this technique for microemulsions and have shown that the high-frequency limit can be extended to more than 20 GHz.

REFERENCES

[1] M. Cazabat and D. Langevin, "Diffusion of interacting particles: Light scattering study of microemulsions," *J. Chem. Phys.*, vol. 74, pp. 3148–3158, 1981.

[2] M. Kotlarcxyk, S. H. Chen, J. S. Huang, and M. W. Kim, "Structure of three-component microemulsions in the critical region determined by small-angle neutron scattering," *Phys. Rev.*, vol. 29A, pp. 2054–2069, 1984.

[3] T. K. Bose, G. Delbos, and M. Merabet, "Dielectric properties of microemulsions by time domain spectroscopy," *J. Phys. Chem.*, vol. 93, 867–872, 1989.

[4] P. G. de Gennes and C. Taupin, "Microemulsions and the flexibility of oil/water interfaces," *J. Phys. Chem.*, vol. 86, 2294–2304, 1982.

[5] W. Widom, "A model microemulsion," *J. Chem. Phys.*, vol. 81, 1030–1046, 1984.

[6] L. M. Prince, *Microemulsions Theory and Practice*, New York: Academic Press, 1977.

[7] T. K. Bose, R. Chahine, M. Merabet, and J. Thoen, "Dielectric study of the liquid crystal compound octylcyanobiphenyl (8CB) using time-domain spectroscopy," *J. Phys.*, vol. 45, pp. 1329–1336, 1984.

[8] T. K. Bose, B. Campbell, S. Yagihara, and J. Thoen, "Dielectric relaxation study of alkylcyanobiphenyl liquid crystals using time-domain spectroscopy," *Phys. Rev. A*, vol. 36, pp. 5767–5773, 1987.

[9] R. H. Cole, J. G. Berberian, S. Mashimo, G. Chryssikos, A. Burns, and E. Tombari, "Time-domain reflection methods for dielectric measurements to 10 GHz," *J. Appl. Phys.*, vol. 66, pp. 793–802, 1989.

[10] R. H. Cole, "Evaluation of dielectric behavior by time-domain spectroscopy. II: Complex permittivity," *J. Phys. Chem.*, vol. 79, pp. 1469–1474, 1975.

[11] R. H. Cole, "Time-domain spectroscopy of dielectric materials," *IEEE Trans. Instrum. Meas.*, vol. IM-25, pp. 371–375, 1976.

[12] S. Mashimo, T. Umehara, T. Ota, S. Kuwabara, N. Shinyashiki, and S. Yagihara, "Evaluation of complex permittivity of aqueous solution by time-domain reflectometry," *J. Mol. Liq.*, vol. 36, pp. 135–151, 1987.

[13] P. Winsor IV and R. H. Cole, "Dielectric properties of electrolyte solutions. I: Sodium iodide in seven solvents at various temperatures," *J. Phys. Chem.*, vol. 86, 2486–2490, 1982.

[14] K. Imamatsu, R. Nozaki, S. Yagihara, and S. Mashimo, "Evaluation of dielectric relaxation spectrum of phospholipids in solution by time domain reflectometry," *J. Chem. Phys.*, vol. 84, pp. 6511–6517, 1986.

[15] A. Ponton, R. Nozaki, and T. K. Bose, "Dielectric relaxation study of oil-continuous microemulsion systems," *J. Chem. Phys.*, vol. 97, 8515–8521, 1992.

[16] R. H. Cole, G. Delbos, P. Winsor IV, T. K. Bose, and J. M. Moreau, "Study of dielectric properties of water/oil and oil/water microemulsions by time-domain and resonance cavity methods," *J. Phys. Chem.*, vol. 89, pp. 3338–3343, 1985.

[17] M. Merabet and T. K. Bose, "Dielectric measurements of water in the radio and microwave frequencies by time domain reflectometry," *J. Phys. Chem.*, vol. 92, pp. 6149–6150, 1988.

[18] J. B. Hastad, *Aqueous Dielectrics*, London: Chapman and Hall, 1973.

6

Satoru Mashimo
Nobuhiro Miura
Tokai University, Kiratsuka, Kanawaga, Japan

Free and Bound Water in Various Matrix Systems Studied by Advanced Microwave Techniques

Abstract. A dielectric relaxation peak due to bound water of globule proteins in aqueous solution was observed by the time-domain reflectometry method. This peak occurs around 100 MHz as well as that of the moist collagen and DNA aqueous solution, and the relaxation strength is proportional to its surface except for trypsin and pepsin of hydrorase. It is suggested that this peak is caused by the orientation of bound water molecules on protein surface. The number of bound water molecules estimated is in good agreement with that obtained by other methods, such as X-ray diffraction measurement. However, the relaxation strength of trypsin or pepsin is larger than that expected. This suggests that the bound water exists in the crack of the protein molecule and may cooperate with a local flactuation of protein. These results suggest that the bound water stabilizes the protein structure and plays an important role in hydrolysis.

6.1. INTRODUCTION

Globule protein has a lot of biological functions in living materials. For example, enzyme works as a catalyst of hydrolysis, oxidation, and reduction. The protein molecules are surrounded by water in most cases. Therefore it is of particular interest to investigate the water structure around the protein. High-resolution X-ray and neutron diffraction measurements in the crystal phase have made clear that there are peculiar water molecules correlating strongly with the globule protein [1].

On the other hand, the sorption measurement of water vapor indicated that the numbers of sorption point of albumin, cytochrome C, and lysozyme obtained are 241, 56, and 60, respectively, which are roughly equal to those of polar residues per protein molecule, which are 375, 65, and 80, respectively [2]. It has been thus suggested that the sorption water is bound to the polar groups [2].

Recently dielectric measurements by time-domain reflectometry (TDR) have revealed two relaxation peaks in the DNA aqueous solution [3] and the moist collagen [4]. The high-frequency peak around 10 GHz is due to the bulk water, judging from its location. The relaxation strength for the low-frequency process around 100 MHz depends on the water content and vanishes for dried collagen. This process is concluded to be due to orientation of strongly bound water to tropocollagen.

In this work, dielectric relaxation measurements are performed on 10 kinds of globule proteins in aqueous solution by TDR to find the bound water [3,4]. High-precision measurements could be carried out over a frequency range from 100 KHz to 10 GHz.

6.2.   EXPERIMENTAL

Samples of cytochrome C (from horse heart), myoglobin (from horse skeletal muscle), hemoglobin (from bovine blood), ribonucrease A (from bovine pancreas), lysozyme (from chicken egg white), ovalbumin (from chicken egg), albumin (from bovine serum), trypsin inhibitor type 1-S (from soybean), trypsin (from porcine pancreas), and pepsin (from porcine stomach mucosa) were purchased from Sigma Co. Ltd. Deionized and distilled water was obtained from Wittaker Bioproducts, Inc.

Dielectric measurements were performed by TDR. The detailed procedure has already been reported [5]. If a certain sample of known permittivity ε_S^* is used as a reference, the permittivity ε_X^* of the unknown sample is

$$\varepsilon_X^*(\omega) = \varepsilon_S^*(\omega) \frac{1 + \{cf_S/j\omega(\gamma d)\varepsilon_S^*(\omega)\}\rho}{1 + \{j\omega(\gamma d)\varepsilon_S^*(\omega)/cf_S\}\rho} \frac{f_X}{f_S}, \tag{6-1}$$

where

$$\rho = \frac{r_S - r_X}{r_S + r_X},$$

$$f_X = Z_X \cot Z_X, \qquad Z_X = (\omega d/c)\varepsilon^*(\omega)^{1/2},$$

and

$$f_S = Z_S \cot Z_S, \qquad Z_S = (\omega d/c)\varepsilon_S^*(\omega)^{1/2},$$

where d is the cell length, γd is the electric length, r_S and r_X are Fourier transforms of the reflected pulse from the reference sample $R_S(t)$ and that from the unknown sample $R_X(t)$, j is the imaginary unit, ω is the angular frequency, and c is the speed of propagation in vacuo.

We chose a water with an appropriate concentration of sodium chloride as the reference sample, which gives the same dc conductivity of the unknown sample. The permittivity of water with sodium chloride is nearly the same as that of pure water except for the dc conductivity. Therefore for the measurement of electrolyte solution like the present solution, we use $\varepsilon_S^*(\omega) - \sigma/j\omega$ instead of ε_S in eq. (6-1), where σ is the dc conductivity of the unknown sample.

6.3.   RESULT

Two relaxation peaks can be seen apparently in the aqueous solution of globule protein if the contribution of dc conductivity is subtracted. As an example, dielectric dispersion and absorption curves for 5 wt% myoglobin solution at 25°C are shown in Figure 6.1.

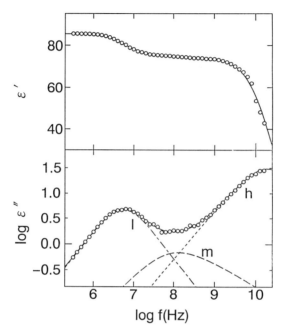

FIGURE 6.1. Frequency dependence of dielectric dispersion and absorption for 5 wt% myoglobin in aqueous solution.

One observed around 20 GHz can be described well by the Cole-Cole representation. If the Cole-Davidson representation is assumed for the low-frequency relaxation, total absorption and dispersion curves are explained relatively well. Nevertheless the absorption and dispersion data around 100 MHz deviate definitely from the sum of those two relaxation processes. Therefore a relaxation process described by the Havriliak-Negami representation is added to the two processes. This representation can be applied to a variety of dielectric relaxations and involves the Cole-Cole and Cole-Davidson representations in it. It can have an extremely wide distribution of relaxation times. The total complex permittivity is explained quite satisfactorily by the sum of three relaxation processes. The highest-frequency process denoted by h is undoubtedly due to the orientation of free water.

The logarithm of the relaxation time τ_1 for the lowest-frequency process is plotted against the logarithm of the molecular weight M for 10 globule proteins in Figure 6.2. All points lie on the same straight line with slope -1. Plots of the logarithm of the relaxation strength $\Delta \varepsilon_m$ for the intermediate process against log M give a straight line with a slope of about $-1/3$ except for trypsin and pepsin of hydrolase, as shown in Figure 6.3.

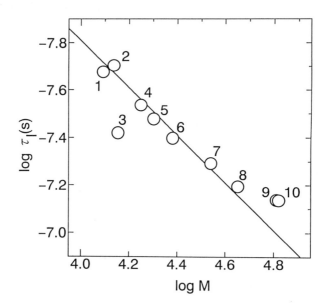

FIGURE 6.2. Plots of logarithm of the relaxation time τ_1 against that of the molecular weight M of globule protein at 25°C. Globule proteins are 1:cytochrome C, 2:ribonuclease A, 3:lysozyme, 4:myoglobin, 5:trypsin inhibitor, 6:trypsin, 7:pepsin, 8:ovalbumin, 9:hemoglobin, and 10:albumin.

6.4. DISCUSSION

If relaxation strength $\Delta \varepsilon_m$ for the intermediate relaxation process is normalized by the number of protein molecules per unit volume, N, we have

$$\Delta \varepsilon / N \propto M \Delta \varepsilon_m \propto M^{2/3} \qquad (6\text{-}2)$$

since $\Delta \varepsilon_m$ observed is proportional to $M^{-1/3}$. This indicates that $\Delta \varepsilon_m$ normalized is proportional to the protein surface. On the other hand, the relaxation time τ_m around 3 ns coincides with that of bound water observed for the DNA aqueous solution [3], moist collagen [4], and moist lysozyme [5].

In general, the relaxation strength depends on the number of dipoles in unit volume and the internal field. If the Debye formula for the field and the value of the square of the dipole moment obtained for the moist collagen, $\langle \mu^2 \rangle = 0.38 \text{ D}^2$ [4], is used to evaluate the square of the effective dipole moment, the number of bound water molecules on one protein surface can be estimated.

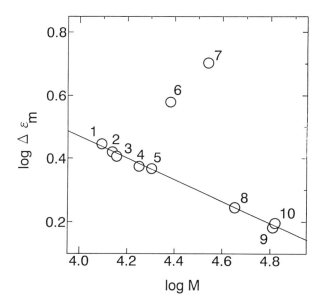

FIGURE 6.3. Plots of logarithm of relaxation strength $\Delta\varepsilon_m$ against that of the molecular weight M of globule protein at 25°C. Globule proteins are 1:cytochrome C, 2:ribonuclease A, 3:lysozyme, 4:myoglobin, 5:trypsin inhibitor, 6:trypsin, 7:pepsin, 8:ovalbumin, 9:hemoglobin, and 10:albumin.

The numbers thus obtained are in good agreement with those of the bound water attached to the main chain and polar groups through hydrogen bonding, obtained by neutron and X-ray diffraction analyses [1] and sorption measurement [2]. For example, the numbers estimated by X-ray and neutron diffractions for lysozyme [6] and myoglobin [7] are 35 and 40 respectively; the numbers obtained in this work are 53 and 58, respectively. The numbers obtained from sorption measurement for albumin, cytochrome C, and lysozyme are 241, 56, and 60, respectively; the numbers from this work are 150, 49, and 53, respectively. These results suggest that the globule protein has the same bound water as the DNA or the tropocollagen and the water molecules produce a short string or a small network on the protein surface around the polar group through hydrogen bonding.

Meanwhile, the relaxation strength $\Delta\varepsilon_m$ of trypsin and pepsin is large, pepsin about twice as large as expected. But the number of polar groups of trypsin or pepsin is not different from that evaluated from other globule proteins. However, pepsin and trypsin of hydrolase have an active site like a crack to site substrate. The pepsin molecule is divided into two domains, and its surface is fairly larger than the spherical molecule. X-ray analysis indicates the existence of a network of bound water in the crack of trypsin [8,9] and in the domain-domain interface of actinidin [1]. Thus, it is thought that fluctuation of the water network cooperates with the movement of active site and plays an important role for hydrolysis.

On the other hand, the relaxation time for overall rotation of the spherical molecule in a viscous medium is described by Debye theory [10], and the relaxation time is proportional to the molecular weight M. Plots of log τ against log M give a straight line with slope -1. This agrees with experimental results in Figure 6.2. Nevertheless, the globule protein is not necessarily a perfect sphere. It is often ellipsoidal. Therefore the relaxation time will have a distribution of relaxation times. This may be why the low-frequency process has a value slightly deviating from unity for α_1.

The large relaxation strength $\Delta\varepsilon_1$ observed especially for cytochrome C or ribonuclease A is interpreted by taking the counterion migration into account. The process always brings about a large relaxation strength.

6.5. CONCLUSION

The dielectric relaxation of globule protein in aqueous solution is observed in three relaxation processes. The high-frequency process observed around 20 GHz is due to orientation of bulk water molecules. The intermediate process around 100 MHz is caused by orientation of bound water molecules on protein molecules, supplemented by fluctuation of polar side groups on the surface. The low-frequency process in the megahertz region reflects an overall rotation of globule protein, which involves the counterion process. The dielectric behavior of globule protein in aqueous solution is thus first clarified by microwave dielectric relaxation measurements. The results obtained here offer very important information on the role of bound water and a key to interpret the function of protein.

REFERENCES

[1] J. T. Edsall and H. A. McKenzie, "Water and proteins. II. The location and dynamics of water in protein systems and its relation to their stability and properties," *Adv. Biophys.*, vol. 16, p. 53, 1983.

[2] M. Levitt and R. Sharon, "Accurate simulation of protein dynamics in solution," *Proc. Natl. Acad. Sci. USA*, vol. 85, p. 7557, October 1988.

[3] T. Umehara, S. Kuwabara, S. Mashimo, and S. Yagihara, "Dielectric study on hydration of A-, B-, and Z-DNA," *Biopolymers*, vol. 30, p. 649, 1990.

[4] N. Shinyashiki, N. Asaka, S. Mashimo, S. Yagihara, and N. Sasaki, "Microwave dielectric study on hydration of moist collagen," *Biopolymers*, vol. 29, p. 1185, 1990.

[5] S. C. Harvey and P. Hoekstra, "Dielectric relaxation spectra of bound water adsorbed on lysozyme," *J. Phys. Chem.*, vol. 76, p. 2987, 1972.

[6] C. C. F. Blake, W. C. A. Pulfford, and P. J. Artymiuk, "X-ray studies of water in crystals of lysozyme," *J. Mol. Biol.*, vol. 167, p. 693, 1983.

[7] B. P. Schoenborn, "A neutron diffraction analysis of myoglobin. III: Hydrogen-deuterium bonding in side chain," in *Cold Spring Harbor Symp. on Quantitative Biology*, vol. 36, p. 569, 1971.

[8] W. Bode and P. Schwager, "The refined crystal structure of bovine β-trypsin at 1.8 Å resolution," *J. Mol. Biol.*, vol. 98, p. 693, 1975.

[9] A. Kossiakoff and S. A. Spencer, "Direct determination of the protonation states of aspartic acid-102 and histidine-57 in the tetrahedral intermediate of the serine proteases: Neutron structure of trypsin," *Biochemistry*, vol. 20, p. 6462, 1981.

[10] N. E. Hill, W. E. Vaughan, A. H. Price, and M. Davies, *Dielectric Properties and Molecular Behavior*, London: Van Nostrand Reinhold, 1969, pp. 90–91.

7

Alexander Brandelik
Gerd Krafft
Institut für Meteorologie und Klimaforschung
Universität Karlsruhe, Karlsruhe, Germany

Measurement of Bound and Free Water in Mixtures

Abstract. We report on the development of a new moisture measurement method. Our work concentrated on the determination of soil moisture, but preliminary data are also available on other materials. The new moisture model has been composed of the recommendation by Birchak et al. [1], a semiempirical soil model by Dobson et al. [2], our new definition of bound water, and our new measurement method based on in situ frost calibration. In the case of soil also swelling and shrinking are taken into account. The method does not require any sampling for the purpose of calibration. The dielectric coefficient of the wet mixture is measured for two different water contents in the natural and in the perfectly frozen states. The information from these four measurements provide four unknown variables, namely the initial water content, the changed water content, the bound water, and a new solid-specific parameter, the solid contribution. The deviation between measurements and accurate expectations is less than 1% for as extreme soils as sand and clay. Investigations have been started to link these soil moisture data to the data obtained from satellites.

7.1. INTRODUCTION

We demonstrate our new method on measurements in moist soils. We use the same method for the determination of water in building stones and in pharmaceutical products as well. In soil physics one (maybe the most) important parameter is the volumetric water content. It breaks down into two portions, free and bound water. We report on the development of a soil model and a device to measure the moisture with an accuracy better than $\pm 1.5\%$. This model is based on the frost-calibrated soil moisture probe [3], but the derivation of the water content from the dielectric coefficients (the so-called mixing rule) is a modification of the mixing rule by Birchak et al. [1], and of the semiempirical model by Dobson et al. [2]. Wang and Schmugge [4] present a comprehensive analysis of the methods used in soil moisture determination. They give a long list of rather different mixing rules. None of them indicates data measured with sufficient accuracy. The experts try to find soil parameters, which help to fit a model to the measurement. De Loor [5] has found that without any fitting parameter used no mixing rule may be accurate enough. Sihvola [6] introduced a new dimensionless parameter related to the phenomenon of percolation. These known mixing rules have in common the need of certain soil-specific supplementary parameters. Our solution provides the accurate actual water content, the characteristic of the bound water, and a further soil-specific parameter without requiring knowledge of any parameter.

7.2. SOIL MODEL

Soil is an extremely complex system. The mutual arrangements of the particles in the soil determine the characteristic of the pore spaces in which water and air are transported or retained. These phases interact strongly with each other. To be able to describe their volumetric proportions, we must consider them as independent constituents [7]. A unit volume is divided into three portions: water W, pure solids S, and air A:

$$W + S + A = 1. \qquad (7\text{-}1)$$

In some mixtures the change of water content also influences the abstract dry density. Soil experts called this phenomenon swelling and shrinking [8]. The volume undergoes a relative change V/V_0, where V_0 is the starting volume:

$$W + S + A = V/V_0. \qquad (7\text{-}2)$$

Clay can change its abstract dry density by up to 15% or by even more. When this correction had not been made, many models failed in the case of clay.

7.3. BOUND WATER

What is bound water? The present situation in this field is rather chaotic. Therefore, we will introduce a new definition. Generally, the transportation of bound water needs more energy than that of free water. Moisture determination by remote sensing can be achieved only by dielectric measurement. There are sufficient arguments in favor of the preference of the dielectric measurement both in the laboratory and in the field. Consequently, the only reasonable definition of bound water must be related to the possible measurement, namely to the change in its dielectric coefficient. Instead of using the former definitions, we call water "bound water" if the real part of its complex dielectric coefficient ϵ is less than that of free water at the same frequency and temperature. Binding of a water molecule is strongest in an ice crystal. So, the dielectric coefficient of ice, ϵ_i, is 3.2 at 250 MHz compared to $\epsilon_f = 80$ at 20°C in free water. In our model, the first water monolayer on a solid surface is as strongly bound as in ice. The binding force diminishes successively, in our model exponentially, with growing distance from the solid surface. At a sufficiently far distance from the solid the water is free, its dielectric coefficient $\epsilon_f = 80$. The continuous function of the dielectric coefficient of water in the neighborhood of a solid, ϵ_w, can be described by the water content W as the free variable (instead of by the distance) [9] (Figure 7.1):

$$\epsilon_w = \epsilon_f - (\epsilon_f - \epsilon_i)\exp(-W/h). \tag{7-3}$$

The slope, h, of the exponential function is characteristic of the surface of the soil particle and of the sum of its binding forces. With our method we get h as one of the results.

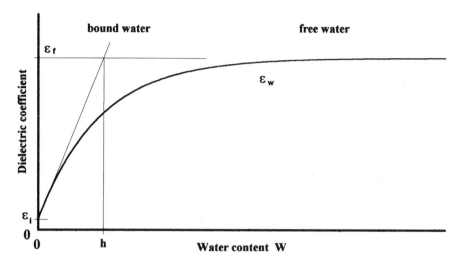

FIGURE 7.1. Diagram of the function ϵ_w.

7.4. MEASURING FREQUENCY

In order to avoid the disturbing influence of the conductivity, one must choose higher frequencies. On the other hand, the depth of penetration decreases with rising frequency. For a representative large measuring volume we need lower frequencies. In the measurements carried out in our laboratory we found the optimal frequency range to lie between approximately 100 and 800 MHz. We report here on measurements made at 250 MHz.

7.5. MIXING RULE

We are convinced [10] that the mixing rule of Birchak, with some proper modifications, gives the best results. These are the modifications we recommend:

1. We prefer the formula by Birchak in the form used by Dobson and Hallikainen which is valid for more than two components. It is

$$\epsilon_m^\alpha = \sum_j V_j \epsilon_j^\alpha, \tag{7-4}$$

where α is a constant and j is the number of the constituents, ϵ_j and ϵ_m are the real parts of the permittivity of the constituents and of the mixture, respectively.

2. The dielectric coefficient of water ϵ_w will be used in the form introduced in eq. (7-3).

3. Also swelling and shrinking have to be used in the mixing rule in the form introduced in eq. (7-2). The new mixing rule is

$$\epsilon_m^\alpha \frac{V}{V_0} = S\epsilon_s^\alpha + \int_0^W \left(\epsilon_f - (\epsilon_f - \epsilon_i)\exp\left(\frac{-w}{h}\right)\right)^\alpha dw + A\epsilon_A^\alpha, \qquad (7\text{-}5)$$

where ϵ_s is the permittivity of the solids and $\epsilon_A = 1$ is that of air. With eq. (7-2),

$$\epsilon_m^\alpha \frac{V}{V_0} = S\epsilon_s^\alpha + \int_0^W \left(\epsilon_f - (\epsilon_f - \epsilon_i)\exp\left(\frac{-w}{h}\right)\right)^\alpha dw + \frac{V}{V_0} - S - W. \qquad (7\text{-}6)$$

Now, we introduce

$$s = S\left(\epsilon_s^\alpha - 1\right), \qquad (7\text{-}7)$$

where s is the new soil-specific parameter, the solid contribution. With eq. (7-7),

$$\frac{V}{V_0}\left(\epsilon_m^\alpha - 1\right) = s + \int_0^W \left(\epsilon_f - (\epsilon_f - \epsilon_i)\exp\left(\frac{-w}{h}\right)\right)^\alpha dw - W. \qquad (7\text{-}8)$$

4. In our patented method it is recommended to make a second measurement of the same mixture at a temperature at which the whole water content W is frozen. The volume of ice is 1.06 times its water volume. This coefficient modificates W in the frozen state. (Consequently, this model is valid up to a water content where 94% of the porosity is filled up.) In frozen state eq. (7-8) reads

$$\frac{V}{V_0}\left(\epsilon_{mc}^\alpha - 1\right) = s + W \cdot 1.06\left(\epsilon_i^\alpha - 1\right), \qquad (7\text{-}9)$$

where ϵ_{mc} is the permittivity of the frozen mixture. The changes of S and ϵ_s are negligible compared to the change of ϵ_f into ϵ_i.

5. The same procedure will be repeated for a water content W_2, where W_2 is not equal to W.

$$\frac{V_2}{V_0}\left(\epsilon_{m2}^\alpha - 1\right) = s + \int_0^{W_2} \left(\epsilon_f - (\epsilon_f - \epsilon_i)\exp\left(\frac{-w}{h}\right)\right)^\alpha dw - W_2, \qquad (7\text{-}10)$$

and for the frozen state

$$\frac{V_2}{V_0}\left(\epsilon_{mc2}^\alpha - 1\right) = s + W_2 \cdot 1.06\left(\epsilon_i^\alpha - 1\right). \qquad (7\text{-}11)$$

We set $\alpha = 0.5$. Now, from eqs. (7-8)–(7-11) we can determine numerically the unknown variables W, W_2, h, and s. The outstanding result of this method is that, unlike with other mixing rules known till now, we do not need any previously known soil parameter. Moreover, the method provides two objective soil parameters besides the water contents. Further measurements under cooling conditions are not needed, not even if we monitor the change of moisture over a long term.

7.6.   APPLICATION TO SEVERAL SOILS

We applied the new mixing rule to several soil types with approximately the same accurate results. Two extreme soils, medium coarse sand and heavy clay (separated in the Winand Staring Centre, Wageningen, The Netherlands), are used as examples. The materials were dried. Then the soil was wetted with controlled volume of water. The moist mixture was filled in an open-end coaxial waveguide (approximately 0.015 m long and 0.03 m outside diameter) layer by layer under a constant pressure of 4 N/cm². Now, we measured the relative volume change and the dielectric coefficient warm and in frozen state after cooling by liquid nitrogen down to approximately −40°C. This procedure was repeated approximately 10 times at different water contents. For every moisture content a new mixture was prepared and filled. Then we picked out a randomly moisture pair and solved the system (7-8)–(7-11). We calculated with the same h and s parameters the functions $\epsilon_m(W)$ and $\epsilon_{mc}(W)$ and compared whether all of the measurements $\epsilon_M(W)$ and $\epsilon_C(W)$ sit on these functions. The new mixing model fitted the measured data extremely well. The relative deviations between the measurements and the calculations were always less than ±1%. Figures 7.2 and 7.3 show these results for clay and sand, respectively.

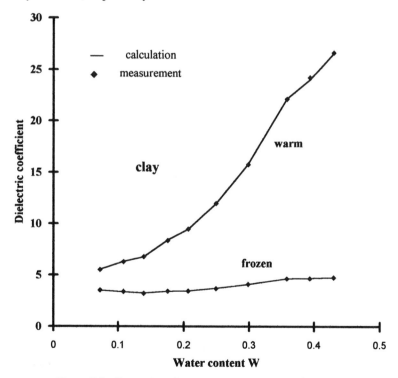

FIGURE 7.2. Comparison between measurements and calculations for clay. The lines are the functions resulting from the new mixing rule and the points are the measured data.

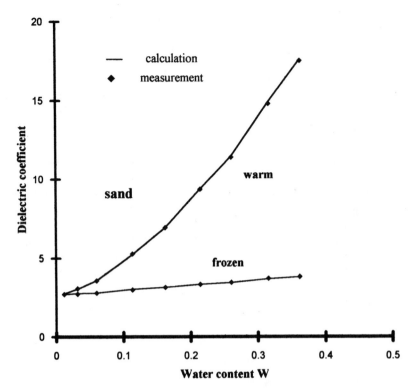

FiGURE 7.3. Comparison between measurements and calculations for sand. The lines are the functions resulting from the new mixing rule and the points are the measured data.

The deformation of the functions of clay is due to shrinking and swelling. The medium coarse sand does not change its volume measurably. The characteristic values of bound water h were 0.0011 and 0.045 for clay and sand, respectively. These are surprising values, which, however, are comprehensible considering the surface chemistry and the complex binding forces of these materials. Notice, h does not characterize the mechanically closed but electrically free water. Further on, we cannot give any information about crystal water. It belongs to the solids. If we remove it, we destroy the material. Every other binding such as capillary, absorption, Coulomb, van der Waals, etc., is compressed in h. We cannot separate them. The solid contributions were 0.806 and 0.635 for clay and sand, respectively.

7.7.   DISCUSSION OF THE POWER α

Several authors tried to find a proper power α to fit their measurements. Birchak [1] and Alharthi and Lange [12] assumed $\alpha = 0.5$. Roth et al. [11] found $\alpha = 0.47$. Dobson et al. [2] prefer $\alpha = 0.65$. A unique result of our work is that we get our measurements with an accurate of $\pm 1\%$ if we only modify the so-called volumetric average of the complex index of refraction by Birchak. The derivation of $\alpha = 0.5$ by Birchak is very simple and correct. Nevertheless, the difference between Dobson et al. and Roth et al. shows a bit of uncertainty. It is remarkable that $\alpha = 0.47$ by Roth et al. is rather near to the theoretically $\alpha = 0.5$. (They did not modify with shrinking and swelling!) We are aware that the simplification by Birchak, namely "assume that the propagation constant is a function of the z coordinate only," is too sensitive. A homogeneous mixture is three-dimensional. In spite of this objection, many results are in the neighborhood of $\alpha = 0.5$ by several authors. For $\alpha = 0.5$ we have a halfway theoretical base; for other values we do not have any physical explanation. So, we use $\alpha = 0.5$.

7.8. SENSOR FOR FIELD MEASUREMENTS

Nine devices were fabricated for field measurements. Figure 7.4 shows one of them. The oldest device has been set up on a hydrological test site since August 1990. Moisture data every 10 min from different depths are furnished. Three other sensors have provided moisture data from the Rhein valley since November 1992, supporting the Regional Klima Project (REKLIP). Two other sensors can be rented from the licencee. These sensors were validated by other meteorological measurements to an accuracy of $\pm 2\%$.

FIGURE 7.4. Cooling of the in situ soil moisture sensor.

If the soil in the field is in a natural state of equilibrium—namely it already has been exposed to some cycles of swelling and shrinking—we do not need to take into account the relative volume change: V/V_0 will be 1. In this case the air gaps in the soil belong to the integral measuring volume. The sensor must have a large enough measuring volume to provide a representative mean value. This is one of the reasons for the large sensor diameter in Figure 7.4.

7.9. CONCLUSION

This study deals with mixtures composed of solids, water, and air. The dielectric moisture determination by electromagnetic waves is in common use, but the optimal measuring frequency has to be chosen more critically than before. Besides the arguments in Section 7.4, two additional conditions limit the measuring frequency range: (1) The wavelength in the mixture and in every constituent has to be much longer than the mechanical dimension of a homogeneous particle or a water film bridge. (2) The mechanical dimension of a particle has to be much larger than a single water molecule (to avoid too much interatomic bound water in a solid molecule). Under these conditions the modified mixing rule by Birchak fits the measurements extremely well. Adopting the recommended measurement after freezing, we do not need any previously known soil parameter for calibration. A repeated warm/cool measurement pair can provide two other solid-specific parameters: the characteristic of bound water (corresponding to the new bound water definition), and the solid contribution, which is proportional to the solid content. Nevertheless, we must not forget that the theoretical derivation of the rule by Birchak is not general enough (restriction of the propagation coefficient in one direction). Further studies are needed to verify the rule of Birchak without restrictions and to derive new rules that fit measurements at least as well as this study provides.

LICENSEE AND PARTNERS

The licensee is Meteolabor AG., CH-8620 Wetzikon, Switzerland. Meteolabor, Kernforschungszentrum Karlsruhe and Ingenieurbüro Roth, Karlsruhe, Germany, are presently cooperating on a project to monitor a waste dumping hill, too. Monitoring of the solid contributions, besides the water content, allows dangerous changes in the compactness and containment to be detected earlier.

ACKNOWLEDGMENTS

The authors gratefully acknowledge the long-term support of the project by Prof. Dr. F. Fiedler, Head of the Institute for Meterology and Climatological Research. We are grateful to the KfK Technology Transfer Coordination Office for granting supplementary financial support. We thank the lectors and especially Dr. A. Kraszewski for critically reviewing this study.

REFERENCES

[1] J. R. Birchak, C. G. Gardner, J. E. Hipp, and J. M. Victor, "High dielectric constant microwave probes for sensing soil moisture," *Proc. IEEE*, vol. 62, no. 1, pp. 93–98, 1974.

[2] M. C. Dobson, F. T. Ulaby, M. T. Hallikainen, and M. A. El-Rayes, "Microwave dielectric behavior of wet soil. II: Dielectric mixing models," *IEEE Trans. Geosci. Remote Sensing*, vol. GE23, no. 1, January 1985.

[3] A. Brandelik and G. Krafft, "Frost calibrated soil moisture measuring," in *IEEE Int. Geosci. Remote Sensing Symp.*, IGARSS'91, Helsinki, 1991.

[4] J. R. Wang and T. J. Schmugge, "An empirical model for the complex dielectric permittivity of soils as a function of water content," *IEEE Trans. Geosci. Remote Sensing*, vol. GE 18, no. 4, October 1980.

[5] G. P. De Loor, "The dielectric properties of wet soils," BCRS Delft, The Netherlands, Report No. 90-13. Project No. AO-2.1, 1990.

[6] A. Sihvola et al., "Dielectric properties of geophysical media: Percolation aspects within mixture theories," in *IEEE Int. Geosci. Remote Sensing Symp.*, IGARSS'92, Houston, 1992.

[7] D. Hillel, *Soil and Water*, New York: Academic Press, 1973.

[8] J. M. Halbertsma and H. G. M. van den Elsen, "Calibration and accuracy analysis of water content measurements of the CAMI TDR unit," Int. Rep. 146. The Winand Staring Centre, Wageningen, The Netherlands, July 1991.

[9] M. A. Berliner, *Feuchtemessung*, Berlin: VEB Verlag Technik, 1965.

[10] A. Brandelik and G. Krafft, "Measurement of soil moisture and its bound water," in *6th Australasian Remote Sensing Conf.*, Wellington, 1992.

[11] K. Roth et al., "Eine Methode zur Messung des volumetrischen Wassergehaltes," *Bull. Bodenkundl. Gesell. Schweiz*, vol. 13, pp. 117–122, 1989.

[12] A. Alharthi and J. Lange, "Soil water saturation," *Water Resour. Res*, vol. 23, no. 4, pp. 591–595, 1987.

8

Ari Sihvola
Helsinki University of Technology, Espoo, Finland

Dielectric Mixture Theories in Permittivity Prediction: Effects of Water on Macroscopic Parameters

Abstract. This paper focuses on the modeling of dielectric mixtures and the question of how the effective permittivity of a mixture can be determined from the component properties and their fractional volumes. A generalized mixing formula is presented which contains several often-used mixing rules as special cases. The effect of water on the dielectric properties of moist substances is emphasized. As an example, snow permittivity is considered. Furthermore, attention is given to the problems of dispersion in the dielectric properties of mixtures and to percolation phenomena.

8.1. INTRODUCTION

An obvious way to model moist substances is to consider them as mixtures with one component being liquid water. As a consequence, the dielectric properties of moist substances are determined by the corresponding properties of the constituent phases, i.e., the permittivities of water and the dry phase of the material. These dielectric properties are essential in the microwave characterization of moist media, because the permittivity (in the full complex form in which losses are included) determines the response of the material as it interacts with electromagnetic fields.

Mixing rules, or formulas, describe quantitatively how the macroscopic permittivity of a mixture depends on the components and the microstructure of the medium. Sometimes (and especially in this article) this large-scale permittivity of the heterogeneous substance is called "effective permittivity" and denoted by ϵ_{eff}.

In the present case, which treats moist substances, the crucial parameters affecting ϵ_{eff} are the permittivity of water and the fractional water volume.

Naturally, the manner in which water is distributed within the structure of the substance is important in the dielectric behavior. In mixing rules this can be taken care of by considering different water inclusion shapes and their effect on the polarization response. However, the exhaustive description of random media requires too many structural parameters to be effective, so certain averaging approaches have to be taken. This makes the analysis approximate.

Several mixing rules have been proposed in the literature for calculating ϵ_{eff} of substances encountered in industrial, geophysical, and materials science applications. A given geometry, however complicated, could be solved from the electromagnetic point of view, at least numerically with a given precision. But in practice, the problem of solving the random medium case is not easy because of the enormous number of degrees of freedom in describing the boundary value problem. Therefore different mixing rules can coexist and possess different areas of application with experimental confirmation.

The theoretical approach to the mixture problem normally starts with a single scatterer in a background medium. This scatterer, or inclusion, is most easily analyzed if its shape is assumed to be ellipsoidal or even spherical. The simplification allows a constant field for the Laplace equation inside such a shape. Therefore the polarizability of the inclusion is expressed rather easily with the parameters of the problem. Polarizability is a crucial parameter in electromagnetic scattering analysis and other mixture-related treatments. For a general description of the dielectric modeling of heterogeneous media, see Refs. 1 and 2, and for a particularly moist–products-oriented discussion, see Ref. 3.

This article gives explicit mixture rules for predicting the effective permittivity of heterogeneous materials in general and moist substances in particular. Many different mixing rules can be combined and shown to emerge from a family of mixing rules that contains a free parameter to select the particular formula. This general mixing formula is used here, and special emphasis is given to two phenomena that are typical of moist materials: dispersion (the dependence of the dielectric properties on the frequency of the operating electromagnetic wave) and percolation (a sudden qualitative change in a macroscopic property of the heterogeneous medium as the composition of the mixture changes).

Although only dielectric mixtures are treated, remember that ϵ is only one electromagnetic material parameter. The permeability μ also affects the electromagnetic response; however, the magnetic parameters will not be treated because most media in microwave applications are nonmagnetic. Using the duality property of Maxwell equations, the dielectric mixing rules apply directly to magnetic mixtures as well. Even more complex media, such as chiral materials (that possess magnetoelectric coupling parameters in addition to permittivity and permeability), can be treated as mixtures. Their macroscopic properties can be shown to obey analogous—although more complicated in form—mixing laws like plain dielectric mixtures [4].

8.2. MIXING RULES

As emphasized in the introduction, there are numerous ways to incorporate the structural shape effects and the weight of different component phases in the electromagnetic modeling of random media. Consider first the simplest conceivable mixture, a discrete two-phase heterogeneity where the guest component is in the form of spheres embedded in the host material. The mixture under study is depicted in Figure 8.1.

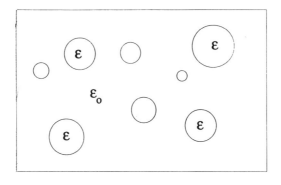

$$\varepsilon_{\text{eff}} = ?$$

FIGURE 8.1. The mixture under study: the inclusions are spheres occupying a volume fraction f of the space filled with background medium.

8.2.1 Lorenz–Lorentz Expression and Maxwell Garnett Mixing Formula

The discrete-scatterer model for the two-phase mixture is inherently nonsymmetric: one of the phases is treated as host with permittivity ϵ_0, and the other as guest with permittivity ϵ. Note here that the permittivities are absolute permittivities and have dimension [A−s/V−m]. Let f denote the volume fraction of the guest material in the mixture.

It is well known that a mixing rule for this mixture can be derived using the polarizability of the dielectric sphere and the Lorenz-Lorentz (sometimes also known as the Clausius-Mossotti formula) refractivity expression [5]:

$$\frac{n^2 - 1}{n^2 + 2} = N\alpha, \tag{8-1}$$

where n is the refractive index of the medium, N is the number density of polarizable molecules, and α is the polarizability of a single molecule. Although originally intended to explain the optical properties of media at the molecular level, this result can be used to derive the microwave properties of mixtures with macroscopically polarizable inclusions. The famous expression for the effective permittivity of the material sample of Figure 8.1 is

$$\epsilon_{\text{eff}} = \epsilon_0 + 3f\epsilon_0 \frac{\epsilon - \epsilon_0}{\epsilon + 2\epsilon_0 - f(\epsilon - \epsilon_0)}. \tag{8-2}$$

This is called the Maxwell Garnett formula [6] or the Rayleigh mixing formula, and is sometimes written

$$\frac{\epsilon_{\text{eff}} - \epsilon_0}{\epsilon_{\text{eff}} + 2\epsilon_0} = f \frac{\epsilon - \epsilon_0}{\epsilon + 2\epsilon_0}. \tag{8-3}$$

The Maxwell Garnett formula (MG) has been applied often with success to predict properties of mixtures for low volume-filling ratios ($f \ll 1$). However, as the volume fraction of the inclusion phase increases, the MG predictions start to fail. This happens especially in the case of high dielectric contrast between the phases composing the mixture, which is, unfortunately, the case for moist substances.

However, one advantage of the Maxwell Garnett formula is frequently overlooked: the fact that MG has been shown to be a lower bound for the effective permittivity of a mixture [7]. In other words, any deviation from its way of taking into account the interactions within a mixture tends to increase the polarization effects. This fact can be formulated as a variational principle for the effective permittivity. Furthermore, the inverse MG (i.e., the case where the host and guest change roles in the mixing treatment) leads to the other limit: the maximum achievable effective permittivity of the mixture. These properties guarantee that the study of MG mixing relation is worthwhile.

The other fact that keeps MG from predicting higher effective permittivity values is the other extremum aspect in eq. (8-2), which is the sphericity assumption for the inclusions. It is well known that spheres have a form producing the smallest dipole moment given the amount of polarizable material [8]. Therefore, it is no wonder that MG has been under attack from experimentally oriented researchers whose objects of interest are composite or geophysical dense media that should be modeled as clusters of spheres or even more complicated shapes.

8.2.2 Generalized Mixing Formula

Because a strong objection has been raised against the Maxwell Garnett mixing rule relating to the unsuccessful regime of high dielectric contrasts ($\epsilon/\epsilon_0 \gg 1$), which, on the other hand, is especially interesting in the case of moist substances, modifications to the MG rule have to be looked for.

It has been shown [9,10] that several mixing rules that have been proposed to explain the dielectric properties of heterogeneous media can be considered to be one family of mixing formulae. This family can be written in the generalized form

$$\frac{\epsilon_{\text{eff}} - \epsilon_0}{\epsilon_{\text{eff}} + 2\epsilon_0 + \nu(\epsilon_{\text{eff}} - \epsilon_0)} = f \frac{\epsilon - \epsilon_0}{\epsilon + 2\epsilon_0 + \nu(\epsilon_{\text{eff}} - \epsilon_0)}. \tag{8-4}$$

In addition to the quantities ϵ, ϵ_0, ϵ_{eff}, and f in the MG rule (8-2), there is an extra parameter ν. This dimensionless parameter determines the nature of the mixing rule. It is easy to see that $\nu = 0$ reproduces the Maxwell Garnett rule, also known as the ATA (average T-matrix approximation) result. Integer values for ν bring forth other known mixing rules, and particularly these mixing models have been used in remote sensing studies in explaining the dielectric behavior of geophysical media.

The value $\nu = 3$ gives the so-called coherent potential formula [11] (CP). In solid-state physics, CP is known also as the GKM (Gyorffy, Korringa, and Mills) rule. Correspondingly, $\nu = 2$ gives the Böttcher mixing rule [12], sometimes referred to as the Polder–Van Santen mixing formula [13],[1] and also known as EMT (effective medium theory) formula. This model, especially dealing with ellipsoidal inclusions, has been applied to snow modeling. The EMT mixing rule is often presented in the form

$$f \frac{\epsilon - \epsilon_{\text{eff}}}{\epsilon + 2\epsilon_{\text{eff}}} + (1 - f) \frac{\epsilon_0 - \epsilon_{\text{eff}}}{\epsilon_0 + 2\epsilon_{\text{eff}}} = 0, \tag{8-5}$$

and in this form it is sometimes called the "Bruggeman symmetric formula" in solid-state studies. Symmetry means that the host and guest phases are treated equally: interchanging ϵ, ϵ_0, f to the triplet ϵ_0, ϵ, $1 - f$ does not change the prediction of the formula. Sometimes the symmetrical properties of mixing laws as such are appreciated, to such an extent that symmetry is considered as a corroboration of the theory.

Note that at least formally, eq. (8-4) is consistent and correct in the dilute and dense extremes. In other words, for any value of ν, eq. (8-4) satisfies the limiting requirements of no guest phase ($f = 0 \rightarrow \epsilon_{\text{eff}} = \epsilon_0$) and vanishing host phase ($f = 1 \rightarrow \epsilon_{\text{eff}} = \epsilon$).

The permittivity of the mixture is now calculable from eq. (8-4). Provided that the mixing rule can be applied to mixtures containing dielectric losses, conductivity behavior emerges from the mixing rule, because in time-harmonic electromagnetic field analysis (frequency domain), the absolute permittivity becomes a complex number:

$$\epsilon = \epsilon' - j\epsilon'' = \epsilon' - j\frac{\sigma}{\omega} \tag{8-6}$$

where σ is the electrical conductivity of the material and ω is the angular frequency. The sign of the imaginary part of ϵ above reveals the assumed convention $\exp(j\omega t)$ for time dependence. The validity of extending mixing rules into the complex domain is discussed elsewhere [15].

As an example of the predictions of the different mixing models, consider the permittivity of dry snow. There exists the model for the dielectric behavior of dry snow, which is based on a large number of measurements [16]. At microwave

---

1. Polder and van Santen derived their formula for ellipsoidal inclusions, and the spherical formula is a special case. However, eq. (8-4) can be generalized also to ellipsoidal mixtures.

frequencies, the dispersion of the dielectric constant of snow is very small, and the real part of the relative permittivity can be expressed as a function of the density of snow ρ (given in units of g/cm^3):

$$\epsilon_r = 1 + 1.7\rho + 0.7\rho^2. \tag{8-7}$$

Now, let us try to predict this with the generalized mixing formula (8-4), treating dry snow as a mixture of air and ice. The volume fraction of ice in the mixture is determined by the density of snow, because the ice density can be taken as $\rho_i = 0.917$ g/cm^3. Then the volume fraction of ice is $f = \rho/\rho_i$. The relative permittivity of ice is 3.15.

Figure 8.2 illustrates the results. It can be seen that all models underestimate the permittivity. This is because of the assumption that the ice phase is in the form of spheres, which naturally is not the case, at least in the case of light snow. Ice particles are rather needles in this range, or even more complicated clusters of grains. Another conclusion from Figure 8.2 is that the Maxwell Garnett model ($\nu = 0$) estimates the smallest permittivity, the coherent potential ($\nu = 3$) predicts the largest, and the Polder–Van Santen model ($\nu = 2$) falls between. Nonetheless, even the CP estimate is smaller than the experimental result.

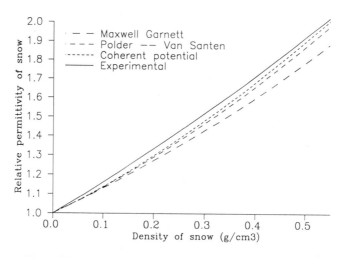

FIGURE 8.2. The dielectric constant of snow: comparison of the experimental model of Ref. 16 with different mixing models. Snow is considered to be a mixture of air and ice, and ice is assumed to be in the form of spheres.

Although only spherical geometry mixtures receive particular attention in this article, the general mixing laws can be written to apply for mixtures with ellipsoidal inclusions [17] and inhomogeneous, layered spherical, and ellipsoidal geometries [18]. With the model of needle- or disk-shaped inclusions as the ice particles in dry snow, the underestimates of Figure 8.2 can be corrected [17]. However, the motivation to focus on the simplest geometry is that from the basic mixing law two special effects present in the dielectric behavior of moist substances can be treated easily: dispersion and percolation.

8.3. DISPERSION IN MOIST SUBSTANCES

The material parameters that appear in the mixing rules have been constants in the analysis above. This implies that the treatment has been carried out at a given frequency. In this section, let us allow the material parameters, i.e., the permittivities to vary with frequency. In more technical terms, the media are temporally dispersive.

Due to the dispersion in mixture constituents, the effective macroscopic permittivity is frequency dependent. This can be anticipated, but what is often overlooked is the strong effect of the mixing process on the global wideband dispersion behavior. To illustrate this, consider a mixture with the presence of water as the guest phase. The simplest mixture is rain; a dilute mixture ($f \ll 1$) where spherical water inclusions are embedded in dispersionless air. As is known [19,20], the permittivity of water follows the Debye-type dispersion:

$$\epsilon(\omega) = \epsilon_\infty + \frac{\epsilon_s - \epsilon_\infty}{1 + j\omega\tau}, \tag{8-8}$$

where ϵ_s and ϵ_∞ are the low-frequency and high-frequency permittivities, and τ is the relaxation time, the inverse of which gives the angular frequency of maximum losses (more exactly, at that frequency the imaginary part of the permittivity is largest). This relaxation frequency is around 9 GHz for water at 0°C. The dispersion is most expressively illustrated in the Cole-Cole diagram in Figure 8.3.

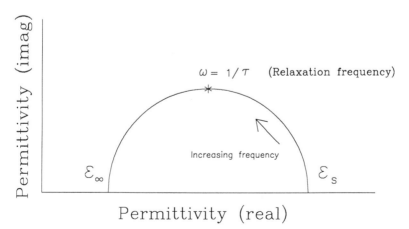

FIGURE 8.3. The Cole-Cole representation of the Debye equation (8-8). The real and imaginary axes for the permittivity are linear. The locus of the permittivity is a semicircle as the frequency increases from static values to optical range.

Substituting this guest dispersive permittivity into the Maxwell Garnett mixing formula (8-2) with the assumption of a low volume fraction of water ($f \ll 1$) yields

$$\epsilon_{\text{eff}} = \epsilon_0 + 3 f \epsilon_0 \frac{\epsilon - \epsilon_0}{\epsilon + 2\epsilon_0}. \tag{8-9}$$

After some algebra, the effective permittivity turns out to follow the same type of Debye spectrum as the water component:

$$\epsilon_{\text{eff}}(\omega) = \epsilon_{\text{eff},\infty} + \frac{\epsilon_{\text{eff},s} - \epsilon_{\text{eff},\infty}}{1 + j\omega\tau_{\text{eff}}} \tag{8-10}$$

with parameters

$$\epsilon_{\text{eff},s} = \epsilon_0 + 3 f \epsilon_0 \frac{\epsilon_s - \epsilon_0}{\epsilon_s + 2\epsilon_0}, \tag{8-11}$$

$$\epsilon_{\text{eff},\infty} = \epsilon_0 + 3 f \epsilon_0 \frac{\epsilon_\infty - \epsilon_0}{\epsilon_\infty + 2\epsilon_0}, \tag{8-12}$$

and the relaxation time for the mixture being

$$\tau_{\text{eff}} = \frac{\epsilon_\infty + 2\epsilon_0}{\epsilon_s + 2\epsilon_0}\tau. \tag{8-13}$$

This means that the overall dispersive behavior of the dielectric mixture remains the same as for the bulk medium. However, the relaxation frequency $f_{\text{rel}} = 1/2\pi\,\tau_{\text{eff}}$ (the frequency of maximum loss) shifts to a higher region; for example, the water-air mixture possesses the relaxation frequency of 114 GHz (at $0°$C, a much higher value than bulk water (9 GHz)). Note that the relaxation frequency is independent of the volume fraction f as long as the low-loading assumption ($f \ll 1$) is valid.

An essential point in this dispersion analysis was the spherical shape of the particles. The inclusion form is crucial in the magnitude of the shift in relaxation frequency. The effect of nonsphericity has been studied [21], and the conclusion was that the spherical shape brings forth an extremum behavior. For ellipsoids deviating from the spherical shape, the relaxation frequency always was lower than for sphere mixtures. For the case of needle- and disk-shaped inclusions, the relaxation frequency would come very close to the relaxation frequency of bulk water.

8.4. PERCOLATION PHENOMENA

The other characteristic feature in moist substances, from the point of view of the effective dielectric behavior, is the appearance of *percolation* effects. This is a consequence of the high dielectric contrast between the phases in the mixture or the large difference in the magnitudes of the permittivities of water and the dry substance.

Percolation and phenomena associated with it have attracted increasing attention in recent years, after the early studies of Hammersley [22]. Percolation theory has indeed been successful and applicable to several phenomena in different fields of physics and other disciplines [23,24]. Among these applications, one may mention ferromagnetism, soil moisture studies, oil penetration in rocks, the spread of epidemics, forest fires, and wafer-scale integration in the manufacture of microchips.

Percolation is inherently connected with heterogeneous and random media. *Percolate* means to flow through. Essential there is the process of flowing and the manner in which the associated flow finds its way. This, naturally, depends on the microstructure of the material. What flows need not be a fluid; it can even be a rumor in the network formed by a group of humans, or, as in the case of moist media treated here, it is the electric flux or current.

Percolation is a nonlinear phenomenon; a very abrupt change in the behavior of certain parameters of a percolating material. The geometry of the matter where percolation takes place is very special; if there are even small changes in the fractions of the components forming the material, the structure behaves totally differently. This fact is characteristic of percolation processes. In percolation studies one often encounters terms like site and bond percolation, mixed and oriented percolation, lattice animals, and clustering. Often the parameters characterizing percolation behavior in the critical region have been enumerated through Monte Carlo simulations.

What is the relevance of percolation to dielectric mixtures and permittivities? As a matter of fact, in many (but not all) of the artificial mixtures that engineers use to design composite materials, percolation needs no special attention. But this phenomenon cannot be overlooked in those cases when the mixtures are composed of phases with strong dielectric contrasts. Percolation is especially conspicuous in metal-insulator mixtures where the electrical properties of the components differ as much as Nature allows. In that case, the electrical properties of the mixture are very sensitive functions of the structure of the mixture [25].

But also the permittivity properties of moist substances clearly show signs of percolation behavior, although the threshold points observed are not as sharp as in metal mixtures and cermet films. The dielectric contrast of water with other material components in moist substances makes the mixture permittivity predictions subject to uncertainties in the vicinity of the volume fractions that correspond to percolating points.

There are two ways to find percolation from mixing rules, analytical and numerical. Looking at the general mixing formula (eq. (8-4)) from the percolation point of view, the straightforward approach is to let the dielectric contrast ϵ/ϵ_0 become large. In this case, the equation is approximately

$$\frac{\epsilon_{\text{eff}} - \epsilon_0}{\epsilon_{\text{eff}} + 2\epsilon_0 + \nu(\epsilon_{\text{eff}} - \epsilon_0)} \simeq f. \tag{8-14}$$

However, eq. (8-14) is only valid as long as $\epsilon_{\text{eff}}/\epsilon_0$ is small, i.e., only up to those volume fraction values where percolation appears. The effective permittivity is in this region

$$\epsilon_{\text{eff}} = \epsilon_0 \frac{1 + f(2 - \nu)}{1 - f(1 + \nu)}. \tag{8-15}$$

Clearly this result breaks down as the denominator reaches the value zero. This very point is interpreted as the percolation threshold point $f = f_c$:

$$f_c = \frac{1}{1 + \nu}. \tag{8-16}$$

This simple result and the values for the percolation threshold f_c can be clearly seen from the numerical experiment of Figure 8.4, where ϵ_{eff} is plotted as the function of guest volume fraction for a high dielectric contrast ($\epsilon/\epsilon_0 = 800$). The percolation threshold of eq. (8-16) is confirmed to be $f_c = 1$ for MG ($\nu = 0$), $f_c = 0.5$ for $\nu = 1$, $f_c = 0.333$ for EMT ($\nu = 2$), and $f_c = 0.25$ for coherent potential ($\nu = 3$). For a more detailed analysis on percolation and mixing formulae, see Ref. 26.

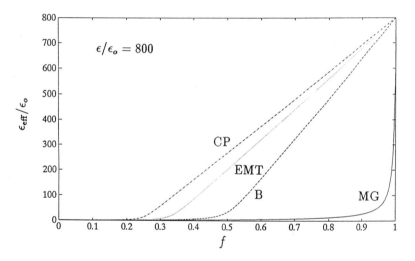

FIGURE 8.4. Illustration of percolation behavior: the effective relative permittivity $\epsilon_{\text{eff}}/\epsilon_0$ of the two-phase mixture with guest permittivity relative to host $\epsilon/\epsilon_0 = 800$ according to the general mixing formula (8-4). The different curves are MG, Maxwell Garnett formula ($\nu = 0$); B, ($\nu = 1$); EMT, effective medium theory ($\nu = 2$); CP, coherent potential ($\nu = 3$).

The analysis done for permittivity is only a little more complicated for the conductivity behavior. Provided that the conductivity of the guest phase is the dominant phenomenon in the mixing process and using eqs. (8-6) in (8-4), the following result applies in the subpercolation regime ($f < f_c$):

$$\epsilon_{\text{eff}}'' = \epsilon_0 \sqrt{\frac{2 - \nu}{1 + \nu} \frac{1 + f(2 - \nu)}{1 - f(1 + \nu)}}. \tag{8-17}$$

Here, the assumption $\epsilon''_{eff} \gg \epsilon'_{eff}$ is made. The earlier reasoning that led to eq. (8-16) is also valid. Looking for the point where the denominator vanishes, the same percolation thresholds of eq. (8-16) apply for the conductivity as for the permittivity mixing.

8.5. CONCLUSION

This article has hopefully shown that quasistatic mixing rules, so often criticized for being too simple to predict the dielectric properties of random media, still are able to explain, even to a quantitative extent, important dielectric characteristics of heterogeneous media. Especially treating moist substances where the effects of water dominate in the macroscopic permittivity behavior, mixing rules are able to analyze the effects of two special phenomena that have been the focal point of this article, dispersion and percolation.

REFERENCES

[1] A. Priou (Ed.), *Progress in Electromagnetics Research*, vol. 6, *Dielectric Properties of Heterogeneous Materials*, New York: Elsevier, 1992.

[2] R. Landauer, "Electrical conductivity in inhomogeneous media," in *American Institute of Physics Conf. Proc.*, no. 40, pp. 2–45, 1978.

[3] S. O. Nelson, "Dielectric properties of agricultural products," *IEEE Trans. Electrical Insul.*, vol. 26, no. 5, pp. 845–869, October 1991.

[4] A. Sihvola and I. Lindell, "Analysis on chiral mixtures," *J. Electromag. Waves Appl.*, vol. 6, no. 5–6, pp. 559–572, 1992.

[5] M. Born and E. Wolf, *Principles of Optics*, 6th ed., Oxford: Pergamon Press, 1980.

[6] J. C. Maxwell Garnett, "Colours in metal glasses and in metal films," *Trans. R. Soc., London*, vol. CCIII, pp. 385–420, 1904.

[7] Z. Hashin and S. Shtrikman, "A variational approach to the theory of the effective magnetic permeability of multiphase materials," *J. Appl. Phys.*, vol. 33, no. 10, pp. 3125–3131, 1962.

[8] A. Sihvola, "Properties of dielectric mixtures with layered spherical inclusions," in *Microwave Radiometric Remote Sensing Applications*, P. Pampaloni, Ed., VSP Press, 1989, pp. 115–123.

[9] A. Sihvola, "Self-consistency aspects of dielectric mixing theories," *IEEE Trans. Geosci. Remote Sensing*, vol. 27, no. 4, pp. 403–415, 1989.

[10] A. H. Sihvola and E. Alanen, "Studies of mixing formulae in the complex plane," *IEEE Trans. Geosci. Remote Sensing,* vol. 29, no. 4, pp. 679–687, 1991.

[11] W. E. Kohler and G. C. Papanicolaou, "Some applications of the coherent potential approximation," in *Multiple Scattering and Waves in Random*

Media, P. L. Chow, W. E. Kohler, and G. C. Papanicolaou, Eds., New York: North-Holland, 1981, pp. 199–223.

[12] C. J. F. Böttcher, *Theory of Electric Polarization,* Amsterdam: Elsevier, 1952.

[13] D. Polder and J. H. van Santen, "The effective permeability of mixtures of solids," *Physica,* vol. XII, no. 5, pp. 257–271, 1946.

[14] D. A. G. Bruggeman, "Berechnung verschiedener physikalischer Konstanten von heterogenen Substanzen. I: Dielektrizitätskonstanten und Leitfähigkeiten der Mischkörper aus isotropen Substanzen," *Ann. Phys.,* 5. Folge, Band 24, pp. 636–664, 1935.

[15] A. H. Sihvola and I. V. Lindell, "Effective permeability of mixtures," in *Progress in Electromagnetics Research,* vol. 6, *Dielectric Properties of Heterogeneous Materials,* A. Priou, Ed., New York: Elsevier, 1992, pp. 153–180.

[16] M. E. Tiuri, A. H. Sihvola, E. G. Nyfors, and M. T. Hallikainen, "The complex dielectric constant of snow at microwave frequencies," *IEEE J. Oceanic Eng.,* vol. OE-9, no. 5, pp. 377–382, 1984.

[17] A. Sihvola and J. A. Kong, "Effective permittivity of dielectric mixtures," *IEEE Trans. Geosci. Remote Sensing,* vol. 26, no. 4, pp. 420–429, 1988. See also Corrections, vol. 27, no. 1, pp. 101–102, Jan. 1989.

[18] A. Sihvola and I. V. Lindell, "Transmission line analogy for calculating the effective permittivity of mixtures with spherical multilayer scatterers," *J. Electromag. Waves Appl.,* vol. 2, no. 8, pp. 741–756, 1988; "Polarizability and effective permittivity of layered and continuously inhomogeneous dielectric spheres," *J. Electromag. Waves Appl.,* vol. 3, no. 1, pp. 37–60, 1989; "Polarizability and effective permittivity of layered and continuously inhomogeneous dielectric ellipsoids," *J. Electromag. Waves Appl.,* vol. 4, no. 1, pp. 1–26, 1990.

[19] J. B. Hasted, *Aqueous Dielectrics,* London: Chapman and Hall, 1973.

[20] E. Nyfors and P. Vainikainen, *Industrial Microwave Sensors,* Norwood, MA: Artech House, 1989.

[21] A. H. Sihvola, "Frequency dependence of absorption attenuation due to hydrometeors," in *Proc. Sixth Int. Conf. on Antennas and Propagation,* ICAP'89, pp. 2:285–288, 1989.

[22] J. M. Hammersley, "Origins of percolation theory," in *Percolation Structures and Processes,* G. Deutscher, R. Zallen, J. Adler, Eds., Annals of the Israel Physical Society, vol. 5, pp. 47–57, 1983.

[23] D. Stauffer, *Introduction to Percolation Theory,* London: Taylor and Francis, 1985.

[24] G. Grimmett, *Percolation,* New York: Springer-Verlag, 1989.

[25] B. J. Last and D. J. Thouless, "Percolation theory and electrical conductivity," *Phys. Rev. Lett.,* vol. 27, no. 25, p. 1719, 1971.

[26] A. Sihvola, S. Saastamoinen, and K. Heiska, "Mixing rules and percolation," *Remote Sens. Rev.,* vol. 9, pp. 39–50, 1994.

9

Sumio Kobayashi
Sumitomo Metal Industries, Ltd., Amagasaki, Japan

Microwave Attenuation in a Wet Layer of Limestone

Abstract. Microwave attenuation in a wet layer of limestone is measured and compared with the calculations based on 12 models selected from literature describing the effective dielectric constant of mixtures. The measurements cover a full range of moisture from a dry state to a water saturation state. Among the models, the index of refraction mixing model (IRMM) predicts best the measured attenuation in a low-moisture range, while the Polder–van Santen model for randomly oriented needle inclusions (PSM-Needle) predicts best with the measured attenuation in a high-moisture range. However, no reasonable explanation is found why both models predict microwave attenuation well. A new model that agrees well the experimental results is proposed, and its physical meaning is discussed.

9.1. INTRODUCTION

The dipolar relaxation of water molecules results in an absorption of microwave energy, which can be used to measure powder moisture. The absorption of microwave energy can be described by the effective dielectric constant of the powder that consists of particle material, water, and air. A number of models have been proposed for describing the effective dielectric constants of mixed materials [1–3], and a number of works comparing the model predictions with the experimental results have been reported. Among the models, the index of refraction mixing model (IRMM) [4] has shown satisfactory predictions for a wide range of mixtures [3–6]. The authors have measured the moisture dependence and the temperature dependence of microwave attenuation in a wet limestone ($CaCO_3$) layer and have shown that the IRMM predicts well the measured results [7].

In this paper, the experimental data from [7] are compared with the predictions based on the models selected from [1–3]. This paper is organized as follows: Section 9.2 presents the selected models to be compared with the experimental data, Section 9.3 describes the experimental method, Section 9.4 presents the

123

results of that comparison, and Section 9.5 discusses the theoretical basis of the IRMM and proposes a new model describing the effective dielectric constant of mixture.

9.2. MODELS FOR EFFECTIVE DIELECTRIC CONSTANT OF MIXTURE

From Sihvola and Kong [1], Sihvola and co-workers [2], and Nelson and co-workers [3], 12 mixing models have been selected for comparison with experimental results. They are listed in Table 9.1 in a form for describing the effective dielectric constant of three-phase mixtures consists of particle material, water, and air.

Models 1 to 3 are classified as exponential [2] and have the general form

$$\epsilon_{\text{eff}}^{\alpha} = \sum_{i=1}^{n} f_i \epsilon_i^{\alpha}, \qquad \text{with } \alpha = 1, \frac{1}{2}, \text{ or } \frac{1}{3}, \tag{9-1}$$

where ϵ_{eff} is the effective dielectric constant of mixture, i denotes the constituent of the mixture, n is the number of phases in the mixture, f_i is the volume fraction of the ith constituent, and ϵ_i is the dielectric constant of the ith constituent. In Table 9.1, the dielectric constant of air ϵ_a is set to 1. Model 4, by Lichtenecker, has a symmetrical form in the constituent-like exponential models.

Models 5 to 12 are classified as structure-dependent and based on a structure that consists of ellipsoidal inclusions [1,2]. The effective dielectric constant of an n-phase mixture, in which all the ellipsoidal inclusions are randomly oriented, can be written as

$$\epsilon_{\text{eff}} = \epsilon + \frac{\dfrac{1}{3}\sum_{i=1}^{n-1} f_i(\epsilon_i - \epsilon)\sum_{k=1}^{3}\dfrac{\epsilon_{\text{ap}}}{\epsilon_{\text{ap}} + N_k(\epsilon_i - \epsilon)}}{1 - \dfrac{1}{3}\sum_{i=1}^{n-1} f_i(\epsilon_i - \epsilon)\sum_{k=1}^{3}\dfrac{N_k}{\epsilon_{\text{ap}} + N_k(\epsilon_i - \epsilon)}}. \tag{9-2}$$

where ϵ is the dielectric constant of the background material that is taken to be air in Table 9.1, f_i is the volume fraction of the ith phase inclusion, ϵ_i is the dielectric constant of the ith phase, $N_k (k = 1, 2, 3)$ are the depolarization factors of inclusion, and ϵ_{ap} is the apparent dielectric constant [1]. In eq. (9-2), it is assumed that the mixture contains the same kind of ellipsoidal form of inclusions independently from the species. In this paper, only the extreme cases of ellipsoids are discussed, namely a sphere with $N = (\frac{1}{3}, \frac{1}{3}, \frac{1}{3})$, a needle with $N = (0, \frac{1}{2}, \frac{1}{2})$, and a disk with $N = (1, 0, 0)$. When the expression

$$\epsilon_{\text{ap}} = \epsilon + \eta(\epsilon_{\text{eff}} - \epsilon), \qquad \text{with } 0 \leq \eta \leq 1, \tag{9-3}$$

is considered, (1) the generalized Maxwell Garnet (MG) formula is obtained with $\eta = 0$ (the Rayleigh formula for the spherical inclusions), (2) the coherent potential approximation (CP) is obtained with $\eta = 1$, and (3) the Polder–van Santen (PS)

TABLE 9.1. Mixing Formulae for the Effective Dielectric Constant of Three-Phase Mixture, ϵ_{eff}, Selected from [1–3]

(1) Volumetric Mixing Model (VMM)

$$\epsilon_{eff} = f_w \epsilon_w + f_m \epsilon_m + f_a$$

(2) Index of Refraction Mixing Model (IRMM)

$$\epsilon_{eff} = [f_w \epsilon_w^{\frac{1}{2}} + f_m \epsilon_m^{\frac{1}{2}} + f_a]^2$$

(3) Landau-Lifshiz-Looyenga Model (LLLM)

$$\epsilon_{eff} = [f_w \epsilon_w^{\frac{1}{3}} + f_m \epsilon_m^{\frac{1}{3}} + f_a]^3$$

(4) Lichtenecker Model (LM)

$$\ln(\epsilon_{eff}) = [f_w \ln(\epsilon_w) + f_m \ln(\epsilon_m)]$$

(5) Generalized Maxwell Garnet Model, Disk (MGM-Disk)

$$\epsilon_{eff} = \frac{1 + \frac{2}{3}[f_w(\epsilon_w - 1) + f_m(\epsilon_m - 1)]}{1 - \frac{1}{3}[f_w(1 - 1/\epsilon_w) + f_m(1 - 1/\epsilon_m)]}$$

(6) Generalized Maxwell Garnet Model, Needle (MGM-Needle)

$$\epsilon_{eff} = \frac{1 + \frac{1}{3}[f_w(\epsilon_w - 1)(\epsilon_w + 3)/(\epsilon_w + 1) + f_m(\epsilon_m - 1)(\epsilon_m + 3)/(\epsilon_m + 1)]}{1 - \frac{2}{3}[f_w(\epsilon_w - 1)/(\epsilon_w + 1) + f_m(\epsilon_m - 1)/(\epsilon_m + 1)]}$$

(7) Generalized Maxwell Garnet Model, Sphere (MGM-Sphere)

$$\epsilon_{eff} = \frac{1 + 2[f_w(\epsilon_w - 1)/(\epsilon_w + 2) + f_m(\epsilon_m - 1)/(\epsilon_m + 2)]}{1 - [f_w(\epsilon_w - 1)/(\epsilon_w + 2) + f_m(\epsilon_m - 1)/(\epsilon_m + 2)]}$$

(8) Polder–van Santen Model, Needle (PSM-Needle)

$$\epsilon_{eff} = 1 + \frac{5}{3}[f_w(\epsilon_w - 1) + f_m(\epsilon_m - 1)] - \frac{4}{3}\left[\frac{f_w \epsilon_w(\epsilon_w - 1)}{\epsilon_{eff} + \epsilon_w} + \frac{f_m \epsilon_m(\epsilon_m - 1)}{\epsilon_{eff} + \epsilon_m}\right]$$

(9) Polder–van Santen Model, Sphere (PSM-Sphere)

$$\epsilon_{eff} = 1 + \frac{3}{2}[f_w(\epsilon_w - 1) + f_m(\epsilon_m - 1)] - \frac{3}{4}\left[\frac{f_w \epsilon_w(\epsilon_w - 1)}{\epsilon_{eff} + \frac{\epsilon_w}{2}} + \frac{f_m \epsilon_m(\epsilon_m - 1)}{\epsilon_{eff} + \frac{\epsilon_m}{2}}\right]$$

(10) Coherent Potential Model, Disk (CPM-Disk)

$$\epsilon_{eff} = 1 + \frac{4}{3}[f_w(\epsilon_w - 1) + f_m(\epsilon_m - 1)]$$

$$- \frac{1}{3}\left[\frac{f_w(\epsilon_w - 1)(2\epsilon_w - 1)}{\epsilon_{eff} + \epsilon_w - 1} + \frac{f_m(\epsilon_m - 1)(2\epsilon_m - 1)}{\epsilon_{eff} + \epsilon_m - 1}\right]$$

(11) Coherent Potential Model, Needle (CPM-Needle)

$$\epsilon_{eff} = 1 + \frac{4}{3}[f_w(\epsilon_w - 1) + f_m(\epsilon_m - 1)]$$

$$- \frac{1}{6}\left[\frac{f_w(\epsilon_w - 1)(3\epsilon_w - 1)}{\epsilon_{eff} + \frac{\epsilon_w}{2} - \frac{1}{2}} + \frac{f_m(\epsilon_m - 1)(3\epsilon_m - 1)}{\epsilon_{eff} + \frac{\epsilon_m}{2} - \frac{1}{2}}\right]$$

(12) Coherent Potential Model, Sphere (CPM-Sphere)

$$\epsilon_{eff} = 1 + \frac{4}{3}[f_w(\epsilon_w - 1) + f_m(\epsilon_m - 1)]$$

$$- \frac{1}{9}\left[\frac{f_w(\epsilon_w - 1)(4\epsilon_w - 1)}{\epsilon_{eff} + \frac{\epsilon_w}{3} - \frac{1}{3}} + \frac{f_m(\epsilon_m - 1)(4\epsilon_m - 1)}{\epsilon_{eff} + \frac{\epsilon_m}{3} - \frac{1}{3}}\right]$$

[a] Subscripts w, m, and a denote water, material, and air, respectively, ϵ_j ($j = w$ and m) is the dielectric constant of the jth constituent, and f_j ($j = w, m, a$) is the volume fraction of the jth constituent.

[b] PSM-Disk is omitted, because it is identical to MGM-Disk.

mixing formula is obtained with $\eta = 1 - N_k$, (the Bötcher formula for spherical inclusions).

9.3. EXPERIMENTS

Limestone was adopted as a test material because it is scarcely soluble in water (solubility 1.5 mg/100 g) and the moisture control in limestone is of practical importance in the iron-making process [7]. A wet layer of limestone under a 5-mm mesh was set between a transmitting antenna and a receiving antenna, and the attenuation of microwave energy in the layer was measured at various moistures and temperatures as shown in Figure 9.1. The microwave frequencies adopted were 4 GHz (the wavelength in vacuum, $\lambda_0 = 7.5$ cm) and 10.525 GHz ($\lambda_0 = 2.85$ cm). The container in Figure 9.1 was made of polyvinyl chloride plates 5 mm thick and had inner dimensions of $0.5 \times 0.5 \times 0.2$ m^3. The limestone

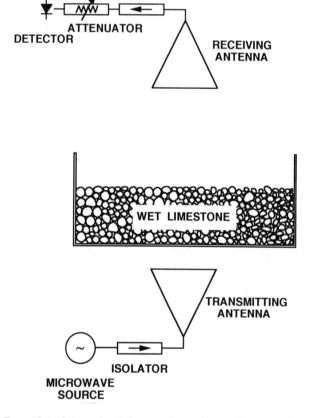

FIGURE 9.1. Schematic of the experimental setup for measuring microwave attenuation in a wet limestone layer.

layer was 13.5 cm thick at 4-GHz measurements or 10.0 cm thick at 10.525 GHz measurements. Optimum pyramidal horns [8] with a gain of 15 dB were used as transmitting and receiving antennas. The separation between antennas was about 40 cm. Microwave attenuation in the sample was determined by adjusting the attenuator in Figure 9.1 to keep the detector output constant. The 0-dB reference of the attenuation was chosen at the received power level when the container was empty. The wet-basis moisture content M was measured with an oven-dry method, where the wet-basis moisture is defined as M = (weight of water contained in the sample)/(weight of the wet sample). The bulk density of the layer was determined by measuring the weight of sample filling the container.

9.4. RESULTS

9.4.1 Bulk Density and Volume Fractions

Figure 9.2 shows the dependence of the bulk density, ρ, of the limestone on the wet-basis moisture content M. The density ρ was almost independent from moisture for $M \leq 0.045$ and increased linearly with moisture for $M \geq 0.045$. The dependence of density on moisture can be expressed as

$$\rho\,(\mathrm{g/cm^3}) = \begin{cases} 1.83 & \text{for } M \leq 0.045, \\ 8M + 1.47 & \text{for } 0.045 \leq M \leq 0.1. \end{cases} \tag{9-4}$$

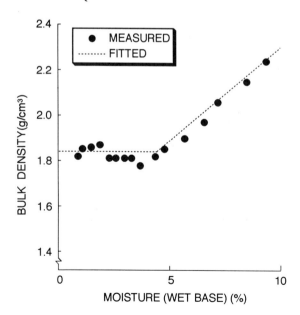

FIGURE 9.2. Bulk density of the moist limestone. $\rho = 1.83$ for $M \leq 0.045$, and $\rho = 8M + 1.47$ for $0.045 \leq M \leq 0.1$.

Figure 9.3 shows the volume fraction of each constituent. The volume fractions were calculated from the equations

$$f_w = (\rho/\rho_w)M,$$

$$f_m = (\rho/\rho_m)(1 - M),$$

and

$$f_a = 1 - (f_w + f_m), \tag{9-5}$$

where subscripts w, m, and a denote water, material, and air, respectively. The adopted data of ρ_w and ρ_m were 1 g/cm^3 and 2.65 g/cm^3, respectively. As shown in Figure 9.3, the sample was saturated with water for $M = 0.1$. Thus the experimental conditions cover a full range of moisture.

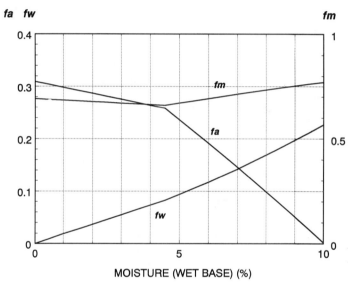

FIGURE 9.3. Dependence of the fractions of constituent f_w, f_m, and f_a on the moisture M.

9.4.2   Microwave Attenuation

Figures 9.4, 9.5, and 9.6 (page 130) compare the measured attenuation with the calculated attenuation. Microwave attenuations were calculated as follows. The complex dielectric constant of the water, ϵ_w, was calculated from the equation

$$\epsilon_w = \epsilon_h + \frac{\epsilon_s - \epsilon_h}{1 + j(\lambda_r/\lambda_0)}, \tag{9-6}$$

with

$$\epsilon_h = 4.5,$$

$$\epsilon_s = 23500/T,$$

$$\lambda_r = (0.618/T)\exp(1980/T) \quad [\text{cm}], \tag{9-7}$$

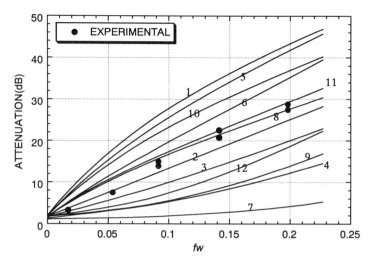

FIGURE 9.4. Dependence of the attenuation on the moisture at 290 K and 4 GHz. Filled circles show the measured attenuation, and curves show the calculated attenuation. Numbers indicate models in Table 9.1.

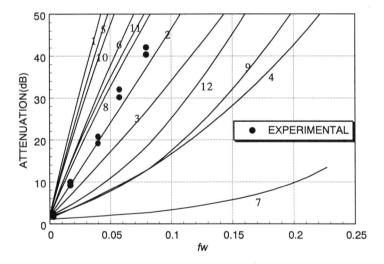

FIGURE 9.5. Dependence of the attenuation on the moisture at 290 K and 10.525 GHz. Filled circles show the measured attenuation, and curves show the calculated attenuation. Numbers indicate models in Table 9.1.

where λ_0 is the wavelength in vacuum, and T is the absolute temperature of the water in kelvins. Equation (9-6) is Debye's relaxation formula [9], and eq. (9-7) was fitted to the experimental data by Grant and co-workers [10]. The adopted data for limestone and air were $\epsilon_m = 10$ and $\epsilon_a = 1$, respectively. (In the previous

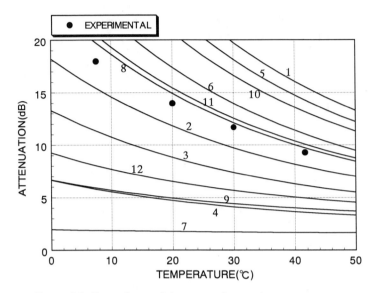

FIGURE 9.6. Dependence of the attenuation on the temperature at 4 GHz. The moisture is 5%. Filled circles show the measured attenuation, and curves show the calculated attenuation. Numbers indicate models in Table 9.1.

paper [7], the authors adopted $\epsilon_m = 4$, which was wrong [11]. But the calculated attenuation was less sensitive to the value of ϵ_m.)

The transmission coefficient of the microwave amplitude through the layer with thickness d and effective dielectric constant ϵ_{eff} is [12]

$$T_A = \frac{p \exp(-j\beta d)}{1 - q \exp(-j\beta d)},$$ (9-8a)

with

$$p = \frac{4\epsilon_{\text{eff}}^{\frac{1}{2}}}{(1 + \epsilon_{\text{eff}}^{\frac{1}{2}})^2},$$ (9-8b)

$$q = \frac{(1 - \epsilon_{\text{eff}}^{\frac{1}{2}})^2}{(1 + \epsilon_{\text{eff}}^{\frac{1}{2}})^2},$$ (9-8c)

and

$$\beta = \frac{4\pi}{\lambda_0}\epsilon_{\text{eff}}^{\frac{1}{2}},$$ (9-8d)

where T_A is the transmission coefficient of microwave amplitude through the layer. When the attenuation through the layer is large, and subsequently the multiple reflection in the layer can be neglected, eq. (9-8a) can be approximated by

$$T_A = p \exp(-j\beta d).$$ (9-9)

Thus, the transmission coefficient of microwave energy through the layer T_E is given by

$$T_E = |T_A|^2 = \frac{16|\epsilon_{\text{eff}}|}{|\epsilon_{\text{eff}}^{\frac{1}{2}} + 1|^4} \exp\left(\frac{4\pi d}{\lambda_0} \operatorname{Im}(\epsilon_{\text{eff}}^{\frac{1}{2}})\right), \tag{9-10}$$

where $\operatorname{Im}(x)$ is the imaginary part of x. The microwave attenuations in Figures 9.4 to 9.6 were calculated by using eq. (9-10).

As shown in Figures 9.4 to 9.6, the calculated attenuations were very much scattered from model to model. Among the exponential models (1–3), the IRMM (2) showed better agreement with the measured attenuation. The VMM (1) overestimated largely the attenuation, while the LLLM (3) underestimated the attenuation. Among the structure-dependent models, the PSM-Needle (8) showed better agreement with the measured attenuation. The difference between the PSM-Needle (8) and the CPM-Needle was small, but the MGM-Needle (6) showed an appreciable overestimation. The disk models (5 and 10) overestimate the attenuation, while the sphere models (7, 9, and 12) underestimate it. The IRMM (2) is better than the PSM-Needle (8) for a low-moisture range ($f_w < 0.08$), while the PSM-Needle is better for a high-moisture range ($f_w > 0.08$).

9.5. DISCUSSION

9.5.1 Mixing Formulae Based on a Stratified Layers Model

The IRMM was proposed by Birchak and co-workers [4] by supposing that plane waves propagate vertically through a stratified medium. In a stratified medium, the electric field E may be described by

$$\frac{d^2 E}{dz^2} + \beta_0^2 \epsilon_r(z) E = 0, \tag{9-11}$$

where z is the direction of propagation and ϵ_r is the dielectric constant of the medium. A WKBJ solution of eq. (9-11) is [13]

$$E = E_0 \exp(-\beta_0 \epsilon_r(z)^{1/2} dz), \tag{9-12}$$

which yields the IRMM. Equation (9-9) is derived on the assumption that

$$\epsilon_r^{\frac{1}{2}} \gg \left|\frac{d\epsilon_r^{\frac{1}{2}}}{dz}\right|; \tag{9-13}$$

that is, the dielectric constant of the medium changes gradually along the propagation path. However, this assumption does not hold for mixtures, in which the dielectric constant changes abruptly at the interfaces. Kraszewski and co-workers [5] also derived the IRMM by considering the propagation of plane waves through stratified layers as shown in Figure 9.7a. They assumed that (1) the thickness of

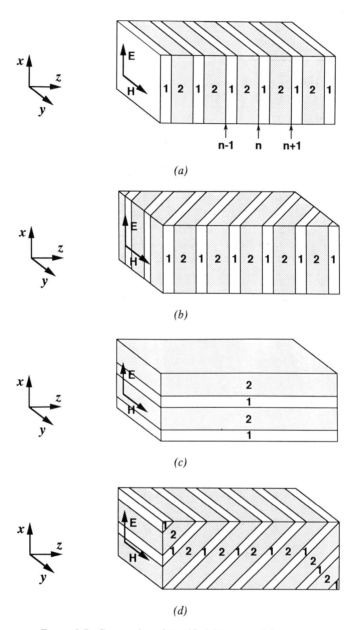

FIGURE 9.7. Geometries of stratified layers model:
(a) propagation of plane waves perpendicular to the layers,
(b) propagation of plane waves in TE mode, general case,
(c) propagation of plane waves parallel to the layers, and
(d) propagation of plane waves in TM mode, general case.

each layer is much smaller than the wavelength, and (2) the interface reflections can be neglected. In this subsection, we deal with this problem more rigorously.

Transmission of Plane Waves Vertical to Stratified Layers. For the stratified two-phase mixture in Figure 9.7a, the transverse fields on the nth interface can be related to the transverse fields on the $(n + 1)$st interface by a four-terminal transmission line matrix, such as

$$\begin{bmatrix} E_n \\ H_n \end{bmatrix} = \begin{bmatrix} A & B \\ C & D \end{bmatrix} \begin{bmatrix} E_{n+1} \\ H_{n+1} \end{bmatrix}. \tag{9-14}$$

The elements in the matrix are

$$A = \cos \theta_1 \cos \theta_2 - (Z_1/Z_2) \sin \theta_1 \sin \theta_2,$$

$$B = jZ_1 \sin \theta_1 \cos \theta_1 + jZ_2 \cos \theta_1 \sin \theta_2,$$

$$C = j(1/Z_1) \sin \theta_1 \cos \theta_2 + j(1/Z_2) \cos \theta_1 \sin \theta_2,$$

$$D = \cos \theta_1 \cos \theta_2 - (Z_2/Z_1) \sin \theta_1 \sin \theta_2, \tag{9-15}$$

where $\theta_i = \beta_i d_i$ $(i = 1, 2)$, β_i is the propagation constant in the ith phase, d_i is the layer thickness of the ith phase, and Z_i is the characteristic impedance of the ith phase. When each layer consists of a dielectric material with dielectric constant ϵ_i, β_i and Z_i are related to the propagation constant in vacuum, β_0, and the characteristic impedance of a vacuum, Z_0, as follows:

$$\beta_i = \beta_0 \epsilon_i^{1/2} \quad \text{and} \quad Z_i = Z_0/\epsilon_i^{1/2} \quad (i = 1, 2). \tag{9-16}$$

When the structure of the stratified medium is periodic, the following equation holds [14]:

$$\begin{bmatrix} E_n \\ H_n \end{bmatrix} = \begin{bmatrix} \exp(j\beta d) & 0 \\ 0 & \exp(j\beta d) \end{bmatrix} \begin{bmatrix} E_{n+1} \\ H_{n+1} \end{bmatrix}, \tag{9-17}$$

where $d = d_1 + d_2$ and β is the effective propagation constant in the stratified medium. From eqs. (9-14), (9-15), and (9-17), we obtain

$$\cos(\beta d) = (A + D)/2$$

$$= \cos \theta_1 \cos \theta_2 - \tfrac{1}{2}(Z_2/Z_1 + Z_1/Z_2) \sin \theta_1 \sin \theta_2. \tag{9-18}$$

When the thickness of each layer is much smaller than the wavelength (i.e., $|\beta_i d_i| \ll 1$), we can, by applying the approximations

$$\cos \theta_i = 1 - \tfrac{1}{2}\theta_i^2 \quad \text{and} \quad \sin \theta_i = \theta_i,$$

get the following equation from eqs. (9-16) and (9-18):

$$\epsilon_{\text{eff}} = \epsilon_1 f_1 + \epsilon_2 f_2, \tag{9-19}$$

where $f_i = d_i/d$ and ϵ_{eff} is defined by

$$\beta = \beta_0 \epsilon_{\text{eff}}^{\frac{1}{2}}. \tag{9-20}$$

Equation (9-19) is the VMM formula.

Since the assumption of periodic structure may be too restrictive, the effects of random distribution of layer thickness were studied by computer simulation. The adopted distribution function was Rosin-Rammler type [15]:

$$p(d_i) = \exp(-d_i/\langle d_i \rangle)/\langle d_i \rangle, \tag{9-21}$$

where d_i is the layer thickness of the ith material ($i = 1, 2$), and $\langle d_i \rangle$ denotes the average thickness. In the simulation, the following values were adopted: $\epsilon_1 = 4.0$, $\epsilon_2 = \epsilon_w$ at 290 K, $\lambda = 7.5$ cm, $\langle d_1 \rangle = 0.1$ cm, $\langle d_2 \rangle = f_2(\langle d_1 \rangle + \langle d_2 \rangle)$, and $f_2 = 0$–0.2. The number of layers was 10^4. The fields at the interface were calculated step by step from eqs. (9-14)–(9-15), and the effective dielectric constant of the medium was determined from the phase delay through the total layers. The results are shown in Figure 9.8. The VMM is also valid for the random distribution case. Those results imply that the interface reflections cannot be neglected even if the layer thickness is much smaller than the wavelength and that the IRMM cannot be derived from the stratified layer model.

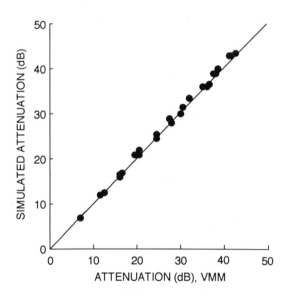

FIGURE 9.8. Comparison of the simulated attenuation in stratified layers with VMM attenuation. The thickness of each layer is randomly distributed.

Furthermore, similar calculation for a more general case as in Figure 9.7b resulted in eq. (9-19). The VMM is valid for the stratified layer model when the electric field is parallel to the layer (i.e., TE mode). Therefore the effective dielectric constant of stratified layers for the TE mode is

$$\epsilon_{TE} = \epsilon_1 f_1 + \epsilon_2 f_2. \tag{9-22}$$

The VMM can also be derived by assuming that the tangential electric field E is uniform in the medium. For both geometries in Figures 9.7a and 9.7b, an average electric flux density $\langle D \rangle$ can be written as

$$\langle D \rangle = \epsilon_{eff} E = (\epsilon_1 f_1 + \epsilon_2 f_2) E. \tag{9-23}$$

Thus, eq. (9-23) yields eq. (9-22).

Transmission of Plane Waves Parallel to Stratified Layers. The geometry in Figure 9.7c is considered. By assuming that the magnetic field has only a y component (i.e., the waves propagate in a TM mode), we derive the equations

$$\frac{d^2 H_y}{dy^2} + (\beta_0^2 \epsilon_i - \beta^2) H_y = 0, \qquad i = 1, 2, \tag{9-24}$$

and

$$E_z = \frac{Z_0}{j\beta_0 \epsilon_i} \frac{d H_y}{dx}, \qquad i = 1, 2, \tag{9-25}$$

where β_0 and Z_0 are the propagation constant and characteristic impedance in a vacuum, respectively, and β is the propagation constant in the z direction. When the structure is periodic, the solutions of eq. (9-24) are

$$H_y = H_1 \cos(\beta_1 x), \qquad \text{for } |x| < a,$$

and

$$H_y = H_2 \cos(\beta_2(x - b)), \qquad \text{for } a < |x| < b,$$

with

$$\beta_i^2 = \beta_0^2 \epsilon_i - \beta^2, \quad i = 1, 2, \quad a = d_1/2, \quad \text{and} \quad b = (d_1 + d_2)/2. \tag{9-26}$$

By applying the continuity conditions for E_z and H_y at $x = a$, we get

$$(\beta_1/\epsilon_1) \tan(\beta_1 a) = (\beta_2/\epsilon_2) \tan(\beta_2(a - b)). \tag{9-27}$$

When the thickness of each layer is much smaller than the wavelength (i.e., $|\beta_i d_i| \ll 1$), the following equation for the effective dielectric constant of the TM mode can be obtained from eq. (9-20) and the approximation of $\tan \theta = \theta$:

$$\frac{1}{\epsilon_{TM}} = \frac{f_1}{\epsilon_1} + \frac{f_2}{\epsilon_2}. \tag{9-28}$$

For the geometry in Figure 9.7d, a mixing formula identical to eq. (9-28) can be obtained by assuming that the tangential electric field is uniform in the medium. In this case, an average electric field in the x direction $\langle E_x \rangle$ is given by

$$\langle E_x \rangle = E_{1x} f_1 + E_{2x} f_2, \tag{9-29}$$

and the electric flux density is

$$D = \epsilon_{TM}\langle E_x\rangle = \epsilon_1 E_{1x} = \epsilon_2 E_{2x}. \qquad (9\text{-}30)$$

The combination of eqs. (9-29) and (9-30) yields eq. (9-28).

The calculation based on eq. (9-27) showed a very small attenuation of less than 1 dB for $M = 0.1$ at 4 GHz. The results show that the water in a wet material absorbs scarcely the microwave energy in the TM mode.

Dual-Mode Mixing Model. If we assume that the wet substance consists of randomly oriented stratified layers, the effective dielectric constant of the substance may be written as

$$\epsilon_{eff} = [\epsilon_{TE} + \epsilon_{TM}]/2. \qquad (9\text{-}31)$$

This model is referred to as the dual-mode mixing model (DMMM). The comparison of the calculated attenuation based on the DMMM with the measured attenuation is shown in Figures 9.9 to 9.11. The DMMM agrees well with the measured attenuation for a high-moisture range as does the PSM-Needle, but the IRMM is better for a low-moisture range.

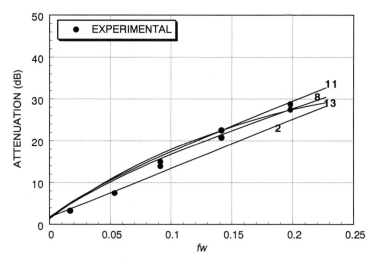

FIGURE 9.9. Dependence of the attenuation on the moisture at 290 K and 4 GHz. Filled circles show the measured attenuation, and curves show the calculated attenuation. Numbers indicate models in Table 9.1. "13" refers to the DMMM model.

The effective constants for the TE and TM modes derived above are closely related to a structure-dependent model for disk-shaped inclusions described in Section 9.2. For uniformly oriented disks, (1) the effective dielectric constant is identical to ϵ_{TE} when the short axis of each disk is perpendicular to the electric field, and (2) the effective dielectric constant is identical to ϵ_{TM} when the short

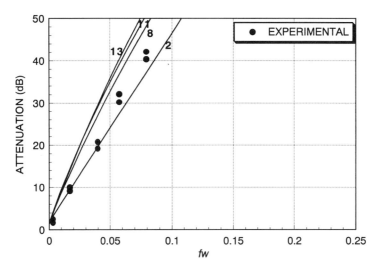

FIGURE 9.10. Dependence on the attenuation on the moisture at 290 K and 10.525 GHz. Filled circles show the measured attenuation, and curves show the calculated attenuation. Numbers indicate models in Table 9.1. "13" refers to the DMMM model.

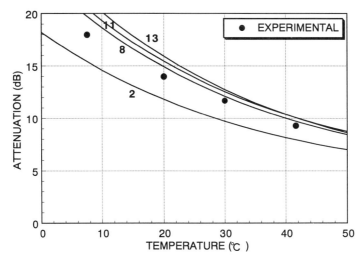

FIGURE 9.11. Dependence of the attenuation on the temperature at 4 GHz. The moisture is 5%. Filled circles show the measured attenuation, and curves show the calculated attenuation. Numbers indicate models in Table 9.1. "13" refers to the DMMM model.

axis of each disk is parallel to the electric field. The comparison of the DMMM with the disk inclusions models, 5 and 10 in Table 9.1, is discussed in the next subsection.

9.5.2   Mixing Formulae Based
on an Ellipsoidal Scatters Model

The structure-dependent models 5–12 in Table 9.1 have clearer physical meanings than the exponential models, and the PSM-Needle showed good agreement with the measured attenuation, as shown in Figures 9.4 to 9.6. Further calculations on the assumptions of spherical limestone and water needle showed the microwave attenuation very close to that of the PSM-Needle. Thus, the model prediction only depends on the shape of water. However, the existence of water needle seems unreal in the powder saturated by water.

Structure-dependent models are derived quite similarly to Onsager's theory on dielectric materials [16]. In Onsager's theory, an inclusion is a constituent molecule having a microscopic length scale (L_m) around 10^{-8} cm, and macroscopic polarization P is given by an average of molecular polarization p in a macroscopic volume $V_M (= L_M^3)$. The macroscopic length scale L_m should be much smaller than the wavelength in the material λ, say $L_M = 0.1\lambda$. The real part of the effective dielectric constant of a wet limestone layer was calculated to be about 9 by the PSM-Needle or the DMMM at $f_w = 0.1$ and $T = 290$ K; i.e., the macroscopic length scale is about 0.25 cm at 4 GHz. Since a very large number of molecules are contained in the macroscopic volume, the macroscopic polarization becomes isotropic for randomly oriented inclusions, even when each inclusion has an anisotropic properties. Thus, the effective dielectric constant can be expressed by the scalar quantity

$$D = \epsilon_{\text{eff}} E = \epsilon E + P. \tag{9-32}$$

Structure-dependent models are valid for $L_m \ll L_M$. On the other hand, for a wet powder, the microscopic length scale is of the order of the particle size that is a 0.5-cm mesh under for the experimental condition; i.e., the macroscopic-length scale is of the same order as the microscopic length scale. In this case, the effective dielectric constant becomes a tensor and is expressed by the following equation instead of eq. (9-31):

$$D = [\epsilon_{\text{eff}}]E = \epsilon E + P, \tag{9-33}$$

where [   ] denotes a tensor. For a disk,

$$[\epsilon_{\text{eff}}] = \epsilon_{\text{TE}} u_1 u_1 + \epsilon_{\text{TM}} u_2 u_2 + \epsilon_{\text{TM}} u_3 u_3, \tag{9-34}$$

where $u_1 (i = 1, 2, 3)$ are unit vectors. For predicting the plane-wave attenuation in the medium, a scalar effective dielectric constant can be defined by

$$\epsilon_{\text{eff}} \langle E \cdot E \rangle = \langle D \cdot E \rangle, \tag{9-35}$$

where $\langle \ \rangle$ denotes the average in the volume with order (illuminated area) × (the macroscopic length scale). The DMMM can be also obtained from eq. (9-34). Therefore, the DMMM can be interpreted such that it gives the effective dielectric constant of randomly oriented disks, sizes of which are comparable to the macroscopic length scale. Since the assumed shape of limestone particle scarcely affects the calculated attenuation, the DMMM is responsible for disk-shaped water or water films. Thus, the DMMM predicts well the microwave attenuation in a high-moisture range where the whole water is film shaped. The DMMM overestimates the microwave attenuation in a low-moisture range where water droplets and water films coexist, because water droplet or spherical water do not attenuate microwaves as much, as shown in Figures 9.4 to 9.6.

9.6. CONCLUSIONS

Microwave attenuation in a wet layer of limestone was measured as a function of the moisture and the temperature and compared with calculations based on 12 models of effective dielectric constant selected from the literature: (1) a group of exponential models consisting of the volumetric mixing model (VMM), the index of refraction mixing model (IRMM), and Landau-Lifshiz-Looyenga model (LLLM), (2) Lihtenecker model (LM), and (3) a group of structure-dependent models consisting of generalized Maxwell Garnet model (MGM), Polder–van Santen model (PSM), and coherent potential model (CPM) for randomly oriented disks, randomly oriented needles, and spheres. The experimental conditions covered almost the full moisture range, from a dry state to a water saturation state. Among the models, the IRMM predicted best the microwave attenuation in a low-moisture range, while the PSM-Needle predicted best the microwave attenuation in a high-moisture range. However, no reasonable explanation is found why both models predict well microwave attenuation. The IRMM has a weak theoretical basis, and the constituent materials of the wet limestone layer does not have needlelike shapes.

A new mixing model called the dual-mode mixing model (DMMM) has been proposed and compared with the measured results. The DMMM predicts well microwave attenuation in the high-moisture range. The theoretical basis of the DMMM is discussed in terms of a stratified-layer geometry as well as a randomly oriented disk geometry. The DMMM is responsible for microwave attenuation by water films when the particle size is not extremely smaller than the wavelength. The behavior of the microwave attenuation in a low-moisture range is also discussed.

ACKNOWLEDGMENTS

The author sincerely thanks Dr. A. Kraszewski, United States Department of Agriculture, for allowing the author to provide the present paper and his papers concerning mixture theory. The author also thanks S. Shibagaki and Y. Sakamoto, Sumitomo Metals. Shibagaki provided the dielectric constant of $CaCO_3$, and Sakamoto assisted the author in numerical calculation.

REFERENCES

[1] A. H. Sihvola and J.-A. Kong, "Effective permittivity of dielectric mixtures," *IEEE Trans. Geosci. Remote Sensing*, vol. 26, no. 4, pp. 420–429, 1988.

[2] A. Sihvola, E. Nyfors, and M. Tiuri, "Mixing formulae and experimental results for the dielectric constant of snow," *J. Glaciology*, vol. 31, no. 108, pp. 163–170, 1985.

[3] S. Nelson, A. Kraszewski, and T. You, "Solid and particulate material permittivity relationships," *J. Microwave Power Electromag. Energy*, vol. 26, no. 1, pp. 45–51, 1991.

[4] J. Birchak, C. Gardner, J. Hipp, and J. Victor, "High dielectric constant microwave probes for sensing soil moisture," *Proc. IEEE*, vol. 62, no. 1, pp. 93–98, 1974.

[5] A. Kraszewski, S. Kulinski, and M. Matuszewski, "Dielectric properties and a model of biphase water suspension at 9.4 GHz," *J. Appl. Phys.*, vol. 47, no. 4, pp. 1275–1277, 1976.

[6] A. Kraszewski, "Prediction of the dielectric properties of two-phase mixtures," *J. Microwave Power*, vol. 12, no. 3, pp. 215–222, 1977.

[7] T. Shiraiwa, S. Kobayashi, A. Koyama, M. Tokuda, and S. Koizumi, "Microwave moisture gauge in limestone," *J. Microwave Power*, vol. 15, no. 4, pp. 255–260, 1980.

[8] W. C. Jakes, Jr., "Horn antennas," in *Antenna Engineering Handbook*, H. Jasik, Ed., New York: McGraw-Hill, ch. 10, p. 0.8, 1961.

[9] H. Frölich, *Theory of Dielectrics, Dielectric Constant and Dielectric Loss*, Oxford: Carendon Press, 1958, ch. 3, pp. 70–90.

[10] E. Grant, J. Buchanan, and H. Cook, "Dielectric behavior of water at microwave frequencies," *J. Chem. Phys.*, vol. 26, no. 1, pp. 156–161, 1957.

[11] S. Shibagaki, private communication.

[12] J. A. Sttraton, *Electromagnetic Theory*, New York: McGraw-Hill, 1941, pp. 511–513.

[13] P. M. Morse and H. Feshbach, *Methods of Theoretical Physics*, Part II, New York: McGraw-Hill, 1953, ch. 9, pp. 1092–1095.

[14] R. E. Collin, *Foundations of Microwave Engineering*, New York: McGraw-Hill, 1966, ch. 8.

[15] T. Allen, *Particle Size Measurement*, London: Chapman and Hall, 1968, ch. 2, p. 31.

[16] H. Frölich, *Theory of Dielectrics, Dielectric Constant and Dielectric Loss*, Oxford: Clarendon Press, 1958, ch. 2, pp. 21–26.

10

Tarek M. Habashy
Pabitra N. Sen
M. Reza Taherian
Schlumberger-Doll Research, Ridgefield, CT

Hydrocarbon and Water Estimation in Reservoirs Using Microwave Methods

Abstract. The pore space of oil field reservoir rocks is filled with water and hydrocarbons. We discuss the problem of estimating hydrocarbon content from microwave measurements of the permittivity and conductivity. We first review some borehole measurement tools and then illustrate the mixing laws as well as the laboratory measurements which are used to interpret the data. The currently available commercial tools employ frequencies ranging from 2 MHz to 1.1 GHz. We highlight features of a device operating at 1.1 GHz and its antenna configuration. The modeling of the electromagnetic wave propagation in a borehole geometry, which is required to estimate permittivity and conductivity from tool measurements, is described. Next, we discuss the application of composite media models to extract the volume fraction of hydrocarbons from the estimated permittivity and conductivity. Complexity of the rock geometry and the presence of mobile counterions in clays render the petrophysical modeling a challenging task. The laboratory data taken over a wide frequency range is used to test the validity and limitations of the models.

10.1. INTRODUCTION

The purpose of this paper is to present some of the current methods of estimating hydrocarbons and water in geological formations using microwave methods. Microwave methods measure the permittivity of the formation, $\epsilon(\omega) = \epsilon'(\omega) - \sigma(\omega)/i\omega\epsilon_0$. Here ω is the angular frequency, $\epsilon'(\omega)$ and $\sigma(\omega)$ are the real part of the permittivity and the conductivity, respectively, and ϵ_0 is the permittivity of vacuum. The real and imaginary parts of the permittivity are related by the well-known Kramers-Kronig relation.

The electrical conductivity has always been the most important logging measurement in estimating hydrocarbon content of the reservoir. A common practice

141

in borehole logging is to estimate the water saturation S_w, which is the fraction of pores occupied by water, from the measured electrical conductivity σ of the rock and the conductivity of the saturating water σ_w. In practice, however, this method often fails because the water conductivity is unknown, or it varies rapidly with depth, or it is too small, or the interpretation schemes based on conductivity fail. It is highly desirable to have a method of estimating hydrocarbons which does not depend on water conductivity. The dielectric measurement is such an avenue [1].

The permittivity of water, ϵ'_w, is about 80 in the gigahertz frequency range while that of rock matrix is between 5 and 9 and of hydrocarbons is about 2. In addition to this large contrast, a weak salinity dependence of ϵ'_w was a motivation for using ϵ to estimate S_w. However, the permittivity of water-saturated rocks shows a complicated frequency-dependent behavior [2]. In the low-frequency range (kHz and below), the permittivity can be as large as 10^8 due to the presence of clays. In the mid-megahertz range there is a strong influence of grain texture. Both of these effects become smaller at higher frequencies in the gigahertz range. We discuss a microwave device that can measure the permittivity and which operates at 1.1 GHz.

10.2. THE ELECTROMAGNETIC PROPAGATION TOOL

The existing Electromagnetic Propagation Tool (EPT trademark of Schlumberger) is a linear array consisting of two-receiver microwave slot antennas sandwiched between two transmitter antennas [3]. Each antenna is a cavity-backed slot antenna. The transmitters are excited by coaxial lines with their inner conductors shorted to the far wall of the cavity. The receivers measure the propagated signal by the same setup where the voltage or current is measured at the input of the coaxial line. These antennas are mounted on a metallic pad which is pressed against the borehole wall. Each transmitter is turned on separately, while phase and amplitude measurements are made at each receiver. In this way the phase shift and attenuation between the two receivers is recorded for signals traveling in opposite directions. Averaging the two readings serves to eliminate imbalances in the two receivers, helps to correct for imperfections on the borehole wall, and symmetrizes the tool response to thin beds. This mode of operation is known as the borehole compensated mode.

Since these slot antennas are small relative to the wavelength, they effectively behave as point magnetic dipoles with moments perpendicular to the axis of the feeding coaxial line. There are two versions of the EPT tool, which differ in the orientations of the magnetic dipoles on the tool pad. In one version, the antennas are mounted so that the dipole moments point end on to each other. This is called the endfire magnetic dipole array. In the other version, the antennas are mounted so that the dipole moments are broadside to each other. This is referred to as the broadside array [4,5].

Because of the high frequency of operation, these antennas are shallow and have a depth of investigation of about 15 cm (6 inches). Thus, the EPT is only sensitive to the part of the formation that has been invaded by the borehole fluid and a thin standoff layer that exists between the tool pad and the formation. For the broadside array, the signal measured at a receiver a distance ρ away from a transmitter is given by

$$S_b = \frac{M}{\pi}\left[\frac{k_1^2}{2\rho}\left(1 + \frac{i}{k_1\rho} - \frac{1}{(k_1\rho)^2}\right)e^{ik_1\rho} + ik_1^2\int_0^\infty dk_\rho \frac{k_\rho}{k_{1z}}J_1'(k_\rho\rho)\frac{R_{12}^{TM}e^{i2k_{1z}h}}{1 - R_{12}^{TM}e^{i2k_{1z}h}}\right.$$

$$\left. - \frac{i}{\rho}\int_0^\infty dk_\rho\, k_{1z}J_1(k_\rho\rho)\frac{R_{12}^{TE}e^{i2k_{1z}h}}{1 + R_{12}^{TE}e^{i2k_{1z}h}}\right].$$

For the endfire array, the receiver response is

$$S_e = \frac{iM}{\pi}\left[-\frac{k_1}{\rho^2}\left(1 + \frac{i}{k_1\rho}\right)e^{ik_1\rho} + \frac{k_1^2}{\rho}\int_0^\infty dk_\rho\frac{1}{k_{1z}}J_1(k_\rho\rho)\frac{R_{12}^{TM}e^{i2k_{1z}h}}{1 - R_{12}^{TM}e^{i2k_{1z}h}}\right.$$

$$\left. - \int_0^\infty dk_\rho\, k_\rho k_{1z}J_1'(k_\rho\rho)\frac{R_{12}^{TE}e^{i2k_{1z}h}}{1 + R_{12}^{TE}e^{i2k_{1z}h}}\right],$$

where M is the transmitter's dipole moment. $J_1(\cdot)$ is the regular Bessel function of the first order, $J_1'(\cdot)$ is its derivative with respect to its argument, and R_{12}^{TE} and R_{12}^{TM} are Fresnel's reflection coefficients at the boundary-separating media (1) and (2) for the transverse-electric (TE) and transverse-magnetic (TM) polarizations, respectively, and are

$$R_{12}^{TE} = \frac{k_{1z} - k_{2z}}{k_{1z} + k_{2z}}, \qquad R_{12}^{TM} = \frac{\epsilon_2 k_{1z} - \epsilon_1 k_{2z}}{\epsilon_2 k_{1z} + \epsilon_1 k_{2z}},$$

$\epsilon_1 = \epsilon_1' - \sigma_1/i\omega\epsilon_0$ and $\epsilon_2 = \epsilon_2' - \sigma_2/i\omega\epsilon_0$ are the permittivities of the standoff layer and the invaded formation, respectively. The thickness of the standoff layer is h, and

$$k_\rho^2 + k_{1z}^2 = k_1^2 = \omega^2\mu_0\epsilon_0\epsilon_1,$$

$$k_\rho^2 + k_{2z}^2 = k_2^2 = \omega^2\mu_0\epsilon_0\epsilon_2.$$

From the phase shift and amplitude ratio measurements, one can derive an effective permittivity and conductivity. These represent an averaged permittivity and conductivity which not only take into consideration the effect of the invaded zone but also any standoff layer that exists between the tool pad and the borehole wall. This standoff layer can be a mudcake layer in front of a permeable sand zone or a layer of borehole mud in front of an impervious shale zone.

Performing an endfire-only or a broadside-only measurement is not enough to obtain the electrical parameters of the invaded zone. This is true since any one set of these measurements provides only two independent measurements, which are not sufficient for the inversion of the two electrical parameters of the invaded zone and those of the standoff layer plus its thickness.

To overcome this problem, we have introduced a cross-dipole antenna that allows the simultaneous endfire and broadside measurements (see Figure 10.1). This provides four independent equations that can be combined with an independent measurement of the standoff thickness (from an ultrasonic device that employs a pulse-echo technique) to invert for the four unknown electrical parameters [6–8].

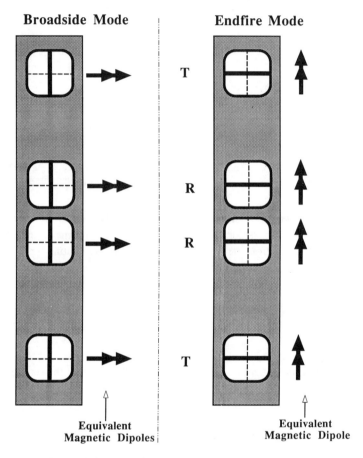

FIGURE 10.1. Schematic of the cross-dipole antenna array showing the endfire and broadside modes of operation.

Figure 10.2 shows the inverted permittivity and conductivity by the endfire mode only, the broadside only, and the cross-dipole mode as a function of the standoff thickness. Also in Figure 10.2, the independently measured formation parameters are depicted by solid lines. It is clear that the endfire only or the broadside only provides the formation permittivity and conductivity with gross errors beyond 0.5-cm standoff, whereas the cross-dipole provides an accurate estimate up to 1.27 cm and even beyond.

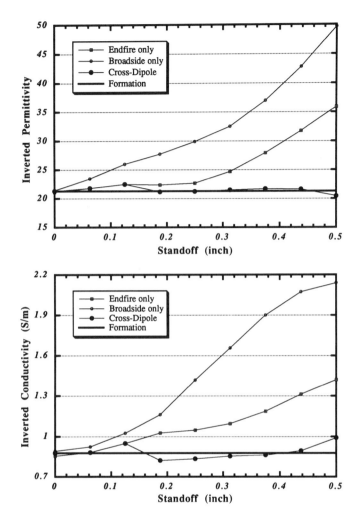

FIGURE 10.2. Inverted permittivity and conductivity using endfire-only, broadside-only, or cross-dipole measurements.

10.3. INTERPRETATION OF THE DATA

In interpreting the dielectric logs, the water saturation is determined by a naive formula which is based on the assertion that the "complex travel time" through a formation is a sum of the corresponding transit times through the components, giving

$$\epsilon^{1/2} = S_w \phi \epsilon_w^{1/2} + (1 - S_w) \phi \epsilon_h^{1/2} + (1 - \phi) \epsilon_m^{1/2}, \qquad (10\text{-}1)$$

where ϕ is the formation porosity and S_w is the water saturation. The dielectric constants ϵ_h and ϵ_m are the permittivities of hydrocarbons and the solid matrix, respectively, which are generally real. The dielectric constant of water $\epsilon_w(\omega) =$

$\epsilon'_w(\omega) - \sigma_w(\omega)/i\omega\epsilon_0$ is complex, and the water conductivity, $\sigma_w(\omega)$, includes any loss due to dipolar rotation.

The above equation often works; however, it would imply that the permittivity should be the same if all quantities on the right-hand side were the same. But experiments show that even when all the quantities on the right-hand side are the same, the permittivity can vary from rock to rock due to textural differences. To demonstrate this, in Figure 10.3 we show the experimentally measured dielectric permittivities of two carbonate rocks as a function of frequency. Both these rock samples have the same porosity of 24% and are fully saturated ($S_w = 1$) with 0.2 ohm-m brine. This figure clearly demonstrates the short comings of eq. (10-1) in describing the permittivity of the rocks.

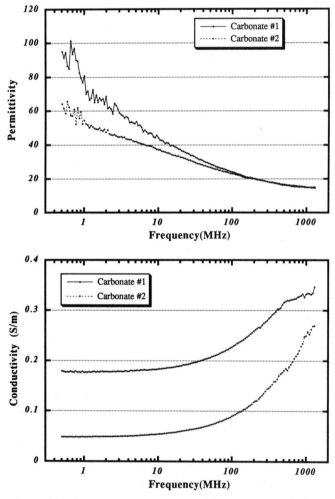

FIGURE 10.3. Laboratory-measured permittivity and conductivity of two carbonate rock samples having the same porosity of 24%.

Microscopic examination of the rock thin sections suggest that rocks with dispersion contain platey grains with one small dimension, and two large dimensions. These platey grains can act as thin capacitors and thereby increase the permittivity of the rock. To model the permittivity of rocks we need to include a distribution of grains having different aspect ratios. Using effective medium theory, Sen [9] has derived the following expression for the change in permittivity of such system arising from a $d\phi$ change in porosity due to an addition of matrix of dielectric constant ϵ_m:

$$-\frac{d\phi}{3\phi} = \frac{d\epsilon}{\epsilon(\epsilon_m - \epsilon)} \frac{1}{\left\langle \dfrac{(1+3L)\epsilon_m + (5-3L)\epsilon}{[L\epsilon_m + (1-L)\epsilon][(1-L)\epsilon_m + (1+L)\epsilon]} \right\rangle}, \qquad (10\text{-}2)$$

where $\epsilon = \epsilon' - \sigma/i\omega\epsilon_0$ is the dielectric constant of the mixture, which is complex. Angular brackets denote an average over the distribution of L-values, and L is the depolarization factor of a spheroid along its principle axis. In deriving eq. (10-2), the volume fraction of solid is built up by the iterative dilute limit method. It involves computing the effective medium properties of a mixture of insulating grains (of an infinitesimal amount) in a conducting host; the resulting medium is used as a host, to which more grains are added. Hydrocarbons are included in an analogous manner [10].

For an arbitrary probability distribution of grains of different aspect ratios, eq. (10-2) cannot be integrated easily. Kenyon [11] showed that the model with only a single value of L did not agree well with experimental results. Thus we need to consider slightly more complicated distribution of L values. The next grain distribution that can be integrated is the bimodal distribution. The grain distribution used is a mixture of spherical grains and platey grains of one aspect ratio. Assuming a value of $L = 1 - \delta$ with a probability p for the platey grains and $L = 1/3$ with a probability $1 - p$ for the spheres in eq. (10-2) gives

$$-\frac{d\phi}{3\phi} = \frac{d\epsilon}{\epsilon(\epsilon_m - \epsilon)} \frac{(\epsilon_m + 2\epsilon)K}{9(1-p)K + p(\epsilon_m + 2\epsilon)[(4-3\delta)\epsilon_m + (2+3\delta)\epsilon]}.$$

where $\qquad\qquad K = [(1-\delta)\epsilon_m + \delta\epsilon][\delta\epsilon_m + (2-\delta)\epsilon].$

Integrating after partial fractioning and using the boundary condition that $\epsilon = \epsilon_w$ (the dielectric constant of water) when $\phi = 1$, we get

$$\phi^{-1/3} = \left(\frac{\epsilon}{\epsilon_w}\right)^{p_1} \left(\frac{\epsilon - \epsilon_m}{\epsilon_w - \epsilon_m}\right)^{p_2} \left(\frac{\epsilon - \epsilon_1}{\epsilon_w - \epsilon_1}\right)^{p_3} \left(\frac{\epsilon - \epsilon_2}{\epsilon_w - \epsilon_2}\right)^{p_4},$$

where the parameters are defined by the following series of equations [11]:

$$p_1 = \frac{\delta(1-\delta)}{9(1-p)\delta(1-\delta) + p(4-3\delta)},$$

$$p_2 = -\frac{1}{3},$$

$$g_1 = 9(1 - p)\delta(2 - \delta) + p(4 + 6\delta),$$

$$g_2 = \epsilon_m[(9(1 - p)(2\delta^2 - 3\delta + 2) + p(10 - 3\delta)],$$

$$g_3 = \epsilon_m^2[9(1 - p)\delta(1 - \delta) + p(4 - 3\delta)],$$

$$w_1 = \left[\frac{g_2^2 - 4g_1g_3}{4g_1^2}\right]^{1/2},$$

$$w_2 = -\frac{g_2}{2g_1},$$

$$\epsilon_1 = w_2 + w_1,$$

$$\epsilon_2 = w_2 - w_1,$$

$$p_3 = \frac{(\epsilon_m + 2\epsilon_1)[(1 - \delta)\epsilon_m + \delta\epsilon_1][\delta\epsilon_m + (2 - \delta)\epsilon_1]}{\epsilon_1 g_1(\epsilon_m - \epsilon_1)(\epsilon_1 - \epsilon_2)},$$

$$p_4 = \frac{(\epsilon_m + 2\epsilon_2)[(1 - \delta)\epsilon_m + \delta\epsilon_2][\delta\epsilon_m + (2 - \delta)\epsilon_2]}{\epsilon_2 g_1(\epsilon_m - \epsilon_2)(\epsilon_2 - \epsilon_1)}.$$

This model is found to explain the frequency dependent ϵ and σ of several clay-free rocks measured in the laboratory over a wide range of frequency and water conductivity. For example, using a nonlinear least-square-fitting routine, we have determined the parameters δ and p for one of the rocks of Figure 10.3. These results, shown in Figure 10.4, lead to a fitted $\delta = 0.112$ and $p = 0.79$. These parameters suggest relatively high fraction of platey grains.

The parameters δ and p can be obtained by applying the model to the experimental results at a single frequency. The values of these parameters when used as input to the model, can reproduce experimentally measured values of ϵ' and σ from 0.5 MHz to 1.3 GHz [12].

In principle, the method is straightforward to generalize the iterated dilute limit technique to allow for a spectrum of grain fractions with different aspect ratios. However, trying to solve the inverse problem of finding the aspect ratio spectrum from dielectric measurements over a wide range of frequency is very difficult. This is because the problem is inherently ill-posed and requires high accuracy in the experimental measurements and water properties. Also, grain-based models may not be so useful for predicting other physical properties of rocks such as fluid permeability which depend strongly on the pore geometry and size rather than grain geometry. Models for dielectric dispersion based on pore geometry consistent with fluid flow properties will be of practical importance.

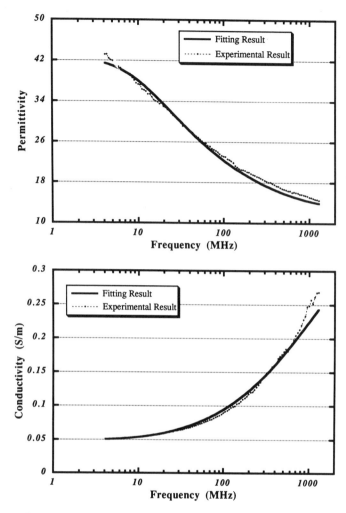

FIGURE 10.4. Fitting the bimodal model to the permittivity and conductivity of carbonate rock 2.

REFERENCES

[1] R. A. Meador and P. T. Cox, "Dielectric constant logging, a salinity independent estimation of formation water volume," SPE paper 5504, 1975.

[2] M. R. Taherian, W. E. Kenyon, and K. A. Safinya, "Measurement of dielectric response of water-saturated rocks," *Geophysics*, vol. 55, pp. 1530–1541, Dec. 1990.

[3] K. A. Safinya, B. Clark, T. M. Habashy, C. Randall, and A. Perez-Falcon, "Experimental and theoretical study of the Electromagnetic Propagation Tool

in layered and homogeneous media," *SPE J. Formation Eval.*, pp. 289–302, Sept. 1987.

[4] R. Gilmore, B. Clark, and D. Best, "Enhanced saturation determination using the EPT-G Endfire antenna array," in *Trans. SPWLA 28th Annual Logging Symp.*, 1987.

[5] B. Anderson, K. A. Safinya, T. M. Habashy, A. Davidson, and R. Gilmore, "The response of the Electromagnetic Propagation Tool to bed boundaries," *SPE J. Formation Eval.*, pp. 458–464, Dec. 1990.

[6] T. M. Habashy, J. A. Beren, and K. A. Safinya, "Electromagnetic logging method and apparatus with scanning magnetic dipole direction," U.S. Patent number 5,210,495, issued May 11, 1993.

[7] T. M. Habashy, M. R. Taherian, A. Dumont, and J. A. Beren, "Electromagnetic logging apparatus and method," filed as a U.S. patent, March 9, 1992, application serial number 848,621.

[8] K. A. Safinya, T. M. Habashy, and J. A. Beren, "Apparatus and method of using slot antenna having two nonparallel elements," U.S. Patent number 5,243,290, issued September 7, 1993.

[9] P. N. Sen, "Grain shape effects on dielectric and electrical properties of rocks," *Geophysics*, vol. 49, pp. 586–587, May 1984.

[10] S. Feng and P. N. Sen, "Geometrical model of conductive and dielectric properties of partially saturated rocks," *J. Appl. Phys.* vol. 58, no. 8, pp. 3236–3249, Oct. 1985.

[11] W. E. Kenyon, "Texture effects in megahertz dielectric properties of calcite rock samples," *J. Appl. Phys.*, vol. 55, pp. 3153–3159, April 1984.

[12] P. L. Baker, W. E. Kenyon, and J. M. Kester, "EPT interpretation using a textural model," paper DD, *SPWLA 26th Logging Symp.*, 1985.

Technical
Applications

Measuring Sensors
and Circuits

R. J. King
J. C. Basuel
M. J. Werner
K. V. King
KDC Technology Corporation, Livermore, CA

11

Material Characterization Using Microwave Open Reflection Resonator Sensors

Abstract. Two-parameter, industrial microwave open reflection resonator sensors have been developed for measuring/monitoring the in situ complex dielectric properties (ϵ', ϵ'') of solids, particulates, and liquids. In turn, the effects of the dielectric properties are relatable to numerous material physical properties such as moisture content and density, chemical reactions, contaminants, porosity, structural characteristics, and others. The sensors can be flush-mounted in chutes, pipes, shakers, hoppers, or glide sleds, or used as a hand-held portable probe. Example results are given for on-line monitoring of moisture and density of grains, processed foods, and wood flakes, absorbed moisture in polymers, and viscosity/curing of polymer resins.

11.1. INTRODUCTION

Two-parameter, industrial microwave open resonator reflection sensors have been developed for continuous measurement/monitoring of many physical properties and/or the complex dielectric properties of solids, particulates, and liquids in situ. When the sensor makes physical contact with the material under test (MUT) its resonant frequency (f_r) and normalized input resistance (r_0) at resonance are influenced by the dielectric constant (ϵ') and loss factor (ϵ'') of the MUT in a quantifiable way. Either of the measured electrical sets $\{f_r, r_0\}$ or $\{\epsilon', \epsilon''\}$ are in turn relatable to numerous physical characteristics of the MUT. Since two independent electrical parameters are measured, it is possible to determine two independent physical parameters simultaneously, e.g., the moisture content (mc) and the oven dry partial density (ρ_d). In this way, density-compensated moisture measurements are made.

These two approaches are depicted in Figure 11.1. In the direct case the *sensor* and the *material* characterizations are combined. Characterization between

153

the independent sets $\{f_r, r_0\}$ and $\{mc, \rho\}$ is empirical. A critical component is establishing and validating suitable mathematical relationships between these two sets. This process can be time-consuming and may involve the use of look-up tables or curve-fitting. Moreover, characterization must be repeated whenever the sensor is adjusted, becomes worn, or is replaced.

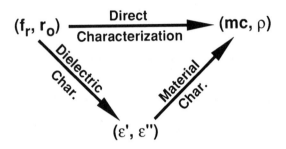

FIGURE 11.1. Two-parameter characterizations.

The indirect approach first involves dielectric characterization of the *sensor* for determining $\{\epsilon', \epsilon''\}$ from the measured $\{f_r, r_0\}$ using one or more "standard" dielectrics. These relationships are analytical functions requiring only two calibration constants. Calibration is straightforward and easily done in a few minutes, usually by the manufacturer. The second step involves characterization of the *material* to relate $\{\epsilon', \epsilon''\}$ to $\{mc, \rho\}$. Like the direct method, these relationships are empirical and usually describable by a mathematical model. This step is also often time consuming, involving laboratory or on-line tests over a desired range of moisture contents. Unlike the direct method, this step is peculiar to the material only and independent of the sensor (at a specified frequency). From the user's viewpoint, the two-step indirect approach is usually preferable because the material needs to be characterized only once. Sensors can easily be adjusted and recharacterized, and worn or damaged sensors can easily be replaced.

11.2. INSTRUMENTATION SYSTEM

Figure 11.2 shows an industrialized electronics system suitable for continuous on-line monitoring of an open reflection resonator sensor. The system functions as a dedicated microwave reflectometer instrument. In operation, the synthesized source steps through an autoscaled frequency span in which the sensor is resonant, with user-specified step sizes equal to multiples of $\Delta f = 0.1$ MHz. Each complete measurement typically takes about 0.5 s. The microwave frequency and frequency span are determined by multiple factors concerning the MUT, such as whether the MUT is a solid, a coarse, medium, or fine particulate, or a liquid, whether the MUT is electronically or ionically conductive, and the MUT homogeneity. Examples are given later.

The reflectometer bridge has a directivity of 40 dB. The reflection coefficient amplitude is detected and passed through a 60-dB logarithmic amplifier to yield

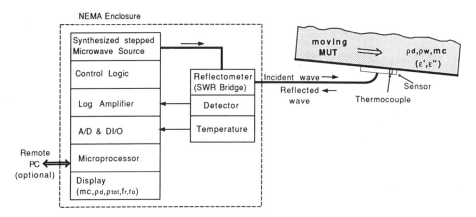

FIGURE 11.2. Industrialized, microprocessor-controlled instrumentation system for on-line measurement of MUT properties using an open reflection resonator sensor.

the return loss in dB. For each frequency scan the spectrum of the return loss is temporarily stored and analyzed by the microprocessor to determine the resonant frequency (f_r) and the return loss (L_r) at f_r. The microprocessor also controls the frequency stepping function, and analyzes the (f_r, L_r) data to yield and display the temperature, date, time, operator ID, lot number, and running averages of the MUT moisture content, wet/dry/total densities, and their statistics. These data can be displayed on a local hand-held control/display terminal or transferred to a satellite PC via an RS-232 communications link. Other forms of data output, e.g., 4–20 mA, can also be provided. Data can be stored in solid-state memory for backup or later retrieval using a floppy disk drive or remote PC. The instrument is entirely digitally controlled from either the hand-held display/control terminal or a satellite PC. For industrial use, the system is packaged in a suitable NEMA enclosure.

11.3. OPEN RESONATOR REFLECTION SENSORS

The forms of resonator sensors are almost unlimited considering the multiplicity of methods for coupling the resonator fields to the MUT and for coupling the microwave source and detector to and from the resonator, respectively [1]. Examples are open-ended coaxial, microstrip, stripline, etc., resonators that are conductively, capacitively, or inductively coupled to the source and wave detector. Of these, two-port transmission sensors are most well known since the resonant frequency f_r and the unloaded Q_0-factor are easily determined by measuring the insertion loss $(|S_{21}|)$ between the input and output ports. The Q_0-factor must generally be large to yield a sharp resonance if f_r is to be measured with high resolution. Unfortunately, for all but low-loss MUTs, a high Q can only result from weak coupling between the resonator and the MUT with a corresponding reduction in sensor sensitivity. As the following shows, one-port reflection resonators offer the advantage of being much more tightly coupled to the MUT and therefore more sensitive to the MUT parameters, while still providing good resolution in the measurement of f_r.

The return loss (L) is defined as the negative magnitude of the reflection coefficient ($|\Gamma|$) at the sensor input port, expressed in dB; i.e., $L = -20 \log |\Gamma|$. For reflection resonators, the reflection coefficient dips to a minimum at f_r. In this case, determining the Q_0-factor is considerably more complicated than for two-port transmission resonators [2,3]. But essentially the same information about the total loss (especially that of the MUT) is contained in the much more easily measured normalized input resistance r_0 at resonance. For this reason, attention is focused on r_0 rather than Q_0. This r_0 is also called the *coupling factor*.

Besides having only one port, an important advantage of reflection resonator sensors is that the sharpness of the return loss near resonance, as expressed in dB, is strongly dependent on the value of r_0. To illustrate, Figure 11.3 gives a typical

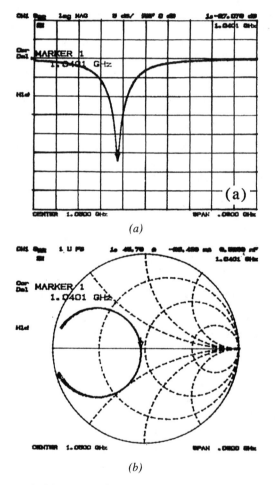

(a)

(b)

FIGURE 11.3. Measured reflection from a typical resonator sensor when contacted with a MUT: (a) return loss or log-magnitude ($|\Gamma|$) display, (b) Smith chart (Γ) display.

log-magnitude return loss display and its corresponding Smith chart presentation for a reasonably well-matched reflection resonator. Since the minimum distance from the Smith chart center to the resonance circle represents the maximum return loss (L_r), it is graphically clear that the log-magnitude plot becomes very sharp as r_0 approaches unity at resonance. Mathematically, the return loss at resonance is related to r_0 as

$$L_r = -20\log\left|\frac{r_0 - 1}{r_0 + 1}\right| \quad \text{dB}. \tag{11-1}$$

It is significant that this logarithmic transformation serves to substantially sharpen the resonance as the resonator becomes well matched ($r_0 \to 1$), as plotted in Figure 11.4. This is important because a sharp resonance permits accurate measurement of the resonant frequency (f_r), even when the Q_0-factor is not large. In many applications f_r plays a very important, if not dominant, role in determining moisture content in the MUT. Thus, for a given Q_0 the sensor feed should be designed or adjusted to achieve a reasonably good match. For example, if $0.7 < r_0 < 1.42$ as shown in Figure 11.4, then the return loss at resonance will be greater than 15 dB, regardless of Q_0.

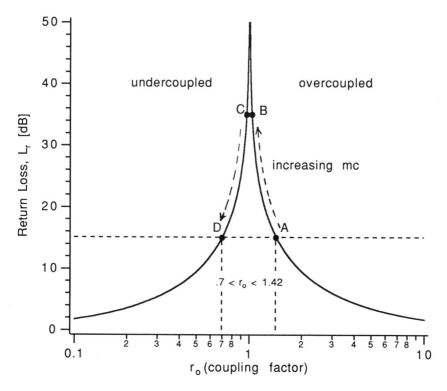

FIGURE 11.4. Return loss (L_r) from a network having a normalized input resistance r_0, as computed from eq. (11-1).

Reflection resonators are said to be overcoupled when $r_0 > 1$, undercoupled when $r_0 < 1$, and critically coupled when $r_0 = 1$ (perfect match). The return loss becomes small when the resonator is highly over- or undercoupled. In either case, accuracy and resolution in the measurement of f_r are lost, regardless of Q_0.

To show the effect of Q_0, the first-order equivalent parallel R_0LC network of a reflection resonator operating near a single-mode resonance [3] is given in Figure 11.5. The normalized input impedance for this network is

$$z = \frac{Z_{in}}{R_c} = \frac{r_0}{1 + jQ_0(y - y^{-1})}, \qquad (11\text{-}2)$$

where $r_0 = R_0/R_c$, Q_0 is the "unloaded" quality factor, $y = \omega/\omega_r$ is the normalized frequency, and $\omega = 2\pi f$ is the angular frequency. The return loss is

$$L = -20\log\left|\frac{z-1}{z+1}\right| \quad \text{dB} \qquad (11\text{-}3)$$

which is plotted in Figure 11.6 for $r_0 = 0.9$ with Q_0 as the parameter. Even though r_0 and Q_0 are not independent (they are both inverse to ϵ'' of the MUT), Figure 11.6 assumes that r_0 can be held constant by design or adjustment of the sensor feed as Q_0 is changed. By definition, the "unloaded" Q_0 is only determined by the dielectric and conductive losses of the MUT and the manner in which the fringing fields of the resonator couple to the MUT, not by the source resistance (R_c) as referred through the network that feeds the resonator. Figure 11.6 demonstrates how increasing the Q_0-factor helps to sharpen the return loss resonance for a given r_0, and Figure 11.4 shows how the resonance is sharpened by r_0 for any Q_0.

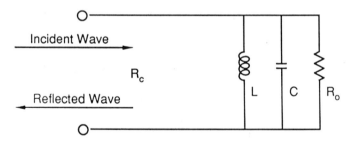

FIGURE 11.5. Equivalent parallel R_0LC network representing the input to a reflection resonator when operating near a single-mode resonance.

This sharpening of the resonance brought about by leveraging the logarithmic function (11-1) or (11-3) is an advantage not enjoyed by two-port transmission resonators where the resonance width is determined exclusively by Q_0. Besides the advantage of having only one port, another advantage of the reflection resonator is that the resonator fields can be more tightly coupled to MUTs having a wide range of loss factors, ϵ''. Moreover, the sensitivities to both ϵ' and ϵ'' are substantially increased.

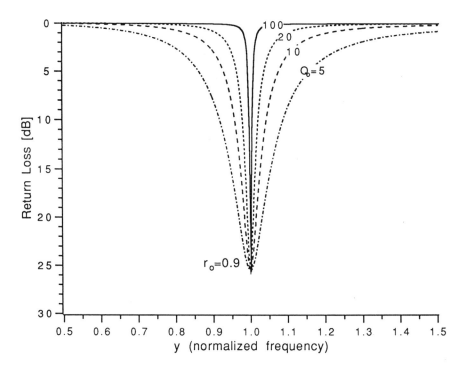

FIGURE 11.6. Return loss for the equivalent network of Figure
11.5, showing how increasing the Q_0-factor helps to sharpen
the resonance for a given r_0.

Examination of (11-1) or Figure 11.4 shows that r_0 is a double-valued function
of the return loss (L_r); i.e.,

$$r_0 \text{ or } r_0^{-1} = \frac{1 - 10^{-L_r/20}}{1 + 10^{-L_r/20}}, \tag{11-4}$$

where the choice of r_0 or r_0^{-1} depends on whether the sensor is over- or undercoupled. Thus, some a priori knowledge is needed to decide the proper solution
to eq. (11-4). This choice is determined by sensor design to a given MUT and
by the range of moisture being measured. For example, it will be seen later that
r_0 is inverse to ϵ'' and mc. Thus, for a given MUT having an expected range of
moisture contents, $mc_1 < mc < mc_2$, the sensor coupling is designed or adjusted
such that the return loss at point A on Figure 11.4 corresponds to mc_1 and point
B corresponds to mc_2, the entire range being on the overcoupled side. Alternatively, the sensor could also be designed or adjusted so that point C corresponds
to mc_1 and D corresponds to mc_2, both being on the undercoupled side. Carrying
this one step further, the sensor might be designed so that points A, B, C, and D
correspond to moisture contents $mc_1 < mc_2 < mc_3 < mc_4$, respectively. These
moisture contents and the spacings between them cannot be arbitrary, but there

does exist some flexibility. Obviously, the range between mc_2 and mc_3 cannot be used because over- or undercoupling is indeterminate from measurement of L_r.

Two types of electromagnetic coupling are subject to design or adjustment. First is the capacitive or inductive fringing of the electromagnetic fields between the resonant structure and the MUT. Adjustment of the fringing factor K (discussed in the next section) can be achieved by changing the physical distance between the resonator and the MUT or by partially shielding the resonator from the MUT using, say, metallic or dielectric shunts or baffles to divert or focus the fringing fields. Adjustment of K directly affects the *sensitivities* of both f_r and r_0 to ϵ' and ϵ'', i.e., $\partial f_r / \partial \epsilon'$ and $\partial r_0 / \partial \epsilon''$. As such, this adjustment *scales* the dynamic ranges of f_r and r_0 as ϵ' and ϵ'' vary, respectively.

The second type of coupling is that between the source feed cable and the resonator; it can be conductive, capacitive, or inductive. In all three cases, the feed network can usually be designed or adjusted to *transform* or *translate* r_0 up or down as desired, akin to translation by a transformer of turns ratio n. Such *translation* together with *scaling* allows positioning the dynamic range of the r_0-factor into a desired range, say $0.7 < r_0 < 1$ or $1 < r_0 < 1.42$, as depicted in Figure 11.4. Obviously, such a sensor is uniquely specific to the particular MUT and the range of moisture it contains.

11.4. DIELECTRIC CHARACTERIZATION

Before proceeding, we must next understand how the measured quantities (f_r, r_0) are related to the complex dielectric properties (ϵ', ϵ'') of the MUT. As depicted in Figure 11.6, a reflection resonator can be regarded as an equivalent parallel R_0LC circuit having resonant frequency $\omega_r = 2\pi f_r = (LC)^{-1/2}$. Since the resonator fields fringe into the MUT, the capacitance (C) can be approximated by the first two terms of a Taylor series:

$$C = C_0[1 + \frac{K}{C_0}(\epsilon' - \epsilon'_{\text{ref}})], \tag{11-5}$$

where C_0 is the capacitance when $\epsilon' = \epsilon'_{\text{ref}}$ (a convenient reference MUT), K is a fringing constant, and $K(\epsilon' - \epsilon'_{\text{ref}})/C_0$ is the small fractional change in C due to variations of $\epsilon' - \epsilon'_{\text{ref}}$.

The calibration constant K/C_0 is found by measuring ω_r and ω_{ref} corresponding to known "standard" MUTs having dielectric constants ϵ' and ϵ'_{ref}, respectively. Then, for MUTs having an unknown ϵ',

$$\epsilon' - \epsilon'_{\text{ref}} = \frac{C_0}{K} \left[\left(\frac{\omega_{\text{ref}}}{\omega_r} \right)^2 - 1 \right] \tag{11-6}$$

Calibration for determining the MUT loss factor ϵ'' is more complicated and is based on the assumption that the MUT losses dominate over radiation and resonator losses. Because the MUT makes physical contact with the sensor, the fringing fields that couple to the MUT are strong and rapidly evanescent with distance from the sensor. Thus, for all except very low loss MUTs, the above assumption holds; it is easy to verify by noting whether r_0 decreases in true inverse proportion to increasing ϵ'', as shown next.

If the fringing capacitance of the equivalent RLC circuit due to the MUT is $K\epsilon'$, then the equivalent fringing conductance is $G = K\omega\epsilon''$ since the fringing fields that couple into the MUT are the same for C and G. Thus, the normalized input resistance of the network in Figure 11.6 at resonance is

$$r_0 = \frac{R_0}{R_c} = \frac{M}{G} = \frac{M}{K\omega_r\epsilon''} = N\frac{[1 + K(\epsilon' - \epsilon'_{\text{ref}})/C_0]^{1/2}}{\epsilon''}, \qquad (11\text{-}7)$$

where M and N are transformation constants that are determined by the coupling between the feed cable and the resonator. Note that because $K(\epsilon' - \epsilon'_{\text{ref}})/C_0$ is small, r_0 only weakly depends on ϵ'.

The calibration constant $N(= M/\omega_{\text{ref}}K)$ in eq. (11-7) can be determined by contacting the sensor with a lossy MUT having known ϵ' and ϵ'' and measuring r_0. This completes the sensor calibration for determining the MUT dielectric properties (ϵ', ϵ'') from measured (f_r, r_0) data.

In some applications absolute calibration is not necessary [4,5]. For example, from eq. (11-6) it is apparent that

$$\epsilon'_M = f_r^{-2} = K_1 + K_2\epsilon', \qquad (11\text{-}8)$$

such that this "modified" dielectric parameter of the MUT is simply a linear transformation of the actual ϵ'. Here, K_1 and K_2 are constants which are not necessarily known. Similarly, a "modified" loss factor

$$\epsilon''_M = (f_r r_0)^{-1} = \epsilon''/f_{\text{ref}}N = K_3\epsilon'' \qquad (11\text{-}9)$$

is linearly proportional to the actual ϵ'', where K_3 is another unknown constant. Typically, f_r is measured in gigahertz.

11.5.   EMPIRICAL MODELING

As noted in Figure 11.1, material modeling at a given frequency relates (ϵ', ϵ'') to the desired MUT parameters, say (mc, ρ). This two-parameter model is empirical and independent of the sensor used; once the modeling coefficients are determined through calibration, this calibration need not be repeated after sensor adjustments or replacement. In the authors' experience, a linear density analytical model has proven able to describe the effects mc and ρ for many material types [6]. In this model, the total density ($\rho_{\text{tot}} = \rho_d + \rho_w$) is separated into its partial densities comprising the oven-dry material ($\rho_d = M_d$/volume) and the contained water ($\rho_w = M_w$/volume). For most materials, ρ_d and ρ_w are linear with (ϵ', ϵ''); i.e.,

$$\epsilon' - A = a_1\rho_d + a_2\rho_w, \qquad \epsilon'' = a_3\rho_d + a_4\rho_w, \qquad (11\text{-}10)$$

where A is a suitable constant chosen as described below, and a_{1-4} are calibration coefficients determined by a least-squares fit to a set of $\{\epsilon', \epsilon'', \rho_d, \rho_w\}$ calibration data. Solving eqs. (11-10) gives

$$\rho_d = \frac{a_4(\epsilon' - A) - a_2\epsilon''}{a_1 a_4 - a_2 a_3}$$

$$\rho_w = \frac{a_3(\epsilon' - A) - a_1\epsilon''}{a_2 a_3 - a_1 a_4}.$$

(11-11)

The absolute moisture content is determined as $\text{mc}_a = \rho_w/\rho_d$ (dry basis), and the relative $\text{mc}_r = \rho_w/\rho_{\text{tot}}$ (wet basis). Note that while $\{\rho_d, \rho_w\}$ are assumed linear with $\{\epsilon', \epsilon''\}$ in this model, neither of these sets are necessarily linear with mc_a or mc_r. In particular, ρ_d may vary with mc due to material shrinkage (or swelling).

The dielectric properties (ϵ', ϵ'') of most materials are temperature dependent, primarily due to the temperature dependence of the contained water. Fortunately, this dependence is linear [1,7] and easily accounted for by letting a_1 become $a_1 + b_1\Delta T$, etc., where $\Delta T = T - T_0$ is the difference between the test and the calibration (T_0) temperatures.

The value of A is chosen to minimize the deviations between the moisture content as determined by a "standard" method (e.g., by weighing and drying samples; see [16]), and by the microwave model (11-11). The best value of A is typically determined by trial and error. For example, first pick an A having a value within the range of ϵ', and calculate the corresponding a_{1-4} coefficients. Then, with the same $\{\epsilon', \epsilon''\}$ data set, use the microwave model (11-11) to compute mc and the standard deviation compared to the "standard" moisture content. Iterate for several assumed values of A until the smallest standard deviation is achieved. In contrast to this optimum value of A, there may also exist a range of A where the model becomes unstable, yielding large uncertainties in mc.

Casting the open reflection resonator interpretive model (11-10) for ϵ' and ϵ'' in terms of the partial water and dry densities ρ_w and ρ_d is empirically based on a similar empirical model for two-parameter transmission moisture meters [7–12]. The similarities lie in the facts that the measured attenuation (ΔA) and phase change $(\Delta\phi)$ for transmission of a microwave beam through a layer of MUT are linearly proportional to the partial basis weights (mass per unit area) of the water and dry material (m_w and m_d), respectively.

The interpretive model (11-10) and its solution (11-11) is called the *density-compensated model* because the moisture content is calculated as the ratio of densities. Note that errors in calculating ρ_d carry forward directly in the calculation of mc_a or mc_r.

Another is the density-independent model which seeks to minimize the effects of density variations at the outset [13,14]. For example, one form of this model is based on the postulated relationships

$$\epsilon' - \epsilon'_{\text{ref}} = \rho_d F(\text{mc}_a) \qquad \text{and} \qquad \epsilon'' = \rho_d G(\text{mc}_a),$$

(11-12)

where F and G are only functions of mc_a. Actually, eq. (11-10) is of this form with $F = a_1 + a_2 \text{mc}_a$ and $G = a_3 + a_4 \text{mc}_a$. But the forms in eq. (11-12) are somewhat more general since they allow for nonlinear F and G. In this model, the ratio

$$\frac{\epsilon' - \epsilon'_{\text{ref}}}{\epsilon''} = \frac{F}{G} = H(\text{mc}_a) \tag{11-13}$$

is independent of ρ_d. The coefficients for this model are found by fitting a (usually quadratic) function (H) to an ensemble set of $\{\epsilon' - \epsilon'_{\text{ref}}, \epsilon'', \text{mc}_a\}$ data. In use, mc_a is determined by solving the thus determined function H for mc_a or via a stored look-up table. A similar model which seeks to minimize the dependence of the total density ($\rho_{\text{tot}} = \rho_d + \rho_w$) can also be postulated and cast into a density-independent function of mc_r [15]. There are undoubtedly other models, the best choice depending on the application.

11.6. MATERIAL CHARACTERIZATION

The bridge between the measured $\{f_r, r_0\}$ parameters and the $\{\text{mc}, \rho, T\}$ parameters is a suitable empirical model, regardless of whether the direct or indirect characterization routes depicted in Figure 11.1 are used. To calibrate this model, procedures must be decided which will permit characterizing $\{f_r, r_0\}$ over appropriate ranges of $\{\text{mc}, \rho, T\}$ for the MUT. Having done so, the model is fitted to these two characterization data sets to compute the model calibration coefficients and some measure of the resulting calibrated model accuracy and robustness.

Procedures for gathering the characterization data are often highly specific to the particular MUT. Great care is required to ensure that the data being gathered truly represent that for real on-line operation. Characterization procedures generally fall within three methods: (1) laboratory characterization by incremented drying of the MUT; (2) near-line characterization using preconditioned MUT samples; and (3) on-line characterization by varying the MUT moisture content.

11.6.1 Laboratory Characterization by Incremented Drying of the MUT

Incremented drying of the MUT usually starts with a single batch of material which is naturally very moist. The batch is weighed and its moisture content determined by, say, oven-drying a few grams according to a standard method for that material [17]. The total density $\rho_{\text{tot}} (= M/V)$ is then determined by pouring a portion of the batch of mass M into a container of known volume V.

The sensor is typically mounted on the side or bottom of a box or fixture that simulates the on-line application. For example, if the on-line sensor is flush mounted on a metal surface, this condition should be replicated in the laboratory box or fixture. The MUT thickness should be sufficient to prevent significant air-MUT interface reflection of the fields at low mc's. When particulates are tested, the density may vary with the manner in which the material is poured into the test fixture and any subsequent vibration or shaking.

The same batch is used throughout the entire measurement sequence (i.e., the total dry mass M_d is held fixed); the batch is poured into the test box or fixture, and (f_r, r_0) are measured at (room) temperature T_0. This process is repeated several times to permit averaging and statistical analysis of the data spread. Successive measurements at reduced mc levels are made after incremental drying of the MUT in steps of $\Delta(\text{mc}) = 1$–2%, with frequent mixing during drying and cooldown to ensure uniform moisture distribution. By monitoring (f_r, r_0) for an hour or so after cooldown, the need for material equilibration can be seen; generally, there is little need for lengthy equilibration because the sensor responds to the bulk mc and is not sensitive to the moisture distribution within macroscopic-sized particulates [16].

At each incremental step, the total density (ρ_{tot}) and batch mass (M) are measured. Then, knowing the starting mc, the mc at each increment can be determined without further drying of small samples; i.e., the decrease in the dry basis moisture content is $\Delta(\text{mc}_a) = \Delta M_w/M_d$, where ΔM_w is the water loss due to incremental drying. Further, the partial densities of the water and dry materials are

Wet basis $\rho_w = \rho_{\text{tot}}(\text{mc}_r)$ $\qquad$ **Dry basis** $\rho_w = \rho_{\text{tot}}\left(\dfrac{\text{mc}_a}{1 + \text{mc}_a}\right)$

$$\rho_d = \rho_{\text{tot}}(1 - \text{mc}_r) \qquad\qquad\qquad \rho_d = \frac{\rho_{\text{tot}}}{1 + \text{mc}_a} \tag{11-14}$$

The final step involves measuring f_r and r_0 as functions of temperature (T) over the range of on-line interest. Data for at least two such temperature runs are needed while holding mc and ρ_{tot} fixed, in order to determine the four temperature coefficients (b_{1-4}). These runs are typically made in the course of incremental drying by sealing the batch in a plastic container or bag and heating to well above the highest temperature of interest. Meanwhile, the test box or fixture should be thermally insulated to reduce the cooling rate when the heated MUT is poured into it. This helps to reduce thermal gradients near the sensor, as a result of heat loss to the environment. The (f_r, r_0) data taken immediately after pouring the heated MUT into the test box or fixture are generally useless because of heat loss to the sensor itself and to gradients. If temperatures below room temperature are needed, the fixture containing the MUT can be placed in a freezer or environmental chamber. Satisfactory (f_r, r_0) data can usually be identified as varying linearly with T.

11.6.2 Near-Line Characterization Using Preconditioned MUT Samples

This method comes closest to replicating true on-line characterization. It is useful for initial calibration or for simple checking of the calibration. The method involves preconditioning several (i.e., six or more) batches of the MUT over the desired range of moisture contents. These batches can often be saved in cold storage for repeated use later. As usual, the precise mc value for each batch is determined by the standard method, and ρ_{tot} is determined by weighing a sample of known volume.

Interfacing the sensor to the MUT usually requires either shutting the production line down or temporarily removing the sensor from the production line. Depending on the particular situation, the sensor is successively exposed to each of the MUTs and (f_r, r_0, T) are measured. Each batch is remixed and remeasured several times to permit data averaging and statistical analysis. Finally, the temperature dependence of (f_r, r_0) is characterized for at least two moisture levels in the same manner as in Section 11.6.1. Throughout the process of handling the MUT, care should be taken to ensure that the density and particle size distribution is not being altered due to handling.

11.6.3 On-Line Characterization by Varying the MUT Moisture Content

Where feasible, this characterization method is the most realistic to actual operating conditions. In some cases it is the *only practical* method. An example is where the physical act of removing the MUT from the production line alters its physical and dielectric properties, as in measurement of extruded foods near the extruder. Another is when measurements must be made in an unusual environment, such as in an oven. Simulating such situations off-line becomes costly, difficult, or even impractical.

During on-line material characterization, $\{f_r, r_0, T\}$ data are recorded while grab samples of the MUT are simultaneously taken to determine ρ_{tot} and mc using a standard method. In some installations it may be possible to vary mc (or even ρ_{tot}) during production, so data can be gathered over the entire range of interest. It may also be possible to vary the temperature (T). In fact, mc, ρ_{tot}, and T are likely to be highly correlated to each other. When these physical parameters are purposely varied, some waste of the output product is likely.

In other on-line operations, mc and T are largely uncontrolled and widely variable, e.g., when monitoring hot foundry sand as it returns from the casting operation before being rewetted and recycled. In these situations, the operator controls the microwave instrument so as to sample and store $\{f_r, r_0, T\}$ at many points over wide ranges of all three of these parameters. Simultaneously, the operator takes many grab samples for immediate determination of mc and ρ_{tot}. As such, data taking is somewhat on a catch-as-catch-can basis.

But in spite of this, the result is a set of $\{f_r, r_0\}$ sensor data which can be least-squares-fitted to the set of $\{mc, \rho_{tot}, T\}$ material data in three dimensions. Many (i.e., 20–30 points) data should be taken to permit averaging and statistical analysis.

11.7. SENSOR CONFIGURATIONS
AND MOUNTING

Open reflection resonator sensors are application-specific, depending on whether the MUT is a solid, a particulate, or a liquid, on the range of (ϵ', ϵ''), on the material abrasiveness, on the desired spatial sampling resolution, and on mounting constraints in the on-line process as depicted in Figure 11.7. They can be made in a variety of shapes—flat or curved for flush mounting to a flat or curved surface, or cylindrical for immersion in a flow stream or for manual insertion into liquids and granular materials.

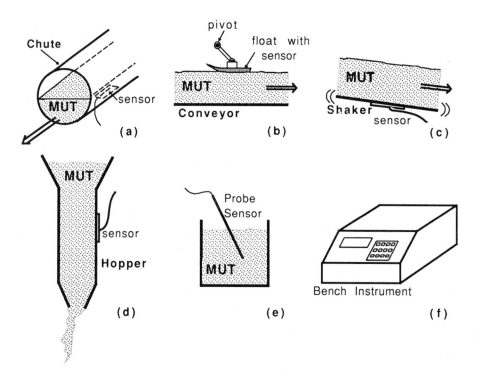

Figure 11.7. Some typical sensor configurations and mounting methods.

The small (2.54-cm diameter) sensors shown in Figure 11.8a typically resonate in the 2–4 GHz range. They are used for liquids and solids. They are also useful for in-mold monitoring of the dynamic viscosity and dielectric cure properties of nonconducting thermoset and thermoplastic polymers [5], and for mapping the spatial dielectric anomalies of composites and dielectric thin films. Sensors having either linearly or circularly polarized field coupling to the MUT have been built. Linear polarization is useful for examining the two-dimensional dielectric tensor of anisotropic MUTs, while circular polarization is effective for averaging such anisotropy.

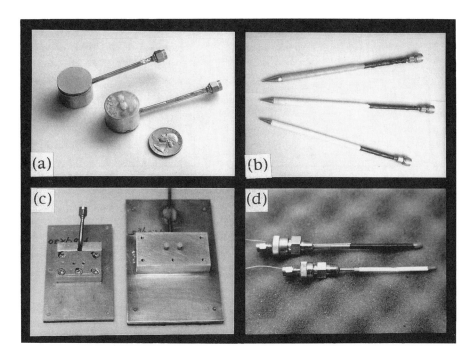

FIGURE 11.8. Photographs of typical sensor designs.

The long (45 cm) cylindrical sensors in Figure 11.8b were developed for hand-held use in measuring the moisture and density of grains, forages, and wood particles, flakes, and chips. Resonance is typically about 500–750 MHz.

The rectangular sensors in Figure 11.8c are designed for monitoring particulates such as grains, dried foods, forages, and wood particles, flakes, and chips. Their larger size (6 to 20 cm length) and low resonant frequency (0.5 to 1 GHz) permit deep field penetration (3 to 6 cm depth).

Finally, the small cylindrical sensors in Figure 11.8d are designed for monitoring liquids in a pipe, drum, etc. These particular designs, which were developed for monitoring the moisture in melted cheese, are about 15 cm long and resonate in the 2–2.5 GHz range.

11.8.   TYPICAL APPLICATION RESULTS

The following results were selected to typify the wide range of sensor applications, particularly related to sensing moisture.

Figure 11.9 correlates the densities (ρ_d, ρ_w, ρ_{tot}) and the moisture content (mc_a; dry basis) of a granular processed (extruded) rice product as determined using the microwave linear density model (11-10) with their counterparts as determined by weighing and drying. Note that this is a low-density product in which the density increases 250% as the moisture content is increased from 9% to 17.5%. The excellent correlations attest to the robustness of the model, in this example,

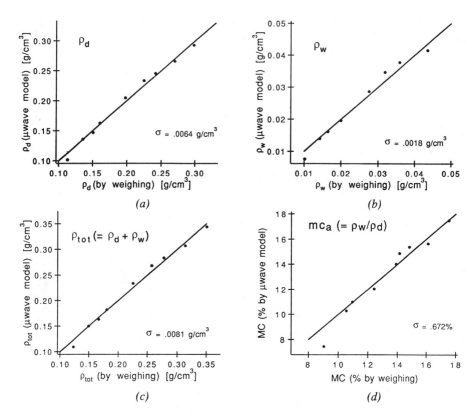

FIGURE 11.9. Correlations of (a) partial dry density (ρ_d), (b) partial moisture density (ρ_w), (c) total density (ρ_{tot}), and (d) absolute moisture content (mc_a; dry basis) as determined by the linear density model (11-10) vs. their counterparts as determined by weighing and drying. Material is a granular processed (extruded) rice product. Nominal $f_r = 2$ GHz.

to track wide swings in mc, ρ_d, and ρ_{tot}. These measurements were made using grab samples taken directly from the production line as the moisture content of the extruded product was varied. Substantial reductions in the standard deviations (σ's) shown are expected when time-averaged measurements are made using an on-line sensor. Also, note that the statistical errors include those due to drying and weighing the samples.

Results for oats tested at 1 GHz using a flat sensor (Figure 11.8c) are shown in Figure 11.10. These data are compensated for nominal variations in temperature (approximate ±3°C) that occurred during laboratory testing. The decrease in ρ_d with increasing moisture due to grain swelling is accurately tracked by the model (11-10) and (11-11). Correlation of mc as determined by the microwave model and by drying and weighing yields a standard deviation of $\sigma = 0.33\%$, including the uncertainty of the drying method. Typically, the uncertainty of determining the mc of grains by drying is 0.2–0.3% [9,10,17].

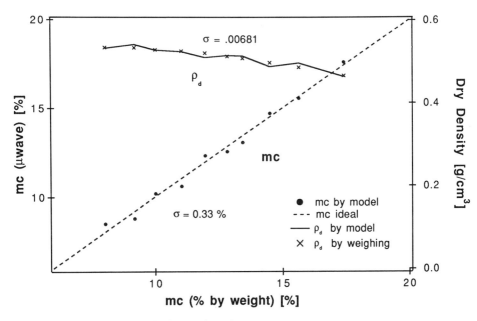

FIGURE 11.10. Correlation of relative moisture content (mc_r; wet basis) and dry density (ρ_d) of oats as determined by the temperature compensated linear density model (11-10) and by drying and weighing. Nominal $f_r = 1$ GHz.

Figure 11.11 shows a nine-day, on-line record of mc_a as recorded for dry rice (with hulls) in a rice mill. The microwave instrument results are compared with those measured using a Motomco 919 (capacitance type) moisture meter. The rice was moving about 1 m/min, and the microwave instrument was programmed to measure mc_a every 10 mins; each output data point is the running average over the previous six measurements (one previous hour). The Motomco measurements were taken about every 2 hr, using grab samples taken a few feet away from the microwave sensor. Hence, variations of mc_a for the two methods are not expected to correspond exactly. The microwave method shows evidence of ±2% moisture variations which are not captured by the sampling method. This is consistent with observations by the mill operators when taking samples 10 mins apart.

Figure 11.12 shows results for Red Alder wood flakes over the range 7 < mc < 40%. For this application, a large flat sensor (Figure 11.8c) which resonates at about 450 MHz was used. Actually, two models are needed (one for mc < 30% and another for 30% < mc) because wood fibers typically saturate at about 26–30% mc. As a result, the slopes of ϵ' and ϵ'' versus mc are slightly different for mc < 30% and mc > 30%. These model results are compensated for nominal temperature (±3°C) variations during laboratory testing.

A critical problem in high-value advanced polymer composites is moisture absorption from the environment. Over a long term (months or even years), absorbed moisture tends to degrade the composite's structural integrity. Classically,

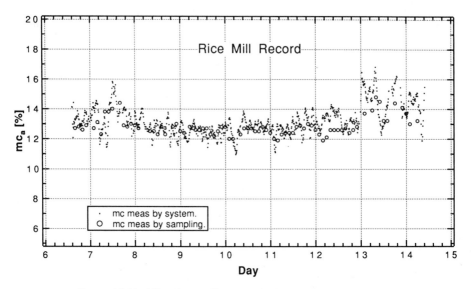

FIGURE 11.11. Nine-day, on-line record of mc_a for dry rice (with hulls) in a rice mill. Nominal $f_r = 1$ GHz.

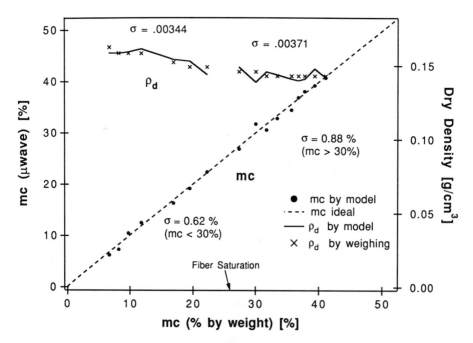

FIGURE 11.12. Correlation of moisture content (mc_a; dry basis) and dry density (ρ_d) of red alder wood flakes as determined by the temperature compensated linear density model (11-10) and by drying and weighing. Nominal $f_r = 450$ MHz.

moisture diffusion begins at the surface and penetrates the composite such that the total accumulated moisture increases linearly with the square root of time, $t^{1/2}[hr^{1/2}]$. At long time (months) the accumulated moisture saturates at a level that depends on the material diffusion constant and on the relative humidity and temperature of the environment.

The sensors shown in Figure 11.8a were successfully used to monitor accumulated moisture in neat (nonreinforced) epoxy resins and fiber-reinforced epoxy composites. The results for a 100 × 100 × 1.6-mm-thick Kevlar fiber-reinforced polyester panel are shown in Figure 11.13. Here, the MUT was soaked in water at 50°C, being removed only momentarily to take measurements. Before each measurement of (f_r, r_0), the free moisture was blotted from the panel surface. Surface moisture was dissipated by ambient air drying for 2 min before the sensor was pressed against the panel surface. Figures 11.13a and 11.13b show that the changes in both $\Delta\epsilon'$ (using eq. (11-6)) and $\Delta\epsilon_M''$ (using eq. (11-9)) track the progressive moisture uptake with time. Initially, the rate of uptake is high, primarily due to the wicking of moisture into the composite via the Kevlar fibers which are directly exposed to the water along the panel edges. After about $t^{1/2} = 2$, the moisture increases linearly in accordance with classical diffusion theory. Saturation had not been reached by the time the experiment was terminated. Figure 11.13c shows both $\Delta\epsilon'$ and $\Delta\epsilon_M''$ versus mc, with time implicit. In general, both $\Delta\epsilon'$ and $\Delta\epsilon_M''$ correlate linearly with accumulated mc, but $\Delta\epsilon_M''$ correlates best. This suggests that the sensor "sees" losses deeper in the material via r_0, while the measured f_r (and hence ϵ') is more weighted by the near-surface moisture. At early times, the near-surface moisture is much greater than that deeper within the panel.

Finally, Figure 11.14 compares the resonant frequency cure profiles for an in-mold sensor (Figure 11.8a) while monitoring the cures of dry and moisture-contaminated prepreg (epoxy that is preimpregnated with fibers prior to curing) composites. Two runs are shown; the "dry" run in which the prepreg had not been exposed to moisture, and the $mc_a = 0.38\%$ (dry basis) run in which the prepreg had been soaked in 20°C water for 3 hr, patted dry and exposed to ambient air for 20 min while being positioned in the mold and curing oven. During both cures, the resin temperature was ramped at about 4°C/min (the actual ramp rates were slightly different for the two runs), then held at about 121°C.

As the resin initially heats, its viscosity drops. This drop in viscosity is closely correlated with f_r [5]. The minimum of f_r provides timing of the point of minimum viscosity (PMV). The earlier occurrence of the PMV for the "wet" prepreg suggests that the moisture acts as a catalyst that accelerates the cure process and lowers the glass transition temperature. When the temperature reaches 100°C the moisture vaporizes, creating erratic variations in f_r. During the upswing in f_r, rapid chemical reaction is occurring, providing information about the rate and degree of cure. As the epoxy transitions from the liquid to the glassy state, the cure rate slows. The difference between the asymptotic levels of the "dry" and $mc_a = 0.38\%$ curves provides information about the degree of residual porosity induced by the contaminating moisture.

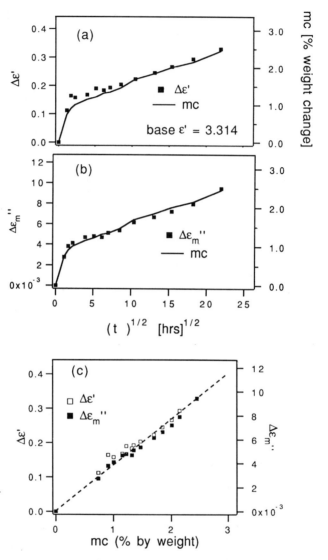

FIGURE 11.13. The open reflection resonator sensor in Figure 11.8a
is used to monitor the progressive diffusion of moisture into a
Kevlar fiber-reinforced polyester panel. In (a) and (b), the initial
rise in $\Delta\epsilon'$ and $\Delta\epsilon_m''$ is due to moisture wicking along the Kevlar
fibers; the subsequent linear rise is due to moisture diffusion into
the polyester matrix. In (c), $\Delta\epsilon'$ and $\Delta\epsilon_m''$ are correlated with the
total accumulated moisture content (dry basis). Base $f_r = 4.5$ GHz.

In curing applications such as this, the raw (f_r, r_0) data are sufficient to
characterize many important dynamic properties. Calibration to quantify the actual
(ϵ', ϵ'') of the MUT is generally unnecessary. Incidentally, although f_r and r_0 are

independent measurables, the information each of them provides about the cure dynamics is very similar for such applications. Practically, only one of them needs to be measured.

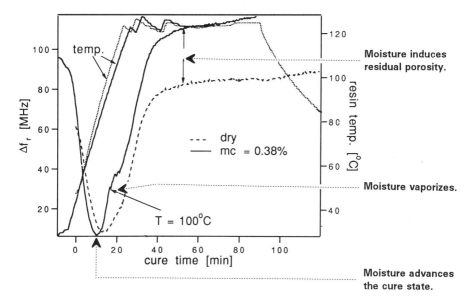

FIGURE 11.14. Comparison of resonant frequency cure records of "dry" and "wet" (mc = 0.38%; dry basis) for fiberglas reinforced epoxy prepreg as monitored by an in-mold sensor (Figure 11.8a). Base f_r = 4.41 GHz.

11.9. CONCLUDING DISCUSSION

Microwave open reflection resonator sensors offer considerable flexibility and versatility for continuous on-line monitoring of industrial products and for process control. Generally, the choice of sensor designs is application-specific, but the number of potential applications is large, as is also the number of economic benefits. Only a few of the applications have been mentioned here.

Compared to well-known noncontacting and nonresonant transmission methods, contacting resonant sensors offer-remarkable (order of magnitude) improvements in sensitivity and resolution for determining small changes in dielectric (ϵ', ϵ'') or physical (e.g., mc, ρ) properties. Moreover, compared to transmission (two-port) resonator sensors, reflection resonator sensors have an additional advantage that measurement of the return loss (in dB) admits to much higher resolution of f_r, especially when the Q is low, provided that the sensor input resistance is well matched to the feed cable.

This paper focused on the varied uses of open reflection resonator sensors and the information they can provide. Space does not permit discussion of the supporting microwave and electronic instrumentation. Costly laboratory scalar or vector microwave network analyzers are not suitable for harsh industrial operation by often unskilled personnel. To meet this need, a line of special industrialized, narrowband, reflectometer analyzers with on-board microprocessors have been developed for instrument control and for data acquisition, processing, display, and output.

REFERENCES

[1] E. Nyfors and P. Vainikainen, *Industrial Microwave Sensors*, Norwood, MA: Artech House, 1989.

[2] E. L. Ginzton, *Microwave Measurements*, New York: McGraw-Hill, 1957, ch. 9.

[3] D. Kajfez and E. J. Hwan, "Q-factor measurement with network analyzer," *IEEE Trans. Microwave Theory Techni.*, vol. MTT-32, no. 7, pp. 666–670, July 1984.

[4] R. J. King, "Microwave sensors for process control. Part II: Open resonator sensors," *SENSORS*, vol. 9, no. 10, pp. 25–29, Oct. 1992.

[5] R. J. King, M. J. Werner, and G. D. Mayorga, "Microwave dynamic dielectric analysis of curing neat resins," *J. Reinforced Plastics Composites*, vol. 12, no. 2, pp. 173–185, 1993.

[6] R. J. King, "Continuous moisture and density measurement of foods using microwaves," in *Proc. Food Processing and Automation III Conf.*, pp. 231–240, 1994.

[7] R. J. King and J. C. Basuel, "Measurement of basis weight and moisture content of composite boards using microwaves," *Forest Products J.*, vol. 43, no. 9, pp. 15–22, Sept. 1993.

[8] A. Kraszewski and S. Kulinski, "An improved microwave method of moisture content measurement and control," *IEEE Trans. Ind. Elect. Control Instrum.*, vol. IECI-23, no. 4, pp. 364–370, Nov. 1976.

[9] A. Kraszewski, S. Kulinski, and Z. Stosio, "A preliminary study on microwave monitoring of moisture content in wheat," *J. Microwave Power*, vol. 12, no. 3, pp. 241–251, 1977.

[10] A. W. Kraszewski, "Microwave aquametry—needs and perspectives," *IEEE Trans. Microwave Theory Techni.*, vol. MTT-39, no. 5, pp. 828–835, May 1991.

[11] A. Kraszewski, S. Kulinski, J. Madziar, and K. Zielkowski, "Microwave on-line moisture content monitoring in low-hydrated organic materials," *J. Microwave Power*, vol. 15, no. 4, pp. 267–275, 1980.

[12] R. J. King, K. V. King, and K. Woo, "Microwave moisture measurement of grains," *IEEE Trans. Inst. Meas.*, vol. IM-41, no. 1, pp. 111–115, Feb. 1992.

[13] W. Meyer and W. Schilz, "Feasibility study of density-independent moisture measurement with microwaves," *IEEE Trans. Microwave Theory Techni.*, vol. MTT-29, no. 7, pp. 732–739, 1981.

[14] E. Kress-Rogers and M. Kent, "Microwave measurement of powder moisture and density," *J. Food Engr.*, vol. 6, pp. 345–376, 1987.

[15] A. Klein, "Microwave determination of moisture in coal: comparison of attenuation and phase measurement," *J. Microwave Power*, vol. 16, no. 3/4, pp. 289–304, 1981.

[16] T. Okabe, M. T. Huang, and S. Okamura, "A new method for measurement of grain moisture content by the use of microwaves," *J. Agric. Engng. Res.*, vol. 18, pp. 59–66, 1973.

[17] ASAE Standards for Moisture Measurement: - S352.2 *Unground Grain and Seeds*; - S358.2 *Forages*; - S353 *Meat and Meat Products*; - S 269.3 *Wafers, Pellets and Crumbles—Definitions and Methods for Determining Density, Durability and Moisture Content*. Available from the American Society of Agricultural Engineers, St. Joseph, MI.

12

Andrzej W. Kraszewski
Stuart O. Nelson
USDA, ARS, Russell Research Center, Athens, GA

Moisture Content Determination in Single Kernels and Seeds with Microwave Resonant Sensors

Abstract. Principles are presented for using microwave resonant cavities to monitor moisture content of individual seeds, grain kernels, etc. Results of research on peanuts, soybeans, and wheat kernels are presented, illustrating the usefulness of the technique for size-independent moisture determination in kernels and seeds of uniform shape.

12.1. INTRODUCTION

Moisture content is one of the most important factors determining quality of food and agricultural products during harvesting, storage, trading, and processing. Because too-high moisture levels will cause product spoilage by rotting, fermentation, and molding, drying of agricultural products is a common practice. Sometimes such destructive action can occur even when the average moisture content of the material is at an appropriate safe level. Often a small percentage of a lot with excessive moisture content, blended with material of appropriate moisture content, is enough to cause spoilage of the whole shipment over time. On the other hand, drying is always an expensive process because of equipment and energy costs. Overdrying of food and agricultural products not only wastes energy but can result in deterioration in texture and nutritional value. Thus, monitoring and control of moisture content during all stages is of significant importance. Not only is the level of moisture content important but its distribution among the lot of material may be of serious interest.

The moisture content of a material, M, expressed in percentage, wet basis, is defined as

$$M = \frac{\text{mass of water}}{\text{mass of wet material}} \times 100 \quad [\%]. \tag{12-1}$$

177

Standard laboratory methods of moisture content determination are based directly on this definition [1]. The mass of water is removed from small samples of wet material (a few to several grams) by evaporation or extraction, and by weighing the sample before and after drying, the required information is obtained. Too often it is assumed that grain and similar materials are homogeneous and uniform, that water is evenly distributed throughout the lot of material, and differences in moisture content determination arise from the uncertainties in weighing. Such ideal lots are the exception, and moisture variation in large-scale industrial production is often a serious problem.

Moisture content of a given grain lot is always taken as a single number, at most averaged for several bulk samples, and is rarely accompanied by data on the distribution or range of values for those samples. The value of moisture content determined by standard methods will be the same for all of the moisture distributions illustrated in Figure 12.1, regardless of the completely different practical steps (technological processes) that need to be taken in each instance. This is especially important for agricultural products, where the situation shown in Figure 12.1c (a sample from a mixture of two lots of different moisture levels) can lead to spoilage of the whole lot of material. To avoid such problems, as many small samples of material as possible, taken from various locations throughout the lot, should be measured. For grain, seeds, and similar materials, testing of individual kernels or seeds could provide the desired information.

Single-kernel moisture meters based on dc conductivity or RF impedance measurements at frequencies of 1 to 5 MHz have been developed [2–4], but we have found [5,6] that microwave resonant cavity measurements provide an interesting alternative, because they are fast, accurate, nondestructive, and do not require physical contact between the kernel and measuring equipment. The purpose of this paper is to summarize the results and experience gained in research on measuring moisture content in single seeds with microwave resonant cavities.

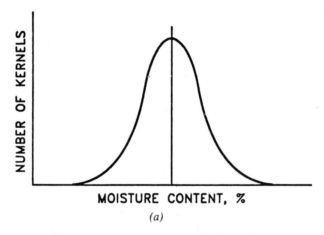

FIGURE 12.1. Examples of various distributions of moisture content for samples of the same size.

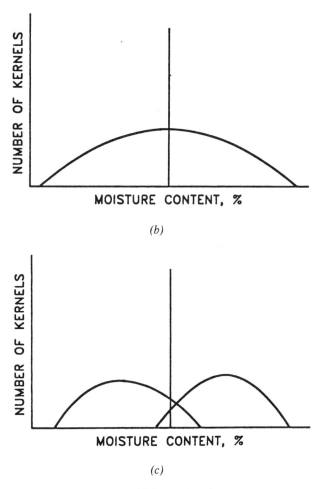

(b)

(c)

Figure 12.1. *continued*

12.2. THEORETICAL BACKGROUND

A dielectric region completely surrounded by a conducting surface constitutes a resonant cavity. A simple cavity can be created with a hollow rectangular or circular waveguide. If a section of such a waveguide is sealed with a short-circuiting conductive wall perpendicular to the direction of wave propagation, the incident and reflected waves are superimposed and create a *standing wave*. The tangential electric field and normal magnetic field are zero at this wall and at distances of integral half-wavelengths from it. In such a nodal plane, a second conductive wall can be located without disturbing the standing wave. Waves reflected by the first short-circuiting wall will be reflected back by the second short-circuiting wall in phase with the incident wave, and a resonant cavity results in which waves are reflected back and forth, and a standing wave persists if energy is supplied to offset losses in the cavity. If the cavity is excited at the proper frequency, through a coupling hole for example, the fields can build up within the cavity in certain restricted configurations or *modes* of operation.

An example of a resonant cavity made from a section of rectangular wave-guide, short-circuited at each end with metal plates containing circular coupling holes and filled with air as the dielectric, is shown in Figure 12.2. The cavity will be resonant at frequencies having an integral number of half-wavelengths in the waveguide equal to the length L; i.e., $L = p\lambda_g/2$, where $p = 1, 2, 3, \ldots$ is an integer. Thus, when a cavity similar to that shown in Figure 12.2 is connected to a swept oscillator and a detector to register transmission as a function of frequency, the general picture at the output is similar to that presented in Figure 12.3. Peaks of transmission through the cavity occur at frequencies for which $p = 3, 4, 5$, and 6 for a given length of the cavity and for the sweep-frequency range indicated in Figure 12.3. These are the resonant frequencies for the different modes of cavity operation designated as the TE_{10p} modes. This indicates that the electric field vector $\mathbf{E}$ is transverse to the z direction of propagation in the waveguide. The first subscript indicates one half-cycle of the E-field pattern in the x direction across the a dimension of the waveguide, the second subscript indicates no E field in the y direction across the b dimension, and the third subscript indicates the value of p, the number of half-wavelengths along the z axis within the length L of the cavity.

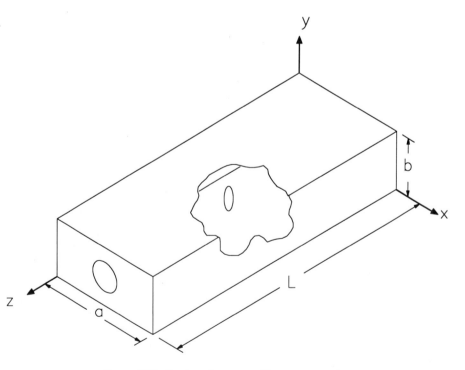

FIGURE 12.2. Rectangular waveguide resonant cavity.

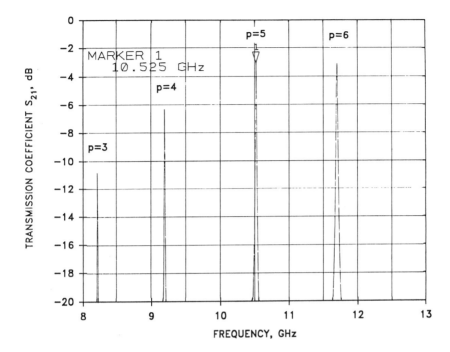

FIGURE 12.3. Multiple resonances in a rectangular waveguide
cavity operating in X-band.

Since the tangential component of the electric field must vanish at the conducting
boundaries, the E field will have a maximum value at the center of the cavity for
the TE_{101} mode or for any TE_{10p} mode for which p is an odd integer.

Resonant cavities have been widely used for determining microwave properties of material samples by measuring the shift in the resonant frequency and the
change in the Q factor of the cavity when a sample is inserted into the cavity [7,8].
Measured parameters depend upon the volume, geometry and mode of operation
of the cavity, as well as the permittivity, shape, dimensions, and location of the
object inside the cavity. For a given cavity and homogeneous sample of regular
(analytical) shape and well-defined dimensions, one can determine the permittivity
of the material. This technique may be demonstrated using an expanded view of
the resonant curve corresponding to $p = 5$ in Figure 12.3 as shown in Figure 12.4.
The resonance of the cavity appears as a peak in transmission through the cavity.
To determine the resonant frequency, the frequency of a signal coupled to the cavity is varied until the maximum transmission is observed. The second parameter
of the resonant curve, as shown in Figure 12.4, is its shape. The apparent Q factor
of the cavity depends upon energy losses in the cavity (walls, coupling, etc.) and
is expressed in the form

$$Q = 2\pi \frac{\text{energy stored}}{\text{energy dissipated per cycle of oscillation}}.$$

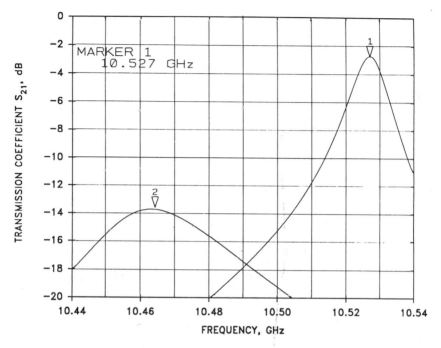

FIGURE 12.4. Expanded view of resonant curve: with (left) and without the perturbing dielectric object.

Thus, when an object is introduced into the cavity, the resonant frequency will decrease and the Q factor will be lowered, causing a broader, flatter resonant curve.

Referring to Figure 12.4, the measured shift of resonant frequency is denoted as $\Delta F = f_0 - f_s$, where subscripts $_0$ and $_s$ refer to the empty cavity and the cavity loaded with an object at the center of the cavity, respectively. Energy dissipated in the object is related to a change in the cavity factor by

$$\frac{1}{Q_s} - \frac{1}{Q_0} = \frac{1}{Q_0}\left(\frac{Q_0}{Q_s} - 1\right) = \frac{1}{Q_0}\left(\frac{V_0}{V_s} - 1\right) = \frac{\Delta T}{Q_0},$$

where V denotes the voltage transmission coefficient at resonance, $\Delta T = 10^k - 1$ is the transmission factor, and $k = 0.05(S_{210} - S_{21s})$, with S_{21} being the measured voltage transmission coefficient at resonance, expressed in decibels. These cavity parameters are related to the dielectric properties of the object by the perturbation equations (see Appendix) [8]

$$\Delta F = 2(\epsilon' - 1)Kf_0\left(\frac{v_s}{v_0}\right), \qquad (12\text{-}2)$$

$$\Delta T = 4\epsilon'' K^2 Q_0\left(\frac{v_s}{v_0}\right), \qquad (12\text{-}3)$$

where $\epsilon^* = \epsilon' - j\epsilon''$ is the material permittivity, v_s and v_0 are the volumes of the sample and the empty cavity, respectively, and K is a factor dependent on object shape, orientation, and permittivity. For example, K has a value of $3/(\epsilon' + 2)$ for spherical objects, and $K = 1$ for thin rods parallel to the electric field. Equations (12-2) and (12-3) can be used for material permittivity measurements when the object is of well-defined dimensions and shape, and the values of f_0, Q_0, and v_0 are known for a given cavity, under the assumption that the volume of the object is much smaller than the volume of the cavity, and the object itself has relatively low loss ($\epsilon'^2 \gg \epsilon''^2$). Both of these conditions are relatively easy to satisfy for solid dielectric materials that can be conveniently machined and measured. However, for biological materials like tissues, seeds, and kernels, these conditions cannot usually be fulfilled satisfactorily, and the resonant cavity techniques are not useful for permittivity measurements of those materials. However, we have found that because both cavity parameters, measured simultaneously for each object located at the center of the cavity, are proportional to the volume of the object and its moisture content, their ratio, X, is independent of object volume:

$$X = \frac{\Delta F}{\Delta T} = \frac{\epsilon' - 1}{\epsilon''} \frac{1}{K} \frac{f_0}{2Q_0} = \varphi(M)\frac{C}{K}, \tag{12-4}$$

where $C = f_0/2Q_0$ is a constant for a given resonant cavity and $\varphi(M) = (\epsilon' - 1)/\epsilon''$ is a permittivity function of the kernel moisture content, and the ratio X is known directly from the microwave cavity measurement.

This ratio then is a function of the object permittivity, which in turn is related to its moisture content, its shape, and a constant describing an empty cavity. Thus, the measurement of the coordinates of the peak of the resonant curves with and without the object inside the cavity provides enough information to determine the value of the permittivity function, describing dielectric properties of the object, related for example to its moisture content. Water content in a material affects its permittivity, but most objects of natural origin (biological) do not have regular, analytical shapes, and they differ in size and shape, even within the same variety or kind. The simplicity of this technique, however, as well as the simplicity of eq. (12-4) for use as a calibration equation for moisture content determination, is so attractive that many attempts have been made to assess its practical usefulness [5,6,9].

Moisture content can be determined from this equation written in a general form:

$$M = \alpha\varphi^{-1}(X) \simeq \frac{a}{X} + b, \tag{12-5}$$

where $\varphi^{-1}(X)$ is the inverse of the permittivity function $\varphi(M)$ and α is a constant. Verification of this relationship with experimental data on several agricultural materials will be discussed. It has proven useful for single soybean seeds [5], single peanut kernels [6], and single kernels of wheat, where slight variations in object shape do not significantly affect the dependence of the ratio upon moisture content. When shape of the objects under test varies significantly, as for corn kernels, two measurements must be taken before and after the object is rotated in the cavity by 90°. Then the ratio of the averaged parameters is a measure of the object moisture content [9].

12.3. MATERIALS AND METHODS

12.3.1 Description of Materials

Samples used to illustrate the usefulness of microwave resonant cavities for measurements reported in this paper include seed lots of soybeans, shelled peanuts, and wheat kernels. The soybeans, *Glycine max* (L.) Merrill, were of the "Wright" cultivar grown in Georgia and stored at 4°C since harvest. Soybeans are seeds of nearly spherical shape, and the ratio of the major to minor diameters for this lot ranged from 1.2 to 1.36, with major diameter between 6.5 and 7.5 mm. Their initial moisture content was about 16%. Microwave measurements were taken, each seed was weighed, permitted to dry under room conditions for various time intervals, and then sealed in glass vials and held at 4°C for at least 24 hr prior to the next microwave measurement to obtain uniform moisture distribution within the seed. In this way each of 18 seeds was measured six times before being dried to determine the dry weight. The same procedure was later applied to 55 seeds used in the validating measurements for moisture determination.

The peanuts, *Arachis hypogaea* L., were of the "Florunner" cultivar grown in Georgia in 1990, shelled and screened into medium and jumbo size classes, and stored at 4°C until needed for measurements. These peanuts had kernels of a roughly prolate spheroidal shape with a major-to-minor-axis ratio between 1.6 and 2.0. The weight of wet kernels ranged from 380 to 940 mg. Initial moisture content was 10% to 12%. The kernels were permitted to dry under ambient conditions for various time intervals between the microwave measurements. After every drying period each kernel was individually sealed in a glass vial and kept at 4°C to equilibrate, usually for two to three days, before it was measured again. In this way microwave measurements and kernel weights were recorded for each kernel three to five times before it was fully oven-dried to determine its dry mass. The same procedure was repeated later on kernels of Runner and Virginia market-types harvested in Georgia in 1991 and measured for verification purposes. The weight of wet kernels ranged from 420 to 1280 mg. Some of these kernels were lightly sprayed with distilled water to increase their moisture content to about 16% from the initial level of 6%.

Kernels of "Coker 9733" soft red winter wheat, *Triticum aestivum* L., grown in Georgia in 1989 and of "Karl" hard red winter wheat grown in Nebraska in 1992 were selected for calibration measurements. Test weight (bulk density under specified conditions) at 24°C for Coker at 12.8% moisture was 795 kg m^{-3}, and for Karl at 12.1% moisture was 755 kg m^{-3}. Samples required to have more than the initial moisture content were conditioned by adding distilled water and sealing them in jars at 4°C for at least a week prior to measurement. Before the measurements, samples were held in jars for up to 16 hr at room temperature (23°C), and after that several kernels were randomly selected and sealed individually in small glass vials just prior to and after the measurements before being weighed and dried. Similar procedures were used during the verification measurements that were carried out on "Stacy" soft red winter wheat grown in Georgia in 1988 and on "Arapahoe" hard red winter wheat harvested in Nebraska in 1992. Respective test weights at 24°C were 786 and 770 kg m^{-3}, recorded at moisture contents of 12.0% and 12.5%.

Moisture contents of individual kernels were determined by forced-air oven drying for time periods and at temperatures prescribed by standards [1,10] for unground bulk samples of these commodities. Soybeans were dried for 16 hr at 105°C, peanut kernels for 6 hr at 130°C, and wheat kernels for 19 hr at 130°C. All samples were cooled in a desiccator over anhydrous $CaSO_4$ upon removal from the oven before being weighed.

12.3.2   Electrical Measurements

Any kind of microwave resonator may be used for moisture content determination in perturbing objects. However, a rectangular waveguide resonant cavity is one of the simplest and easiest to build. Principles of general perturbation theory [11] can be used to establish some general rules of resonant cavity applications for agricultural products. In standard rectangular waveguide cavities, the object-to-cavity volume ratio should be $< 10^{-3}$. Figure 12.5 shows how operating frequency f_0 is related to the volume of the empty cavity operating in the TE_{105} mode with $f_c/f_0 = 0.65$, where f_c is the cutoff frequency. Volumes of cavities used for some kernels and seeds are indicated, as well as middle frequencies of some standard rectangular waveguides.

Each of the resonant cavities used in this study consisted of a section of standard rectangular waveguide coupled with external waveguides through two

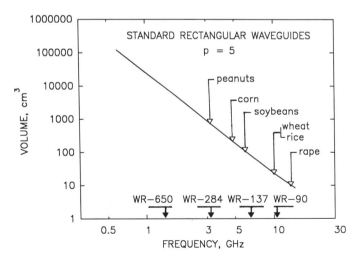

FIGURE 12.5. Required cavity volume as a function of resonant frequency. WR refers to rectangular waveguides according to the EIA designation.

identical coupling holes at each end of the cavity. A dielectric tube was installed in the center of each cavity, as shown in Figure 12.6, to facilitate positioning the object at the center of the cavity. Parameters of the cavities used in the study are listed in Table 12.1. Several rectangular waveguide resonant cavities are shown in Figure 12.7. Each cavity is located between two waveguide-to-coaxial transitions which allowed it to be connected to a computer-controlled automatic network analyzer calibrated in the transmission mode. The analyzer generated 801 discrete frequencies within a selected range for given cavity resonance that varied from 16 MHz for peanuts to 96 MHz for wheat kernels. In this way, measurement of the transmission through the cavity was possible in increments of 20 kHz to 0.12 MHz, respectively, by reading the coordinates of the marker on the test-set CRT display. A "marker-to-maximum" command automatically accomplished the determination of the coordinates of the peak of the resonant curve (resonant frequency, f, and transmission loss, S_{21}, in Figure 12.4), without and with an object located in the cavity, providing all information necessary to determine the two measured quantities ΔF and ΔT.

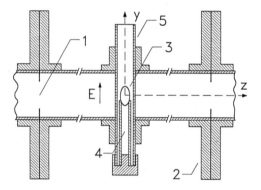

FIGURE 12.6. Cross section of a rectangular waveguide resonant cavity showing the coupling irises (1) held between circular waveguide flanges (2) at each end of the cavity and the object (3) supported at the center by a plastic tube (4) inside the plastic tube (5) within the cavity.

TABLE 12.1. Comparison of Parameters for Rectangular Waveguide Resonant Cavities Operating in TE_{105} (H_{105}) Mode Used in This Study

PARAMETER	S-BAND	C-BAND	X-BAND
Cross section [cm]	7.214×3.404	4.755×2.215	2.283×1.016
Length	30.5 cm	20.3 cm	9.14 cm
Volume	749 cm^3	214 cm^3	21.0 cm^3
Resonant frequency	3,175.9 MHz	4,832.9 MHz	10,526.5 MHz
Q factor	865	1280	900
3-dB bandwidth	3.67 MHz	3.75 MHz	11.64 MHz
Coupling hole diam.	20.6 mm	13.0 mm	6.68 mm

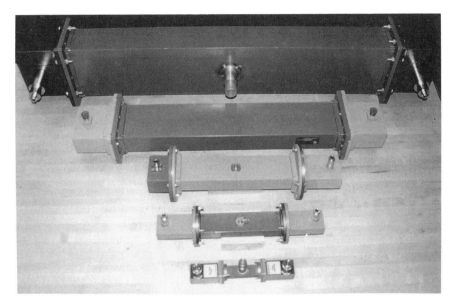

FIGURE 12.7. External view of several rectangular waveguide resonant cavities consisting of standard rectangular waveguides: WR-650, WR-430, WR-284, WR-187, and WR-90, as seen from top to bottom.

12.4. EXPERIMENTAL RESULTS

12.4.1 Peanut Kernels

The microwave measurements were taken first at high moisture contents and then at intervals of a few days as the kernels dried under ambient conditions. The resonant frequency shift, ΔF, and the transmission factor, ΔT, as a function of moisture content for several kernels of different dry mass, are presented in Figures 12.8 and 12.9, respectively. Strong dependence of both of these variables on size of peanut kernels may be easily observed. The ratio of these two variables, X, as a function of moisture content is shown in Figure 12.10 for more than 500 measured data points on more than 100 peanuts.

To study the effects of kernel shape and size, the measured shift in resonant frequency and the change in the transmission factor for several peanut kernels of different shape factors, expressed as a ratio of their lengths to their diameters, are presented in Figures 12.11 and 12.12, respectively, as a function of the mass of water in the kernels. Three kernels each of three different dry-mass groups, averaging 0.39, 0.65, and 0.91 g were selected. The dry masses of the three kernels in each mass group were within $\pm$ 0.012 g. The three kernels in each mass group differed in their length L to diameter D ratios (L/D), which are identified in Figure 12.12. The kernel shape affects the transmission factor to a much greater extent than it does the shift of the resonant frequency. For small or relatively dry kernels

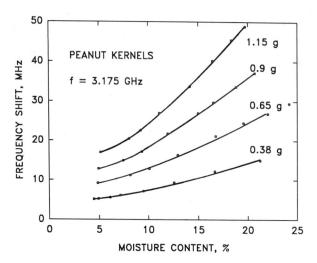

FIGURE 12.8. Resonant frequency shift, ΔF, as a function of moisture content for several peanut kernels of indicated dry mass.

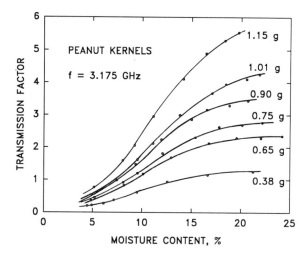

FIGURE 12.9. Transmission factor, ΔT, as a function of moisture content for several peanut kernels of indicated dry mass.

both relationships are linear, as might be expected from the cavity perturbation theory. When moisture content increases, the transmission factor deviates from linearity much faster than the resonant frequency shift, indicating that the effect of losses in the material is stronger than the effect of the size of the object.

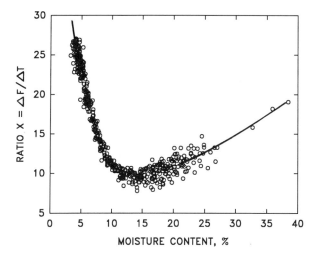

FIGURE 12.10. Ratio $X = \Delta F / \Delta T$ as a function of moisture content for peanut kernels (extended range of moisture).

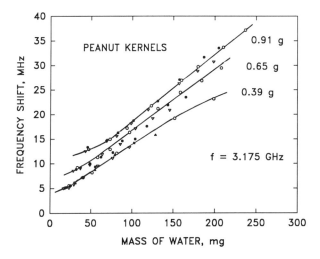

FIGURE 12.11. Resonant frequency shift as a function of mass of water in several peanut kernels of indicated dry mass and different shape factors (see Figure 12.12 for identification of L/D ratios).

For the 46 peanut kernels measured at various moisture levels in the S-band rectangular resonant cavity (total of 184 data points), the ratio X as a function of moisture content is shown in Figure 12.13. The experimental data were fitted by the equation

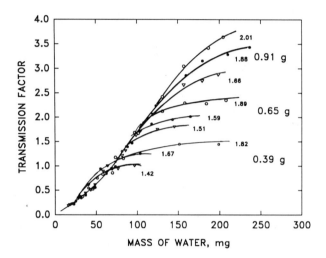

FIGURE 12.12. Transmission factor as a function of mass of water in several peanut kernels of indicated dry mass and different shape factors (values of L/D indicated).

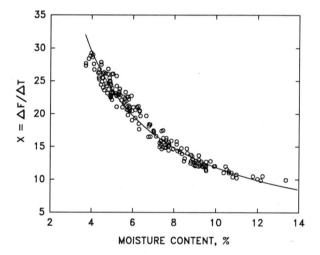

FIGURE 12.13. Dependence of the ratio X on moisture content in single peanut kernels harvested in 1990—total of 184 data points.

$$\frac{1}{X} = 0.008437M + 0.000159, \qquad r = 0.9843, \qquad (12\text{-}6)$$

where r is the correlation coefficient. Solving for M provides the calibration equation for the kernel moisture content in percent, wet basis, equivalent to eq. (12-5), as

$$M = \frac{118.5}{X} - 0.02. \tag{12-7}$$

To verify the above regression constants, another set of 91 peanuts from different lots was tested in the same resonator. A total of 318 data points from 4% to 14% was collected. The moisture contents of the kernels were determined by the standard oven method and compared with those calculated from eq. (12-7). The standard error of performance (SEP) (the standard deviation of the differences) for the method, determined as

$$SEP = \left\{ \frac{1}{N-1} \sum_{i=1}^{N} (\Delta M_i - \Delta M)^2 \right\}^{1/2}, \tag{12-8}$$

where ΔM_i is the difference in moisture content between the value predicted from eq. (12-7) and that of the air-oven determination for the ith kernel, and N is the number of kernels observed, was 0.466% moisture content. The mean value of the differences (bias)

$$\Delta M = \frac{1}{N} \sum_{i=1}^{N} \Delta M_i \tag{12-9}$$

was −0.21% moisture.

Moisture content values predicted from eq. (12-7) are compared with the reference values in Figure 12.14. The distribution of differences between oven moisture and calculated values is shown in Figure 12.15. The predicted values of peanut moisture content agree well with those determined by the standard method, which provides a proof of practical usefulness of the technique.

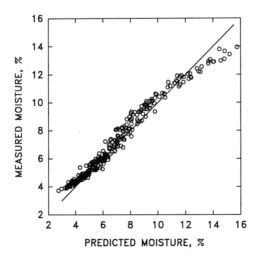

FIGURE 12.14. Predicted moisture content of mixed medium and jumbo sized peanut kernels versus moisture content determined by standard oven method.

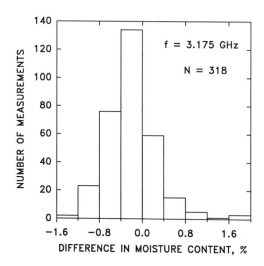

FIGURE 12.15. Distribution of differences between oven moisture content and moisture content from the calibration equation (eq. 12-7) for 318 data points on peanuts harvested in 1991.

12.4.2   Soybean Seeds

Experimental results for 18 seeds, measured at various moisture levels in the C-band resonant cavity providing a total of 108 data points, are shown in Figure 12.16. Procedures identical to those already described were applied, and results for the calibration equation and validation are listed in Table 12.2.

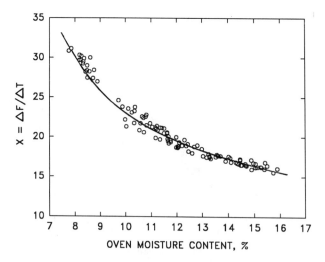

FIGURE 12.16. Dependence of the ratio X on moisture content in single soybean seeds—108 data points.

TABLE 12.2. Comparison of Calibration and Verification or Validation Information for Three Commodities Used in This Study

PARAMETER	PEANUTS	SOYBEANS	WHEAT
Resonant frequency	3.176 GHz	4.833 GHz	10.527 GHz
Constant C (eq. (12-4))	1.836	1.888	5.848
Constant a (eq. (12-5))	118.5	244.0	288.1
Constant b (eq. (12-5))	−0.02	−0.7	4.76
r (eq. (12-6))	0.9843	0.9782	0.9452
N (eqs. (12-8), (12-9))	318	55	534
SEP (eq. (12-8))	0.466%	0.450%	0.983%
Bias (eq. (12-9))	−0.21%	0.13%	−0.036%

12.4.3 Wheat Kernels

Experimental results for 474 wheat kernels measured at various moisture levels at 24°C in the X-band resonant cavity are shown in Figure 12.17. The variables in the calibration equation are listed in Table 12.2, and moisture content values predicted for 534 wheat kernels of different varieties by the calibration equation are compared with the reference values in Figure 12.18. The distribution of differences between oven moisture content and calculated values is shown in Figure 12.19. The predicted values of wheat kernel moisture content, when compared with those determined by the standard method, have a bias of −0.036% moisture and standard deviation of differences of 0.983% moisture content.

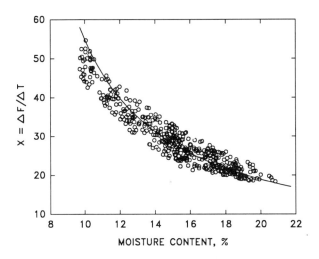

FIGURE 12.17. Dependence of the ratio X on moisture content in single kernels of wheat—total of 474 kernels of soft and hard red winter wheat.

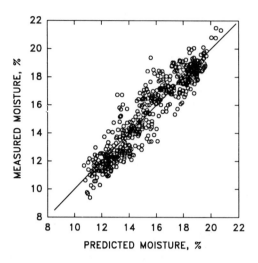

FIGURE 12.18. Predicted moisture contents of 534 single wheat kernels versus moisture content determined by standard oven method.

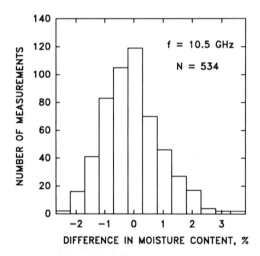

FIGURE 12.19. Distribution of differences between oven moisture content and moisture content from the calibration equation for 534 single kernels of wheat.

12.5. UNCERTAINTY ANALYSIS

The precision of system calibration is affected by an uncertainty of the measuring system (σ_M) consisting of the repeatability of the results for the same kernel of a given moisture content and the repeatability of the results provided by the standard oven method used for the system calibration (σ_c). Since both of these magnitudes are of random character, the uncertainty in using the microwave resonant cavity for moisture content determination in a single kernel may be defined as

$$\sigma_s = \pm\sqrt{\sigma_M^2 + \sigma_c^2}. \tag{12-10}$$

The uncertainty in the measuring system can be determined by differentiating eq. (12-5):

$$\Delta M = \sigma_M = -\left(\frac{M}{X} + \left|\frac{b}{X}\right|\right)\Delta X, \tag{12-11}$$

where

$$\Delta X = X\left(\frac{\delta f_0 + \delta f_s}{f_0 - f_s} + B\frac{\delta S_{210} + \delta S_{21s}}{10^k - 1}\right) = X\left(\frac{n_1\delta f}{\Delta F} + B\frac{n_2\delta S}{\Delta T}\right), \tag{12-12}$$

with

$$B = \frac{2.301}{20}10^k = 0.115(\Delta T + 1),$$

where

δf = discrete elementary error in resonant frequency measurement

δS = elementary error in transmission coefficient measurement (dB)

n_1, n_2 = integers equal to 0, 1, 2, ..., n

The discrete character of the measurements causes a similar distribution of errors. Because there is no continuous spectrum of readings available, the error in the resonant frequency measurement, δf, may have a value of $\delta f_i/2$ or any multiple thereof, where δf_i is the increment of the frequency change in the measuring system and is listed for three cavities in Table 12.3. The incremental errors in the transmission coefficient measurement, δS_{21}, were determined experimentally for each system, and these values are listed in Table 12.3.

An expression for the uncertainty in the moisture content determination by the reference method can be obtained by differentiating eq. (12-1), which is the definition of the relative moisture content determined on the wet basis. The expression has the form

$$\sigma_c = \frac{100}{m_m^2}(m_m\Delta m_d + m_d\Delta m_m) = \frac{100\Delta m}{m_m^2}(m_m + m_d), \tag{12-13}$$

where Δm_m and Δm_d denote errors in weighing moist and dry kernels, respectively. The uncertainty σ_c for $\Delta m_m = \Delta m_d = \Delta m = 0.05$ mg is 0.24% to 0.29% moisture for large and small wheat kernels, respectively. Corresponding uncertainties provided for peanut kernels and soybean seeds are 0.17% and 0.22% moisture, respectively. These values are also listed in Table 12.3.

TABLE 12.3. Comparison of Measuring Conditions and Expected Uncertainties for Kernels and Seeds in the Rectangular Waveguide Cavities Used in This Study

Parameter	S-band	C-band	X-band
Frequency span	16 MHz	25 MHz	96 MHz
Average ΔF	11 MHz	15 MHz	60 MHz
Average ΔT	0.715	0.868	2.32
δf_i	0.02 MHz	0.032 MHz	0.12 MHz
δS_i	0.02 dB	0.02 dB	0.02 dB
a in eq. (12-5)	118.5	244.0	288.1
b in eq. (12-5)	−0.02	−0.07	4.76
B in eq. (12-12)	0.1972	0.2148	0.3818
Aver. M	7.6%	12.0%	14.8%
Aver. X	15.5	17.3	26.0
ΔX from eq. (12-12)			
$n_1 = 1, n_2 = 0$	0.0282	0.0369	0.052
$n_2 = 1, n_1 = 0$	0.0855	0.0856	0.086
$(M + b)/X$	0.4916	0.6977	0.753
σ_M from eq. (12-11)	−0.056%	−0.086%	−0.103%
σ_c from eq. (12-13)	0.17%	0.22%	0.27%
σ_s from eq. (12-10)	$\pm \begin{matrix} 0.427\% \\ 0.483\% \end{matrix}$	$\pm \begin{matrix} 0.483\% \\ 0.561\% \end{matrix}$	$\pm \begin{matrix} 0.970\% \\ 1.058\% \end{matrix}$
SEP from eq. (12-8)	± 0.466%	± 0.450%	±0.983%
Real value $n_1 + n_2$	7–8	5–6	9–10

12.6.   DISCUSSION AND CONCLUSIONS

The microwave resonant cavity technique has proven to be a useful tool for moisture content determination in single grain kernels and seeds. The expected uncertainty in moisture content determination using the calibration equations developed, in the range of practical interest, can be less than 1% moisture at the 95% confidence level. The measuring circuit can be simple and requires only commercially available devices. Changes in resonant frequency and the transmission coefficient of the cavity when loaded with a grain kernel are the only measured values. Thus, long-term stability of the measuring system is not required, because checking of the reference values for an empty cavity may be performed as often as necessary. Absolute values of either of these parameters are of no interest in routine measurements.

The measurement uncertainty results mainly from variations in kernel shape which varies significantly even within samples from one variety. Numerical examples calculated from expressions developed (see Appendix) show that for wheat kernels of the same moisture content (assuming identical kernel densities, meaning identical values of the permittivity function) and with L/D ratios changing from 1.42 to 2.01, ΔF will increase by 1.193 while the increase in ΔT will be 1.44. The authors have noted that changes in ΔT are always the square of those for ΔF, within the accuracy of the measurement error.

From data presented in Figure 12.8 and Figure 12.9, both ΔF and ΔT depend upon the mass and moisture content of peanut kernels, although in a different nonlinear way. As for ΔF, the change of slope with moisture content evident at about 6% to 8% moisture may be related to a change in the character and properties of bound water in the kernel. Above that level, the increase in ΔF with moisture content is nearly linear for kernels of all sizes. The slope of the transmission factor ΔT versus moisture content changes in the same moisture range, but the slope changes again at higher moisture levels, which most likely is related to the character of the perturbed resonator. When losses in the kernel increase above a certain level, the response of the resonant cavity is no longer linear. Although this is not a limitation for the applications considered in this paper, it would be of interest to determine this limit quantitatively. The physical basis for this limit was established [11] in the form

$$\frac{\Delta F}{f_0} \approx \frac{\Delta T}{4Q_0}.$$

It is the permissible limit of the change in the Q factor which determines the applicability of the perturbation theory generalized to high-loss materials for accurate permittivity determination.

The nonlinear character of both measured variables, ΔF and ΔT, is reflected in the particular shape of the plot of ratio X, which is a size-independent function, and, as shown in Figure 12.10, has a broad flat extremum. Although peanuts cannot be stored when their moisture content is higher than 10.5%, detection of higher moisture kernels is of practical interest. However, the cavity used to obtain data illustrated in Figure 12.10 would be limited to measuring moisture contents lower than $\sim$ 14%. In case kernels of moisture higher than 14% are present, the criterion of $\Delta T < 1.5$ could be useful for excluding such kernels (Figure 12.9).

For objects of a given material and range of shapes and sizes, there is a certain optimum cavity volume which can be obtained from Figure 12.5, which will provide useful relationships between the measured variables and moisture content. However, for all material parameters constant, only the cavity constant C (eq. (12-4)) can affect the value of the ratio X and in turn the slope of $\varphi^{-1}(X)$ (eq. (12-5)). When the slope of $\partial X/\partial M$ is not steep enough for practical use (expected uncertainty of the inverse function is too high), then increasing f_0 and/or decreasing the cavity Q factor may be a desired solution. However, for a resonant cavity, increasing the resonant frequency is equivalent to using a smaller cavity and to raising the v_s/v_0 ratio which may lead to a strongly nonlinear behavior. A reduction of the Q factor is then the only acceptable solution, and it may be accomplished by increasing the cavity coupling factor. Enlargement of the diameter of the coupling holes will provide a rapid reduction of the Q factor with no significant change in the resonant frequency f_0. In any case, the value of Q_0 should not be reduced below about 1000 to attain other advantages of the resonant cavity method.

It may be noted from Figure 12.14 that the accuracy of moisture content prediction for single peanut kernels is lower at both ends of the moisture content range (for very dry and very moist kernels). This is probably a result of developing the calibration equation for the moisture content range of 4% to 12% (Figure 12.13) and the verification being performed for the wider, 3.5% to 14% moisture content range. But for both peanut and wheat kernels, the distribution of differences between measured and predicted moisture contents appears to have a normal distribution (Figures 12.15 and 12.19), which would suggest a random character of measurement error. In the case of soybean seeds that have very uniform shapes, the error distribution had a nonuniform character [5] because of the limited number of seeds used in the validation procedure (55). The broader spread in results for wheat kernels (Figures 12.17 and 12.18) than for the other two commodities reflects the effect of a much larger eccentricity ratio (L/D) among the wheat kernels than among the other commodities tested.

As mentioned, the discrete character of the measurement creates values of resonant frequency and transmission coefficient differing from their "true" values by $n_1 \delta f$ and $n_2 \delta S$, respectively. That in turn gives discrete values of moisture content calculated from the calibration equation that differ by $n \Delta X$ from the "real" value. If a set of such quantized values is subtracted from a uniformly distributed set of moisture content values determined by the reference method, the discrete character of differences may become obvious, as long as the number of samples (kernels) remains low in a statistical sense. While the uncertainty in ΔX related to the frequency measurement increases for higher frequencies (see Table 12.3), that related to the transmission coefficient measurement remains virtually the same for S-band, C-band, and X-band cavities. The number of discrete steps n_1 and n_2 by which the measured variables differed from their respective "real" values, during the reference measurement on the empty cavity *and* on the cavity loaded with an object, may be considered a measure of system uncertainty. The fact that this number is higher for nonuniformly shaped wheat kernels than for almost identically shaped soybean seeds may be related to the shape variations among the objects rather than to poorer stability of one measuring system compared to the other. However, each commodity was measured in a different cavity, of different resonant frequency and Q factor, and those factors also have to be taken into consideration. More study on different commodities in similar cavities needs to be conducted before conclusive statements can be presented.

Although it is too early to speculate on the physical and/or biological meaning and significance of the numerical factors of the calibration equations for particular commodities, simple comparisons may be appropriate. The inverse of the size-independent function for the three commodities tested in this study is shown as a function of the moisture content in Figure 12.20. This form of presentation was recommended by Kent and Meyer [12] as the function is linearly related to the material moisture content. This is also true for all three commodities used in our

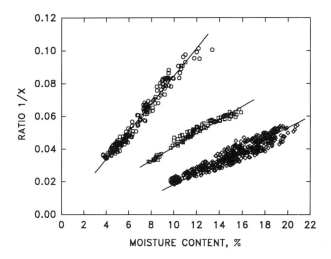

Figure 12.20. Inverse of the ratio X as a function of moisture content for three groups of objects of biological origin: circles—peanut, squares—soybean, and diamonds—wheat.

tests. Combining eqs. (12-4) and (12-5), one obtains

$$\frac{\varphi(M)}{K} = \left(\frac{a}{M-b}\right)\frac{2Q_0}{f_0}.$$

For parameters of the three cavities listed in Table 12.1 and Table 12.2, and values of M close to the equilibrium moisture content, which is 4.2%, 8.5%, and 11.3% moisture for peanuts, soybeans, and wheat, respectively, the value of the permittivity function divided by the shape factor K is 15.29, 15.08, and 7.53 for these commodities, respectively. Biological similarities of peanuts and soybeans include high oil content, which is not the case for wheat, but, again, more study is needed before certain conclusions can be drawn.

Although results of static measurements were reported in the paper, seeds or grain kernels could be introduced individually into the cavity, measured and released automatically at a high rate, determined only by the characteristics of a transport system and time needed for measurement. The measured variables—two coordinates of the resonant curve, with and without the object—can be determined rapidly with existing microprocessor tracking routines for finding the maximum value, in time periods ranging from several to 100 ms. As the differences between those two variables are needed for computations, their absolute values are of no importance, and long-term stability of the measuring system is not required. Thus, the measuring circuit could be simple and requires only commercially available devices. Standard microprocessors could also govern kernel delivery and release from the cavity and provide the computations required for moisture content determination. A printout of individual values and/or values averaged over the number of kernels being tested and the distribution of values for a given lot could be easily provided.

By applying the high sensitivity of microwave resonant sensors, a fast, accurate, and nondestructive method of moisture content determination has been developed for grain kernels and seeds. An accuracy of 0.5% moisture content may be obtained for individual kernels and seeds without need for contact between these objects and the measuring system, thus providing an effective tool for determining the distribution of moisture content among kernels and detecting mixed lots of grain of different moisture levels.

APPENDIX

When a dielectric object is placed in an electric field E_0, polarization P is induced [13]:

$$P = \frac{\epsilon - 1}{1 + A(\epsilon - 1)} \epsilon_0 E_0 = (\epsilon - 1)\epsilon_0 K E_0, \qquad (12\text{-}A1)$$

where ϵ_0 is the permittivity of free space and $\epsilon = \epsilon' - j\epsilon'' = \epsilon' - j(\sigma/\omega_0\epsilon_0)$ is the complex permittivity of the material, with σ being the material conductivity and ω_0 the operating angular frequency; K is the complex shape factor used previously [8], and A is the depolarization factor which depends upon the shape of the object and which is well determined for simple geometries such as a sphere and a thin, infinite rod. Other shapes found in nature have to be approximated, e.g., by prolate spheroids [14]. For a prolate spheroid of radii a, b, and c ($a > b = c$), the appropriate expressions for the depolarization factor in the direction of the respective radii have the form

$$A_a = \frac{1 - e^2}{2e^3} \left(\ln\frac{1 + e}{1 - e} - 2e \right), \qquad (12\text{-}A2)$$

$$A_b = A_c = \tfrac{1}{2}(1 - A_a), \qquad (12\text{-}A3)$$

where the eccentricity is

$$e = \sqrt{1 - (L/D)^2}.$$

For peanut kernels, L is the length and D is the diameter of the kernel.

If an object, small compared to the wavelength, is inserted into the cavity, small changes in the resonant frequency and Q factor occur. These changes are expressed as a complex frequency shift, which is related to a small change in the energy stored in the cavity and is given by [15]

$$\frac{\delta\Omega}{\omega_0} = -\frac{P \cdot E_0^* + M \cdot H_0^*}{2W}, \qquad (12\text{-}A4)$$

where P and M are the total induced electric and magnetic moments of the object. E_0 and H_0 are the unperturbed electric and magnetic fields in the location of the object, and W is the energy stored in the cavity, which is given by

$$W = \frac{1}{2} \int_{v_0} \{\epsilon_0 |E|^2 + \mu_0 |H|^2\} \, dv, \qquad (12\text{-}A5)$$

where μ_0 is the magnetic permeability of free space, and the integration is over the volume of the cavity. The complex angular frequency Ω is related to the resonant frequency and the Q factor by

$$\Omega_0 = \omega_0(1 + j/2Q_0).$$

When the object is placed in the maximum electric field, the magnetic field H at this point is zero, thus simplifying eq. (12-A4). Using eqs. (12-A1)–(12-A5) and performing the integration for the TE_{10p} cavity, one obtains

$$
\frac{\delta\Omega}{\omega_0} = -2\left(\frac{v_s}{v_0}\right)\frac{\epsilon - 1}{1 + A(\epsilon - 1)}
$$

$$
= \frac{\omega_s - \omega_0}{\omega_0} + j\frac{1}{2}\left(\frac{1}{Q_s} - \frac{1}{Q_0}\right) = -\frac{\Delta F}{f_0} + j\frac{\Delta T}{2Q_0},
$$

(12-A6)

where v_s and v_0 are the object volume and the cavity volume, respectively. Separating eq.(12-A6) into the real and imaginary parts yields

$$
\Delta F = 2\left(1 - \frac{1 + A(\epsilon' - 1)}{[1 + A(\epsilon' - 1)]^2 + [A\epsilon'']^2}\right)f_0\left(\frac{v_s}{v_0}\right),
$$

(12-A7)

$$
\Delta T = \frac{4Q_0\epsilon''}{[1 + A(\epsilon' - 1)]^2 + [A\epsilon'']^2}\left(\frac{v_s}{v_0}\right).
$$

(12-A8)

When it can be assumed that the material is not very lossy, $\epsilon'^2 \gg \epsilon''^2$, these two equations can be written in much simpler form:

$$
\Delta F \approx \frac{2(\epsilon' - 1)f_0}{1 + A(\epsilon' - 1)}\left(\frac{v_s}{v_0}\right),
$$

$$
\Delta T \approx \frac{4Q_0\epsilon''}{[1 + A(\epsilon' - 1)]^2}\left(\frac{v_s}{v_0}\right),
$$

(12-A9)

which are identical with eqs. (12-2) and (12-3), where $K = 1/[1+A(\epsilon'-1)]$ is the shape factor, which is dependent upon object shape, orientation, and permittivitty. If $A = \frac{1}{3}$, eq. (12-A9) provides the relations for a spherical object with $K = 3/(\epsilon' + 2)$. If $A = 1$, the shape factor $K = 1/\epsilon'$ for a thin disk perpendicular to the E_0 vector, while for $A = 0$, $K = 1$ for a thin long rod parallel to the E_0 vector.

To illustrate quantitatively the relationship between the two measured variables ΔF and ΔT and the dimension ratio L/D, some simple calculations were performed. The relationship between the L/D ratio and the depolarization factor along the L dimension, being parallel to the E-field vector, as calculated from eq. (12-A2), is shown in Figure 12.21. Dependence of the variables ΔF and ΔT upon the L/D ratio for peanut kernels of $m_d = 0.65$ g, as shown in Figures 12.11 and 12.12 (constant volume and $\epsilon = 6 - j2$), is presented in Figure 12.22. For

L/D ranging from 1.51 to 1.89, the shift of the resonant frequency increases by 1.122, and the transmission factor increases by 1.26. The ratio X changes by 1.123 from 13.04 to 11.61, which would cause a change in moisture content determination by eq. (12-7) from 9.07 to 10.19% moisture only because of differences in peanut shape. Similar changes calculated for wheat kernels are even greater, which is reflected in the spread of the experimental results presented in Figures 12.17–12.19.

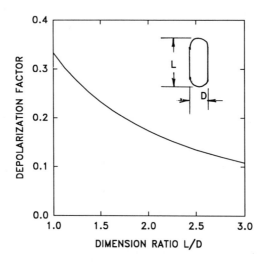

FIGURE 12.21. Relationship between the dimension ratio L/D and the depolarization factor A for prolate spheroids.

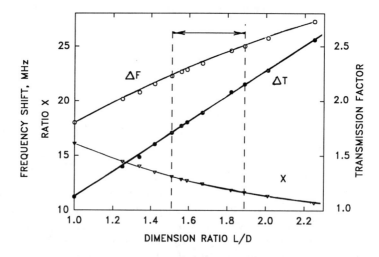

FIGURE 12.22. Effect of the shape of the object upon two variables measured in the S-band cavity and their ratio. Numerical values in indicated range correspond to peanuts of $m_d = 0.65$ g as shown in Figures 12.11 and 12.12.

REFERENCES

[1] ASAE Standards: ASAE S352.2 "Moisture measurement—unground grain and seeds," ASAE S410.1 "Moisture measurement—Peanuts," St. Joseph, MI: American Society of Agricultural Engineers, 1992.

[2] C. A. Watson, W. T. Greenaway, G. Davis, and R. J. McGinty, "Rapid proximate method for determining moisture content of single kernels of corn," *Cereal Chem.*, vol. 56, pp. 137–140, 1979.

[3] S. O. Nelson and K. C. Lawrence, "Evaluation of a crushing-roller conductance instrument for single-kernel corn moisture measurement," *Trans. ASAE*, vol. 32, pp. 737–743, 1989.

[4] C. V. K. Kandala, S. O. Nelson, and K. C. Lawrence, "Non-destructive electrical measurement of moisture content in single kernels of corn," *J. Agric. Eng. Res.*, vol. 44, pp. 125–132, 1989.

[5] A. W. Kraszewski, T.-S. You, and S. O. Nelson, "Microwave resonator technique for moisture content determination in single soybean seeds," *IEEE Trans.*, vol. IM-38, pp. 79–84, Feb. 1989.

[6] A. W. Kraszewski and S. O. Nelson, "Moisture content determination in single peanut kernels with a microwave resonator," *Peanut Sci.*, vol. 20, pp. 27–31, 1993.

[7] H. Altschuler, "Dielectric constant," in *Handbook of Microwave Measurements*, M. Sucher and J. Fox, Eds., New York: Polytechnic Press, 1963, pp. 530–536.

[8] A. W. Kraszewski and S. O. Nelson, "Observations on resonant cavity perturbation by dielectric objects," *IEEE Trans.*, vol. MTT-40, pp. 151–155, Jan. 1992.

[9] A. W. Kraszewski, S. O. Nelson, and T. S. You, "Moisture content determination in single corn kernels by microwave resonator techniques," *J. Agric. Eng. Res.*, vol. 48, pp. 77–87, 1991.

[10] ISTA, "International rules for seed testing. 9: Determination of moisture content," *Proc. Int. Seed Test. Assoc.*, vol. 31, pp. 128–139, 1966.

[11] A. Bonincontro and C. Cametti, "On the application of the cavity perturbation method to high-loss dielectrics," *J. Phys. E: Sci. Instrum.*, vol. 10, pp. 1232–1233, 1977.

[12] M. Kent and W. Meyer, "A density independent microwave moisture meter for heterogeneous foodstuffs," *J. Food Eng.*, vol. 1, pp. 31–42, 1982.

[13] C. Kittel, *Introduction to Solid State Physics*, New York: Wiley, 1956.

[14] A. H. Sihvola and I. V. Lindell, "Effective permeability of mixtures," in *Dielectric Properties of Heterogeneous Materials*, A. Priou, Ed., New York: Elsevier, 1992, p. 158.

[15] R. A. Waldron, *The Theory of Waveguides and Cavities*, London: Maclaran, 1967.

Seichi Okamura
Takaaki Masuda
Shizuoka University, Hamamatsu, Japan

13

A New Moisture Content Measurement Method by a Dielectric Ring Resonator

Abstract. This paper describes a new moisture content measurement method for small or random size material from the resonant frequency shift of a dielectric ring resonator caused by the moisture contained in the material which is placed in the hollow core of the ring. The theoretical resonant frequency shift of the resonator is obtained from a new analysis using the mode matching method. The accuracy of the theory is confirmed by the close agreement between the calculated and measured values for the disk-shaped and different size samples whose dielectric constants are known. The theoretical result shows that the measurement of resonant frequency shift per unit weight of a sample is very effective for moisture content measurements and this is shown in experiments on cardboard samples.

13.1. INTRODUCTION

Microwaves become important for the moisture content measurements of industrial or agricultural products such as paper, textile, green tea, etc., and are gradually replacing the conventional measurement methods [1,2]. In these measurements the microwave attentuation or phase shift caused by the water contained in such materials is used to determine the moisture content. For reliable measurements it is necessary for the samples to have a larger cross section than the area of the microwave propagating region and to have a uniform density.

So far, it has been difficult to measure the moisture content of single and small size samples. Wooden chips, for example, used in production of paper have small and irregular shapes. Therefore their moisture content is measured by the conventional drying method.

The use of a dielectric ring resonator has been proposed for those measurements [3]. In this literature the quality factor Q of the resonator was used to measure the moisture contents, but it is complicated because the frequency and amplitude of microwaves must be measured.

In this paper, we introduce a new moisture measurement method in which the resonant frequency of a dielectric ring resonator only is used. A sample of a material to be measured is placed in the center of the hollow core of the dielectric ring resonator. The resonant frequency of the resonator with an inserted sample is reported [4], but that analysis neglected the air gap between the dielectric ring and the sample. The rigorous analysis considering the air gap had not yet been reported.

13.2. RESONANT FREQUENCY ANALYSIS

The dielectric ring resonator for the new moisture measurement method is shown in Figure 13.1a. It is composed of a Teflon ring, whose relative permittivity is 2.08, and two conducting plates. The resonant mode is TE_{011}; the size of the dielectric ring is decided in accordance with the resonant mode chart of dielectric rod resonators [5]. The resonant frequency changes with the permittivity of the sample, which is influenced by the moisture content of the material. We consider the influence only of the real part of permittivity, because the influence of the imaginary part can be neglected when the quality factor of a resonator is higher than 353 [4,6]. In experiments, the resonator of this paper has factors of about 1000 or more.

Figure 13.1b shows the cross section of the resonator used in the resonant frequency analysis. Region I is a sample of a material to be measured, region IV is a dielectric ring, regions II, III, and V are air. The outer and inner radii of the dielectric ring and the radius of the sample are indicated by a, b, and c, and the half-height of the dielectric ring and of the sample are represented by L and h, respectively. The shaded portions show two metal plates attached to the ends of the dielectric ring.

In general, the electromagnetic fields ψ satisfies the wave equation (13-1) in cylindrical coordinates (r, θ, z), derived from Maxwell's equations [7]:

$$\frac{\partial}{r\partial r}\left(r\frac{\partial\psi}{\partial r}\right) + \frac{\partial^2\psi}{r^2\partial\theta^2} + \frac{\partial^2\psi}{\partial z^2} + k^2\psi = 0. \tag{13-1}$$

Since the relative permittivity ϵ_{rn} of the resonator is different in each region, as in eq. (13-2), the mode matching method of considering the higher-order resonance modes is used in the analysis to obtain the resonant frequency.

$$\epsilon_{rn} = \begin{cases} \epsilon_1 & \text{(sample)} & n = 1, \\ 1 & \text{(air)} & n = 2, \\ 1 & \text{(air)} & n = 3, \\ \epsilon_D & \text{(dielectric ring)} & n = 4, \\ 1 & \text{(air)} & n = 5. \end{cases} \tag{13-2}$$

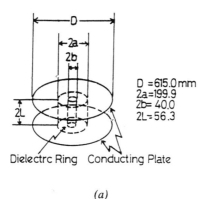

D =615.0 mm
2a=199.9
2b= 40.0
2L= 56.3

Dielectrc Ring   Conducting Plate

(a)

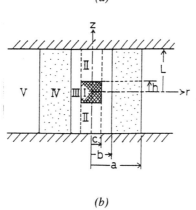

(b)

FIGURE 13.1. Dielectric ring resonator for moisture content measurements: (a) resonator used in this paper; (b) cross section of the resonator when a sample is set in the center of the hollow core.

Hence, the magnetic field component of the resonator is given in eq. (13-3) as a sum of the components of H_{nz}, which is expressed by the dominant and higher-order mode of the pth TE_{01p} ($p = 1, 2, 3, \ldots$). The subscript n is a number from 1 to 5 corresponding to the region from I to V.

$$H_{nz} = -\frac{j\omega}{k_0^2} \sum_{p=1}^{\infty} R_{np}(r)\lambda_{np}Z_{np}(z), \qquad (13\text{-}3)$$

where ω and k_0 are the angular frequency and the propagation constant of the electromagnetic wave in a vacuum. $R_{np}(r)$ and $Z_{np}(z)$ are radial and axial functions shown later in eqs. (13-7) and (13-10), respectively, and

$$\lambda_{np} = k_0^2\epsilon_{rn} - \beta_{np}^2, \qquad (13\text{-}4)$$

where β_{np} is the propagation constant in the z direction.

The other related electric and magnetic fields are

$$E_{nr} = E_{nz} = 0,$$

$$E_{n\theta} = \frac{1}{\epsilon_0} \sum_{p=1}^{\infty} R'_{np}(r) Z_{np}(z), \tag{13-5}$$

$$H_{nr} = -\frac{j\omega}{k_0^2} \sum_{p=1}^{\infty} R'_{np}(r) Z'_{np}(z), \tag{13-6}$$

$$H_{n\theta} = 0,$$

where the prime means the derivative with respect to the variable inside the parentheses, the Bessel function $R_{np}(r)$ is

$$R_{np}(r) = \begin{cases} A_p P_{1p}(r), & n = 1, \\ B_p P_{2p}(r), & n = 2, \\ C_p P_{3p}(r) + D_p Q_{3p}(r), & n = 3, \\ E_p P_{4p}(r) + F_p Q_{4p}(r), & n = 4, \\ G_p W_{5p}(r), & n = 5, \end{cases} \tag{13-7}$$

where A_p, B_p, C_p, D_p, E_p, F_p, and G_p are amplitude coefficients of the pth higher-order-mode wave. The functions $P_{np}(r)$, $Q_{np}(r)$, and $W_{np}(r)$ are represented by the Bessel function of the first kind, $J_0(x)$, or the second kind, $Y_0(x)$, and the modified Bessel function of the first kind, $I_0(x)$, or the second kind, $K_0(x)$, using real numbers ξ_{np} and η_{np} according to the sign of λ_{np} as follows:

$$(\lambda_{np})^{1/2} = \begin{cases} \xi_{np}, & \lambda_{np} \geq 0, \\ j\eta_{np}, & \lambda_{np} < 0, \end{cases} \tag{13-8}$$

$$P_{np}(r) = \begin{cases} J_0(\xi_{np}r), \\ I_0(\eta_{np}r), \end{cases}$$

$$Q_{np}(r) = \begin{cases} Y_0(\xi_{np}r), \\ K_0(\eta_{np}r), \end{cases} \tag{13-9}$$

$$W_{np}(r) = K_0(\eta_{np}r).$$

The function $Z_{np}(z)$ is given in each region as follows:

$$Z_{np}(z) = \begin{cases} \cos(\beta_{1p}), & n = 1, \\ \sin(\beta_{2p}(z - L)), & n = 2, \\ \cos(\beta_{3p}z), & n = 3, \\ \cos(\beta_{4p}z), & n = 4, \\ \cos(\beta_{5p}z), & n = 5. \end{cases} \tag{13-10}$$

The propagation constants of β_{1p} and β_{2p} are taken from the conditions for magnetic fields at the boundary between regions I and II as follows:

$$\beta_{1p} \tan(\beta_{1p}h) = \beta_{2p} \cot(\beta_{2p}(L - h)),$$
$$\beta_{1p}^2 - \beta_{2p}^2 = (2\pi/\lambda_0)^2(\epsilon_1 - 1). \tag{13-11}$$

The propagation constants of β_{3p}, β_{4p}, and β_{5p} are obtained from the boundary condition of zero tangential electric component on the conducting plates.

$$\beta_{3p} = \beta_{4p} = \beta_{5p} = (2p - 1)\pi/2L \qquad (p = 1, 2, \ldots). \tag{13-12}$$

Furthermore, the amplitude coefficients of A_p and B_p must satisfy eq. (13-13) from the continuity of the tangential electric field components at the boundary between regions I and II:

$$\frac{B_p}{A_p} = -\frac{\cos(\beta_{1p}h)}{\sin(\beta_{2p}(L - h))}. \tag{13-13}$$

Equations (13-14) to (13-16) are obtained from the boundary conditions with respect to the radial direction. The continuity of the electric and magnetic fields at the boundaries of regions IV and V ($r = a$), III and IV ($r = b$), and I and III ($r = a$) produces the following equations:

$$\lambda_{4q}(E_q P_{4q}(a) + F_q Q_{4q}(a)) = \lambda_{5q} G_q W'_{5q}(a),$$
$$E_q P'_{4q}(a) + F_q Q'_{4q}(a) = G_q W'_{5q}(a), \tag{13-14}$$

$$\lambda_{3q}(C_q P_{3q}(b) + D_q Q_{3q}(b)) = \lambda_{4q}(E_q P_{4q}(b) + F_q Q_{4q}(b)),$$
$$C_q P'_{3q}(b) + D_q Q'_{3q}(b) = E_q P'_{4q}(b) + F_q Q'_{4q}(b), \tag{13-15}$$

$$\sum_{p=1}^{\infty} \lambda_{1p} A_p P_{1p}(c) \cos(\beta_{1p}z) \qquad (0 < z < h),$$

$$\sum_{p=1}^{\infty} \lambda_{1p} B_p P_{1p}(c) \sin(\beta_{2p}(z - 1)) \qquad (h < z < L),$$

$$= \sum_{q=1}^{\infty} \lambda_{3q}(C_q P_{3q}(c) + D_q Q_{3q}(c)) \cos(\beta_{3q}z),$$

$$\sum_{p=1}^{\infty} A_p P'_{1p}(c) \cos(\beta_{1p}z) \qquad (0 < z < h),$$

$$\sum_{p=1}^{\infty} B_p P'_{1p}(c) \sin(\beta_{2p}(z - L)) \qquad (h < z < L),$$

$$= \sum_{q=1}^{\infty} (C_q P'_{3q}(c) + D_q Q'_{3q}(c)) \cos(\beta_{3q}z),$$

$$\tag{13-16}$$

where the higher-order mode numbers in regions I and II are indicated by subscript p, and those in regions III–V are indicated by subscript q.

Eliminating the other amplitude coefficients exclusive of A_p from eqs. (13-13) through (13-16), eq. (13-17) with respect to A_p is obtained for all q ($q = 1, 2, 3, \ldots$):

$$\sum_{p=1}^{\infty} \left[\left(S_{pq} - \frac{\cos(\beta_{1p}h)}{\sin(\beta_{2p}(L-h))} T_{pq} \right) \left(P'_{1p}(c) - \frac{\lambda_{1p}}{\lambda_{3q}} L_q P_{1p}(c) \right) \right] A_p = 0, \qquad (13\text{-}17)$$

where

$$S_{pq} = \int_0^h \cos(\beta_{1p}z) \cos(\beta_{3q}z) \, dz,$$

$$T_{pq} = \int_h^L \sin(\beta_{2p}(z-L)) \cos(\beta_{3q}z) \, dz, \qquad (13\text{-}18)$$

$$L_q = \frac{P'_{3q}(c) + M_q Q'_{3q}(c)}{P_{3q}(c) + M_q Q_{3q}(c)}.$$

Moreover,

$$M_q = \frac{\lambda_{3q} P_{3q}(b)V_q - \lambda_{4q} P'_{3q}(b)U_q}{\lambda_{4q} Q'_{3q}(b)U_q - \lambda_{3q} Q_{3q}(b)V_q},$$

$$U_q = P_{4q}(b) + N_q Q_{4q}(b),$$

$$V_q = P'_{4q}(b) + N_q Q'_{4q}(b), \qquad (13\text{-}19)$$

$$N_q = \frac{\lambda_{4q} P_{4q}(a)W'_{5q}(a) - \lambda_{5q} P'_{4q}(a)W_{5q}(a)}{\lambda_{5q} Q'_{4q}(a)W_{5q}(a) - \lambda_{4q} Q_{4q}(a)W'_{5q}(a)}.$$

A nontrivial solution of A_p can exist in eq. (13-17) when the determinant of the coefficients for A_p is zero. The determinant is called a characteristic equation, which can be used to calculate a theoretical resonant frequency of a dielectric ring resonator containing a sample to be measured.

The value of the equation is more accurate when it is calculated for larger numbers of p and q. In this paper we consider the higher-order modes up to the 25th mode because the values converge well enough up to the 16th mode.

13.3.   EVALUATION OF CHARACTERISTIC EQUATION

To evaluate the characteristic equation we compared the theoretical values with those measured from dielectric samples with known permittivities. The measurement was performed with the system shown in Figure 13.2. The dielectric ring resonator is excited by a sweep oscillator. The waves detected with coupler 2 are fed to the scalar network analyzer and the resonant frequencies are measured. The frequency is about 2120 MHz when no sample is inserted. The theoretical resonance value is 2117.4 MHz.

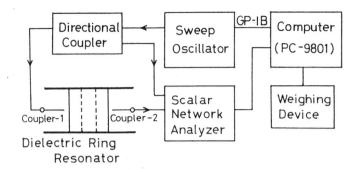

FIGURE 13.2. Block diagram of the resonant frequency measurement system for a dielectric ring resonator which is used as a new moisture content measurement system.

The resonant frequency changes by several megahertz due to fluctuations in room temperature. Therefore, the difference of the resonant frequencies before and after insertion of a sample in the resonator is used as the resonant frequency shift. In this way the effect of temperature fluctuation is minimized. The value of the shift is negative in experiments, and the absolute value of the shift is shown as ΔF in Figure 13.3.

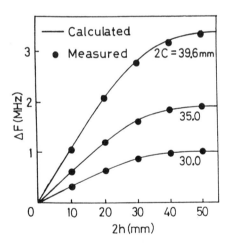

FIGURE 13.3. Resonant frequency shift of ΔF as a function of sample thickness.

The data in Figure 13.3 are obtained from samples of Teflon with different diameters ($2c$) and thicknesses ($2h$). The solid lines represent theoretical values based on the characteristic equation, and the dark circles indicate measured values. It can be seen that the theoretical values agree well with the measured ones. Therefore, the characteristic equation accurately predicts the resonant frequencies of the resonator.

The new moisture content measurement method utilizing a dielectric ring resonator is discussed next.

13.4. RESONANT FREQUENCY SHIFT AND MOISTURE CONTENTS

In low-moisture contents the relative permittivity ϵ_s of a moistened sample is approximated as follows [8]:

$$\epsilon_s = \epsilon_{s0} + \eta M_d, \tag{13-20}$$

where ϵ_{s0} is the permittivity of the sample having no moisture, η is a coefficient for the permittivity depending on the sample density, and M_d is the moisture content at dry basis.

A sample with a different moisture content can be replaced by a sample with a different permittivity as shown in eq. (13-20). Figure 13.4 shows the theoretical resonant frequency shift of the samples of different permittivities as a function of their thicknesses. It is evident that the resonant frequency shift ΔF is proportional to the thickness when ΔF is less than about 3 MHz. As the thickness is proportional to the sample weight, one should expect that the resonant frequency shift per unit weight ΔF_w is independent of the sample's weight, that is, the quantity.

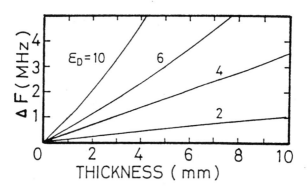

FIGURE 13.4. Theoretical resonant frequency shift of ΔF as a function of sample thickness.

Figure 13.5 shows the experimental results of various moisture contents taken from four cardboard samples having different thicknesses. In this experiment it was possible to measure the moisture contents up to 118% at dry basis for various sample thicknesses. For high-moisture-content measurements thin samples were used, whereas thick samples were used for low-moisture measurements. The data of all samples with different thicknesses are shown by the dotted line.

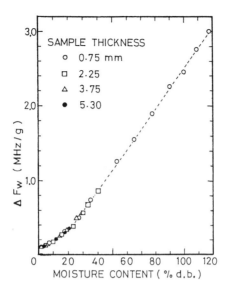

FIGURE 13.5. Resonant frequency shift per unit weight (ΔF_w) of a sample as a function of moisture content in four cardboard samples (sample diameter = 35 mm).

As is clearly indicated, one can measure the moisture contents of cardboard from the measurement of ΔF_w.

13.5. CONCLUSIONS

A new method for moisture content measurement is presented. The method is based on the shift of the resonant frequency of a dielectric ring resonator when a sample to be measured is placed in the hollow core of the ring. The theoretical resonant frequency is calculated using the characteristic equation obtained from the analysis represented in this paper. There is a good agreement between theoretical calculations and experimental results.

It is confirmed from examining the theoretical resonant frequencies that the resonant frequency shift by a sample per the unit weight is determined only from the moisture content of the sample. This is nicely shown in the measurements for many cardboard samples having various sizes and thicknesses.

For samples have a random shape, however, further study is necessary to determine the accuracy of the method.

REFERENCES

[1] E. Nyfors and P. Vainikainen, *Industrial Microwave Sensors*, Norwood, MA: Artech House, 1989, pp. 201–230.

[2] P. V. Vainikainen, E. G. Nyfors, and M. T. Fischer, "Radiowave sensor for measuring the properties of dielectric sheets: Application to veneer moisture content and mass per unit area measurement," *IEEE Trans. Instrum. Meas.*, vol. IM-36, pp. 1036–1039, 1987.

[3] S. Okamura and H. Ohishi, "Moisture content measurement with a microwave dielectric resonator," *IEICE Trans.* (*c*), vol. J70-C, pp. 1523–1528, 1987.

[4] S. Okamura and M. Sone, "Resonant frequency analysis of a dielectric ring resonator for a new method of moisture content measurement," *IEICE Trans.* (*c*), vol. J75-C, pp. 164–170, 1992.

[5] Y. Kobayashi and S. Tanaka, "Resonant modes of a dielectric rod resonator short-circuited at both ends by parallel conducting plates," *IEEE Trans. Microwave Theory Tech.*, vol. MTT-28, pp. 1077–1085, 1980.

[6] M. Nakajima, *Microwave Engineering*, Morikita, 1984, pp. 152–153.

[7] D. Kajfez and P. Guillon, *Dielectric Resonators*, Norwood, MA: Artech House, 1986, pp. 72–73.

[8] S. Oka and O. Nakada, *Solid-State Dielectric Theory*, Iwanami, 1966, pp. 284–285.

G. Biffi Gentili
G. F. Avitabile
G. F. Manes
Laboratorio di Microelettronica, Università di Firenze
Firenze, Italy

14

An Integrated Microwave Moisture Sensor

Abstract. An innovative moisture sensor is introduced and experimented in this paper. The sensing element is an open-ended transmission line inserted in the feedback loop of a self-oscillating patch antenna. The perturbation induced by the material under test results in a frequency shift of the waveform transmitted by the antenna. Moisture content can be then directly measured and remotely read out, resulting in a simple test set with on-field operation capability.

14.1. INTRODUCTION

Local monitoring and control of the moisture content in surface layers is of great importance to the disciplines of agriculture, industrial applications, and environment sensing.

Microwave dielectrometry represents a very promising method for moisture content estimation of porous and matrix systems.

Techniques based on the measurement of the complex reflection coefficient of an open-ended transmission line or waveguide perturbed by the material under test are well established [1,2]. In most applications a coaxial cable represents the transmission line, while the reflection coefficient measurement is usually performed on a network analyzer.

The same technique can be extended to moisture content measurements by properly modeling the complex dielectric permittivity of substances as a function of water content. In particular, in [3] it has been theoretically and experimentally demonstrated that a good estimation of the moisture content of a porous material of known density can be simply obtained through the measurement of the reflection coefficient phase of an open-ended coaxial termination embedded in the medium itself.

A major drawback of this approach is represented by the cable connecting the sensing element to the network analyzer. The length of this cable should be kept as short as possible in terms of wavelengths to reduce parametric sensitivity. Furthermore, the need of the network analyzer itself makes the test set unsuitable for in-field applications.

An innovative moisture sensor based on the previously described principle is introduced and discussed in this paper. The sensor exhibits the unique capability to perform the required moisture measurement and, at the same time, to transmit the associated information to a remote receiver.

The sensor is configured as an active patch antenna. The sensing element is an open-ended transmission line embodied in the feedback loop of a microwave oscillator which employs the patch antenna as a resonator. The moisture content of the medium under test is, thus, directly transformed to an oscillating frequency through the sensing element reflection coefficient variation. The resulting wave-form is radiated by the patch and remotely read out using an identical receiving patch antenna.

A simple, low-cost moisture sensor is thus realized, whose attractive features include operational flexibility, absence of artefacts, and multipoint operation capability.

14.2. SENSOR OPERATION PRINCIPLE

The schematic of the new sensor is represented in Figure 14.1.

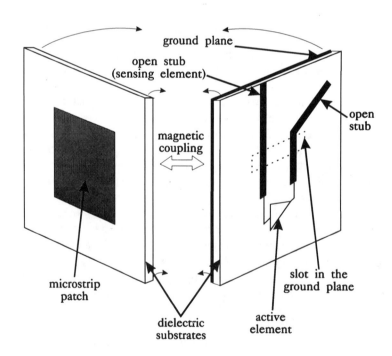

FIGURE 14.1. Active integrated antenna structure.

The active integrated antenna (AIA), already introduced [4], is a multilayer structure based on a patch antenna fed by a pair of microstrips through a non-resonant slot. The patch antenna represents the frequency-selective element of a phase-shift transistor oscillator. The AIA frequency of operation is determined both by the patch geometry and the electrical loads connected at the two microstrip ends. A convenient termination is normally represented by two open-ended stubs whose electrical length is $\lambda/4$ at the patch resonant frequency [4].

In order to operate as a moisture sensor, one of the open stubs is replaced by the open-ended coaxial transmission-line which represents the sensing element. Alternatively, a microstrip or coplanar waveguide element can be used for noninvasive sensing operation.

An equivalent lumped-element circuit of the proposed configuration is presented in Figure 14.2. The parallel resonant circuit represents the patch antenna, and the transformer represents the coupling between the antenna and the oscillator circuit. The sensing element is modeled by a variable reflectance, $\Gamma(\varepsilon)$, connected to the amplifier input port through a patch feeding microstrip. The other port is terminated with a load.

A noteworthy feature of the configuration is represented by the inherent capability of automatic compensation of frequency drifts determined by medium- and long-term parametric variations. This can be obtained by switching the amplifier input port between the sensing element (the variable reflectance in Figure 14.2) and

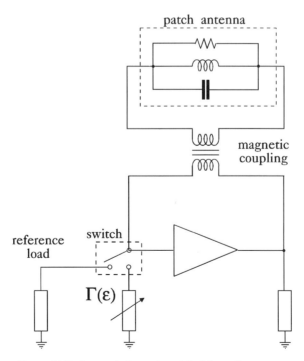

FIGURE 14.2. Lumped-element model of the active sensor.

a reference load at a fixed rate. This results in a two-tone FSK-modulated waveform, where the first tone frequency is determined by the reference load while the second is determined by the sensing element loaded by the medium under test. Assuming that both frequencies are affected by the same parametric variations, frequency offset between the two tones directly gives a compensated moisture measurement.

Multipoint measurements can be also attained by frequency multiplexing of a number of individual battery-powered sensors.

The configuration is planar and inherently low cost, being implemented on plastic substrates and allowing standard bipolar transistors to be used. Operation up to X-band can be easily achieved.

14.3. SENSOR OPERATION DEMONSTRATION

A prototype of the active sensor has been fabricated and tested. The AIA was implemented using a $\varepsilon_r = 2.55$, 30-mil-thick CuClad substrate. The sensor was designed to operate at a 2.450-GHz reference frequency (ISM band). An 18-mm open-ended UT-141 semirigid coaxial cable was used as sensing element. The layout of the realized prototype is shown in Figure 14.3. The active sensor operates according to the differential mode previously described, using a microstrip stub as reference load. The measurements will be referred in the following as frequency shifts, Δf, with respect to the reference frequency obtained connecting the reference load to the amplifier input in the oscillator feedback loop.

The active sensor was calibrated using the following procedure:

1. The frequency shift was first measured using the active sensor and a set of low-loss (tan δ < 0.04) materials of known dielectric characteristics [5,6]. The set of materials and the corresponding measured frequency shifts are reported in Table 14.1.

2. Numerical evaluation of reflection coefficient with reference to the open-ended termination of the sensing element was then performed for the same set of materials, according to the method described in [2]. Amplitude ($|\Gamma|$) and phase ($\angle\Gamma$) of the reflection coefficient are reported in Table 14.1.

Figure 14.4 shows the calibration curve of the active sensor obtained by combining the measurements with the theoretical calculations. This curve can be used to evaluate the water content, Wc, of a moistened medium, under the following assumptions. First, the dependence of $\angle\Gamma$ with Wc for the specific sensing element is known; second, $|\Gamma|$ is close to unity (low dielectric losses in the material).

Figure 14.5 shows the measured AIA frequency offset versus the volumetric water content of a 0.6-porosity sand, as an example of the measuring capabilities of the proposed integrated sensor.

With the aid of a CAD model of the sensor and by using the method described in [3], it can be easily demonstrated that the value of $|\Gamma|$ only slightly affects the

Laboratorio di Microelettronica
D.I.E - Firenze

C.C

FIGURE 14.3. Layout of the 2450-GHz prototype.

TABLE 14.1. Measured Frequency Offset and Calculated
Reflection Coefficient of the Sensing Element for Various
Low-Loss Materials (2450-MHz reference frequency)

MATERIAL	$|\Gamma|$	$\angle\Gamma(°)$	ΔF (MHz)
Heptane	0.999	−5.55	37
Carbon tetrachloride	0.999	−6.43	36
Chloroform	0.991	−13.37	32
1.2 - Dichloroethane	0.989	−27.7	26
Aceton	0.975	−51.92	16
Acetonitrile	0.959	−79.88	9
Water	0.962	−123.17	4

Wc estimation for a wide class of moistened porous materials and a wide variation
of the ionic content of the water.

A total error less than 3% has been theoretically and experimentally verified
in the moisture measurements of porous media with $0 \leq \text{Wc} \leq 0.8$ and NaCl
water solution up to 0.1 molality (porosity ranging from 0.3 to 0.8).

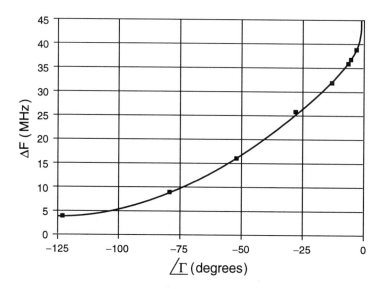

FIGURE 14.4. Calibration curve of the active sensor.

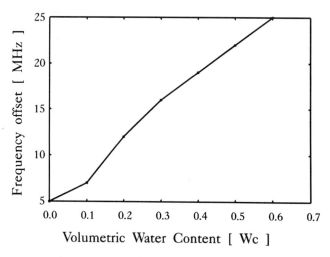

FIGURE 14.5. Measured frequency offset vs volumetric water content for a 0.6-porosity sand.

14.4.   CONCLUSION

A simple, low-cost moisture sensor based on an active integrated antenna has been discussed and demonstrated in this paper. The sensing element acts as tuning element of the oscillating frequency of the AIA.

The moisture content can be remotely evaluated through the AIA frequency-shift measurements with respect to the simultaneously transmitted reference frequency. This dramatically reduces the cost and size of the measurement test set while ensuring in-field operations free from medium- and long-term parametric variations of the sensor environment.

Other unique features of the innovative configuration are remote readout capability, multipoint operation, and a virtual absence of artefacts. Invasive or noninvasive operation can be easily obtained by simply changing the sensing element structure.

ISM-band operation has been demonstrated using a simple prototype; a relationship between frequency offsets and reflection coefficient phase changes of the sensing element induced by the water content variation of a porous material has been derived, in qualitative agreement with theoretical calculations.

Future work will be devoted to the optimization of the microwave circuit and to an accurate characterization of the sensor through laboratory and in-field measurements.

Studies oriented to extend the applicability of the integrated sensor to permittivity measurements of high-loss substances through an independent but simultaneous estimation of the low-frequency conductance of the sensing element are also in progress.

REFERENCES

[1] E. Tanabe and W. J. Joines, "A non-destructive method for measuring the complex permittivity of dielectric materials at microwave frequencies using an open transmission line resonator," *IEEE Trans. Instrum. Meas.*, vol IM-25, no. 3, Sept. 1976.

[2] J. R. Mosig, J. C. E. Bessons, M. Gex-Fabry, and F. E. Gardiol, "Reflection of an open-ended coaxial line and application to nondestructive measurement of materials," *IEEE Trans. Instrum. Meas.*, vol. IM-30, no. 1, May 1981.

[3] G. Biffi Gentili and F. Gori, "Open end coaxial probe for soil mixture measurement," *Int. Microwave Symp.*, Rio de Janeiro, 1987.

[4] G. F. Avitabile, S. Maci, G. Biffi Gentili, L. Roselli, and G. F. Manes, "A two-port active coupled microstrip antenna," *Electron. Letts.*, vol. 28, no. 25, Dec. 1992.

[5] R. H. Cole, J. G. Berberian, S. Mashimo, G. Chryssikos, A. Burns, and E. Tombari, "Time domain reflection methods for dielectric measurements to 10 GHz," *J. Appl. Phys.*, vol. 66, p. 793, 1984.

[6] U. Kaatze, R. Pottel, and K. Schäfer, *J. Phys. Chem.*, vol. 93, p. 5623, 1989.

15

Ferenc Völgyi
Technical University of Budapest, Budapest, Hungary

Integrated Microwave Moisture Sensors for Automatic Process Control

Abstract. The basic definitions and relations for calculating the main characteristics of a transmission-type microwave moisture measurement using two antennas in the radiating near field are introduced. Making known the preliminary considerations for the design of integrated microwave moisture sensors, a block diagram is discussed. A number of measurements (transmission and return loss, temperature effects) were carried out, including the compensation of moisture content of yellow dent field corn.

15.1. INTRODUCTION

New businesses are important in the economic situation of Hungary. Such firms should manufacture, and the farmers should produce, cheaper marketable goods and crops, using the benefits presented by modern technology. One of the most important technological processes is drying corn to the moisture content specified for storage. For automatic operation of dryers, moisture sensors are needed.

Our versatile microwave moisture sensors [1] were made with microwave hybrid integrated circuits (MHIC) on plastic substrates [2]. Recently, we have developed copper thick-film MHICs on alumina ceramic which are cheaper and have better characteristics than gold-film circuits. These were integrated with microstrip antennas (active array), creating integrated microwave moisture sensors.

Using the classification of microwave sensors being used for moisture content measurement in [3], our instrument is transmission type, aperiodic, open, and constructed with two antennas. For such an instrument, used for agricultural and industrial automatic process control, high reliability, stability, simplicity, small size, and low cost (if possible) are the major requirements.

This paper introduces the basic definitions and principles of the moisture measurement using two antennas, reviews our integrated microwave moisture sensor family, and shows some useful measurement results relating to our near-field circuit stability and quality, and temperature-effect examinations.

15.2. BASIC DEFINITIONS AND PRINCIPLES

Assuming a plane wave (TEM) propagating in lossless free space, the attenuation between a transmitting antenna with gain G_t and receiving antenna with gain G_r at a distance R is

$$A_f = 20 \log \ 4\pi R/\lambda - G_t - G_r \quad \text{[dB]}, \qquad (15\text{-}1)$$

where λ, the free-space wavelength, is

$$\lambda = c/f; \qquad \lambda \ \text{[cm]} \cong 30/f \ \text{[GHz]}. \qquad (15\text{-}2)$$

A lossy dielectric material ℓ cm thick attenuates the TEM wave as follows:

$$A = \alpha\ell = (0.91\sqrt{\epsilon'}(\tan\delta)f\ell \quad \text{[dB]}, \qquad (15\text{-}3)$$

where α is the attenuation factor (dB/cm), f is the frequency (GHz), and ϵ' is the real part of the complex relative permittivity:

$$\epsilon = \epsilon' - j\epsilon''. \qquad (15\text{-}4)$$

The power dissipation is indicated by the loss factor ϵ'', which is proportional to the tangent of the loss angle δ:

$$\tan\delta = \epsilon''/\epsilon'. \qquad (15\text{-}5)$$

The field strength of the microwave field penetrating the lossy material decreases exponentially. Skin depth is defined as the distance from the surface where the power density is reduced by a factor of $1/e$ ($e = 2.7142$) and is written

$$d = \lambda/2\pi\sqrt{\epsilon'}\tan\delta. \qquad (15\text{-}6)$$

The characteristic impedance of a plane wave in a lossy space is

$$Z_0 = 377[1 - \frac{3(\tan\delta)^2}{8} + \frac{j\tan\delta}{2} \ \frac{1}{\sqrt{\epsilon'}}. \qquad (15\text{-}7)$$

The power dissipated in a volume V of homogeneous material is

$$P_d = 5.56 \times 10^{-11} f\epsilon'(\tan\delta)|E|^2 V \quad \text{[W/m}^3\text{]}, \qquad (15\text{-}8)$$

where E is the effective value of the electric field strength (V/m).

After calculating a few characteristic values referring to water in the 0.4–10 GHz frequency range, we make the following observations:

1. The power dissipated in water rapidly increases with frequency and is proportional to $f\epsilon' \tan \delta$; this product varies from 0.6 to 297, skin depth varies from 67.6 to 0.12 cm, and the attenuation factor varies from 0.06 to 36.4 dB/cm.

2. The free-space attenuation at 10 GHz between two antennas with 5-dB gain 10 cm apart is 22.4 dB. With small microstrip antennas, it is easy to achieve a 12-dB gain; hence the base attenuation at calibration can be expected to be no more than 8.4 dB.

3. Water as a lossy material exhibits an impedance with approximately a 45-ohm real part; therefore, surface reflections are expected.

There are reflections from other parts of the sample, too, depending on the geometry of the sample. This was examined [4] with an approximate approach based on geometrical optics, and it was determined that minimum thickness and transverse dimensions of the sample are needed, depending on the moisture content. To analyze these problems, we must know the exact solution of the electromagnetic fields between the antennas used in moisture content measurement, which is difficult.

In moisture measurement using two antennas, the receiving antenna is usually located in the radiating near field (Fresnel region) of the transmitting aperture, and the expected attenuation at calibration (empty cell) will be higher than in eq. (15-1), because G_r and G_t are the so-called far-field gains (measured in the Fraunhofer region).

The limit of the Fresnel region is

$$x = \frac{R}{2L^2/\lambda} = 1, \tag{15-9}$$

where L is the larger dimension of the radiating aperture. It is useful to examine quantitatively the radiating near-field variation, especially along the axis, for such applications as personnel radiation hazards and short-distance illumination of moist substances. For a tapered circular aperture, the normalized on-axis power density [5] is

$$p(x) = 26.1 \left[1 - \frac{16x}{\pi} \sin \frac{\pi}{8x} + \frac{128x^2}{\pi^2} \left(1 - \cos \frac{\pi}{8x} \right) \right], \tag{15-10}$$

which is an oscillating function, and if $x = 1$ then $p = 1$. The peak power density occurs at about $x = 0.1$ and is nearly 42 (or 16.2 dB). The asymptotic value for a small distance is 26.1 (14.2 dB). The on-axis power density at $R = 2L^2/\lambda$ is

$$p = 3\pi P/64L^2, \tag{15-11}$$

where P is the radiated power and L is the aperture diameter.

Directivity is reduced and the main beam is broadened in the near field. The relative reduction of gain is

$$g(x) = G(x)/G_0 = p(x)x^2, \tag{15-12}$$

where $p(x)$ is the normalized on-axis power density given by eq. (15-10), and G_0 is the gain at $x = 1$. The beam-broadening factor is the square root of the reciprocal of the directivity reduction factor; i.e.,

$$\vartheta(x) = \frac{\theta(x)}{\theta_0} = \frac{1}{\sqrt{g(x)}} = \frac{1}{x} \frac{1}{\sqrt{p(x)}}, \qquad (15\text{-}13)$$

where θ_0 is the beamwidth at $x = 1$. Assume $P = 10$ mW, $f = 10$ GHz, and $L = 2.5$ cm (tapered illumination). Then the calculated on-axis power density (0.42 cm from the aperture), from eqs. (15-9)–(15-12), is $p = 10$ mW/cm^2, which is the personnel safety limit.

15.3. DESIGN OF INTEGRATED MICROWAVE MOISTURE SENSORS

Preliminary considerations:

1. It is useful to select the frequency of the moisture sensor in the X-band, because the sensitivity of the microwave attenuation related to moisture content is dominant.

2. For safety, no more than 10 mW of radiated power is suggested.

3. Using Figure 15.1, the measured microwave attenuation of wet corn having 28% moisture content (at 10 GHz) is approximately 40 dB if the thickness of the sample is 4.9 cm.

4. We must apply a minimal input signal level of −40 dBm, using a simple Schottky detector, to achieve a sufficient signal-to-noise ratio at the output of our equipment.

5. A distance of 10 cm between antennas is suitable for various applications.

From the above considerations, the free-space attenuation will be no more than 10 dB at calibration, and from eq. (15-1) comes the minimal value of 11.2 dB of gain needed for such antennas. To get reliable information about the moisture content of wet substances, a minimal value of illuminated cross section is needed. It is important from a practical point of view that we have only limited fluctuation of the received signal during calibration or during measurement of quasi-dry substances, if the range of the antennas has a little variation (because of dryer vibration or, e.g., textile swinging). We summarize the main specifications and our measurement results for such antennas in Table 15.1.

Before designing the moisture sensors for different purposes, preliminary measurements [1] were conducted on lots of granulated materials and grains to obtain reference attenuation data [6]. With given density and temperature of sampled materials, the microwave attenuation is characteristic of the moisture content if a constant sample thickness is maintained with a proper closed sensor structure [7]. Based on measured data in the X-band, it can be noted that in the typical

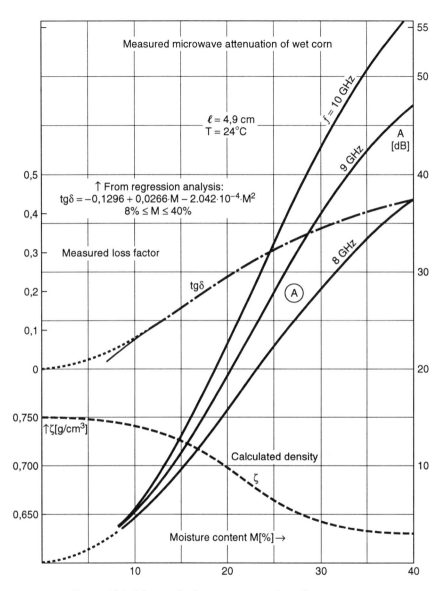

Figure 15.1. Measured microwave attenuation of wet corn as a function of moisture content.

moisture range characteristic attenuation values are 1–50 dB for corn, 2–20 dB for wheat, 2–20 dB for leather, and 0.5–12 dB for textiles. Figure 15.1 shows the microwave attenuation of wet corn as a function of moisture content, the average density value taken from the literature, and the loss factor calculated from the measurement together with its regression expression.

TABLE 15.1. Parameters and Measured Values of Tested Antenna Models

MODEL TYPE	G_0 (dB)	ILLUMINATED SHAPE (APPROX.)	A (dM^2)	θ_3 (DEG)	$R = 10$ CM $A_m \pm \Delta A$ [dB]	ΔG (DB)	F (GHz)
A	7.4	$D = 10$ cm, circular	0.8	76	18.5 ± 0.12	-0.20	10.5
B	6.1	$D = 11.5$ cm, circ.	1.0	88	21.0 ± 3.50	-0.15	10.5
C	12.6	7.2×7.2 cm squ.	0.5	42	9.0 ± 0.15	-0.70	10.5
D	12.0	7.5×7.5 cm square	0.6	45	10.9 ± 0.60	-1.0	10.5
E	13.0	12.5×10 cm rectang.	1.2	56/28	8.6 ± 0.65	-2.1	8.0
F	18.6	11.3×8.7 cm rect.	1.0	19/17.5	6.4 ± 1.20	-5.3	10.5

G_0 = far-field gain in dB
A = "illuminated" surface (approx.) at $2x(-3$ dB), taking into account beam broadening
θ_3 = 3-dB beamwidth
A_m = measured transmission loss (dB) at $R = 10$ cm
ΔA = transmission loss variation near the distance of 10 cm
ΔG = measured gain reduction in the near field $R \cong 10$ cm

The examined antenna models are:
A = circular primary feed, $D_i = 21$ mm, $D_0 = 48$ mm, four grooves
B = open circular waveguide (C-120), using extreme large flange $D_i = 17.5$ mm, $D_0 = 60$ mm
C = circularly polarized microstrip antenna using four elements
D = linearly polarized microstrip antenna with four elements
E = circularly polarized microstrip antenna with eight elements
F = optimal pyramidal horn, $A \times B \times L = 110 \times 84 \times 102$ mm

In our equipment, instead of two-parameter (attenuation and phase) measurements, which are accurate and well suited for laboratory conditions, we measure only the attenuation (it simplifies the receiver), and the errors are decreased by other means.

The block diagram of the integrated moisture sensor for grain is shown in Figure 15.2. The transmitter and receiver are built together and located outside the sample space. A square-wave-modulated (30 kHz) system was developed, and thus a reduced oscillator noise and simplified signal processing on the receiver side can be achieved. The transmitter oscillator (dielectric resonator oscillator (DRO) working around 8 GHz, with simple on-off modulation) is integrated with the miniature circulator (reducing the load-pull effect) and microstrip antenna array. In the receiver, a simple microstrip (planar) detector is used, integrated onto the receiver microstrip antenna (1.6 mm thick) protected by a plastic planar radome and located in the flowing moist grain. Because of the high dynamic range requirements, a logarithmic amplifier is used in the receiver. The $U_0(M)$ output signal of the receiver is obtained after detection and dc amplification. During processing of calibration data, temperature (measured by a thermistor temperature sensor) compensation is organized by a microprocessor. The low variation of corn density in the sample space (surrounded by 40×40 cm^2 metal tube) was provided by the mechanical means of directing sheets. In practice, an optimal control was found based on 1 sample/min with moving averaging.

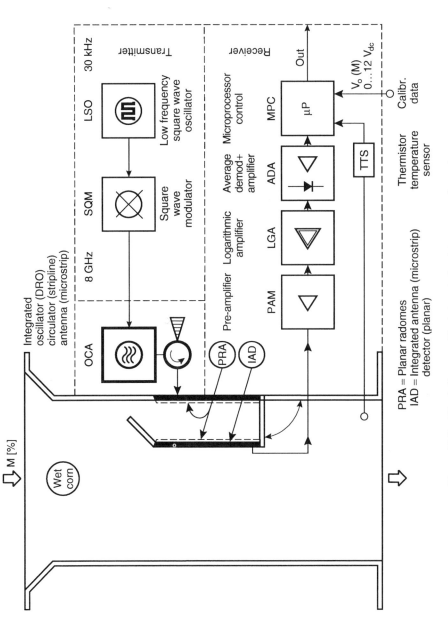

FIGURE 15.2. Block diagram of the integrated moisture sensor for grain.

229

15.4. EXPERIMENTAL RESULTS

15.4.1 Integrated Moisture Sensor for Grain (MSG)

Our instruments (shown in Figure 15.2) were mounted at the input and output of a corn dryer with 80-ton capacity and 15–20 ton/hr drying speed. A Burrows Model-700 digital moisture meter was used for calibration (see Figure 15.3). The built-in microprocessor selects automatically the appropriate A or B characteristic, so the sensitivity at the linear range of M (16–29%) will be as high as $\Delta V / \Delta M = 1.119V/\%$. For yellow dent field corn, the simple linear temperature compensation $\Delta M = -0.14(T - 22.5)$ was used, which was sufficient in practice. The measured uncertainty of the moisture content determination was $\Delta M = 0.41\%$ (20 replicates) at $M = 23.7\%$. Using the numerical coefficients and the equation for one-parameter measurement in [7], the calculated relative fluctuation of the material density in the measuring cell was $\Delta \rho / \rho = 0.105$, which is a real value for this simple mechanical solution using directing sheets.

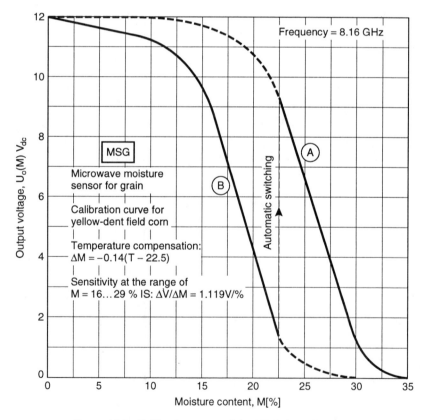

FIGURE 15.3. Calibration curve of the MSG for yellow dent field corn, showing sensitivity ranges A and B. Temperature compensation and sensitivity values are also given.

15.4.2 Quality and Stability of Microwave Circuits Used in Our Moisture Sensor

To characterize the quality and stability of our copper thick-film microwave circuits used in the MSG, the unloaded quality factor Q_0 (dashed line in Figure 15.4) and resonant frequency measurements of ring resonators were executed in a broad temperature range (in comparison with our high-temperature superconductor microstrip resonators). At 300 K, $Q_0 = 145$ at the first resonant frequency $f_1 = 2.86$ GHz, and the frequency-temperature stability factor is $\Delta f (f \Delta T)^{-1} = -93$ ppm/K. The attenuation constant for microstrip line having $Z_0 = 50$-ohm characteristic impedance is $\alpha = 5.1$ dB/m and the frequency dependence is $\alpha(f) = 2.97\sqrt{f} + 0.017 f$ [GHz]. The measured results on the third resonance (around 9 GHz) are $Q_0 = 250$, $\alpha = 9.1$ dB/m. These parameters are better than the measured values for gold thick-film resonators realized on the same alumina substrate.

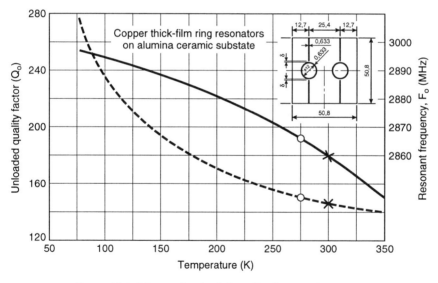

FIGURE 15.4. Measured unloaded quality factor and resonant frequency of copper thick-film ring resonator on alumina substrate as a function of temperature.

In our MSG instruments, small-dimension and low-profile microstrip antennas were used as radiating elements. To check the temperature dependence for these resonant arrays, a relatively narrowband four-element rectangular microstrip antenna for 4.4 GHz was made. The results for the temperature-dependent input return loss versus frequency measurement are shown in Figure 15.5. The relative frequency-temperature stability of the resonant array is 86 ppm/K. Because our microstrip antennas used in the MSG have 2.3 times higher relative bandwidth at 8 GHz, this problem is practically negligible.

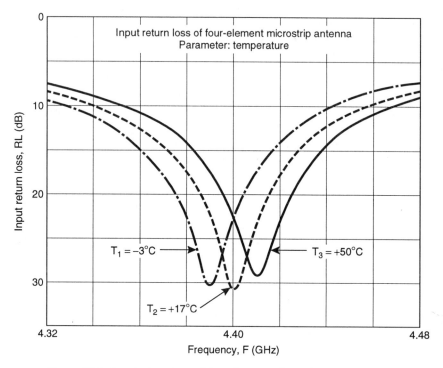

FIGURE 15.5. Input return loss of four-element microstrip antenna as a function of frequency. The parameter of the curves is the temperature.

15.4.3 Radiating Near-Field Experiments

Transmission Measurement (Calibration).   Different types of antennas are used for microwave moisture measurement. Low transmission loss (at calibration, moist substance out), matched input impedance, given value of "illuminated" cross section, and relatively smooth loss-distance characteristic are the main requirements. Because closed formulas are known only for simple apertures (e.g., for tapered circular aperture: eq. (15-1)), we examined this problem experimentally.

Pyramidal horns are frequently used in practice. The received relative power of an optimal pyramidal horn (model F, see Table 15.1) as a function of the range between antennas is shown in Figure 15.6. At $R = 10$ cm, the relative power is -6.4 dB; the rapid variation near this distance is ± 1.2 dB. Far-field gain is $G_0 = 18.6$ dB, and from eq. (15-1) we find that the reduced gain in the near field at $R = 10$ cm is only 13.3 dB.

Figure 15.7 shows the relative axial power densities of receiving antennas (models A, D, and C) as a function of distance from the transmitting loop. The extrapolated "gain" of the loop from measurements was about -15 dB. These curves are in good agreement with theoretical predictions; at 6–14 cm models A

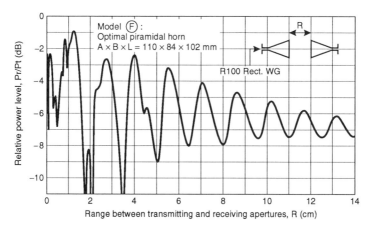

FIGURE 15.6. Received relative power level of optimal pyramidal horn as a function of the range between antennas.

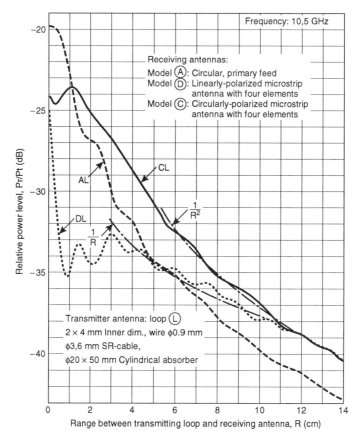

FIGURE 15.7. Relative axial power levels of receiving antennas as a function of distance from transmitting loop.

and C are in the far-field region ($1/R^2$ function), and model D at 4–10 cm is in the radiating near-field (Fresnel) region. Transmission loss (at moisture measurement calibration) is shown in Figure 15.8 for models A, C, and E. At 10 cm, models C and E are in the $1/R$ region and the attenuation is about 9 dB. Model A is in the $1/R^2$ region, with a relative level of received power of −18.5 dB.

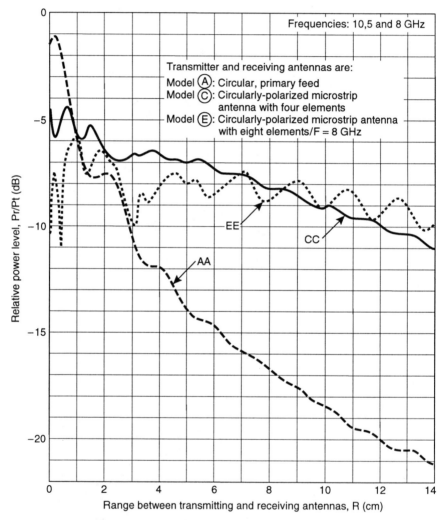

Figure 15.8. Transmission loss as a function of the range between antennas.

The radiating properties of an antenna (measured in the near field) are changed considerably when placing the radiator into a metal box. Modeling this problem, an open circular waveguide with an extremely large flange (this is the front wall of the cylindrical metal box) was measured. The result is shown in Figure 15.9 (curve

BB) where a compressed vertical scale (5 dB/div) is used. A very large periodical variation of the received power is observable, while it has a smooth function at model A (curve AA).

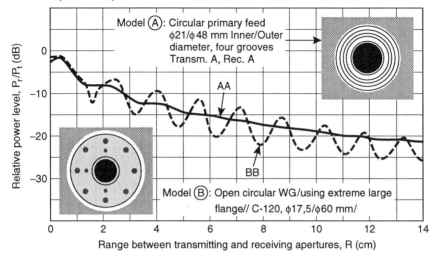

FIGURE 15.9. Relative power levels in 5 dB/div compressed vertical scale of models A and B, as a function of range between transmitting and receiving apertures.

The above-mentioned problem is avoidable by using circular polarization. We have made four-element microstrip antennas with circular polarization (well known from the literature). Locating the input microstrip lines in the same and opposite directions, the received relative power levels as a function of range between transmitting and receiving antennas are shown in Figures 15.10 and 15.11. The best results are given by the last configuration, because of the high received level (−9 dB) and low variation (±0.15 dB). In our original solution, the microstrip power splitter network between radiating elements had quasi double symmetry [1].

Input Return Loss Measurement (with Moist Substance). According to eq. (15-7), the input return loss of the transmitting antenna will degrade, locating moist substance near the radiating aperture. This effect was measured at 2.45 GHz with a motorized automatic input return loss measurement setup using two types of microstrip antennas. Figure 15.12 shows the input return loss of the transmitting rectangular microstrip antenna, loading it with different moist materials at a distance R [cm]. The nearest optimum position is $R = 4$ cm, where the minimal return loss is $RL = 12.5$ dB, which means VSWR = 1.6; the reflection coefficient $|\Gamma| = 0.24$, and the mismatch loss $ML = 0.25$ dB. Using a broadbanded microstrip antenna [8] at the nearest optimum (also $R = 4$ cm), $RL = 15$ dB, VSWR = 1.4, $|\Gamma| = 0.18$, and $ML = 0.14$ dB. Sometimes the moist substance gives greater reflection than the water tank itself. On the other hand, it is possible to obtain "matching" by the dry sample (Figure 15.13, curve 2, at $R = 5$ cm),

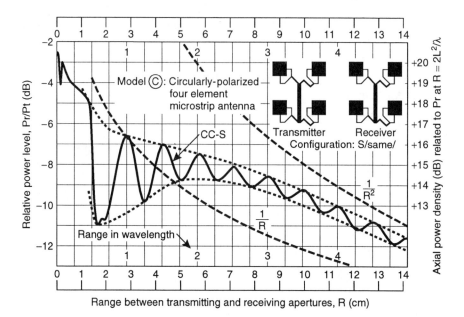

FIGURE 15.10. Received relative power level as a function of range between transmitting and receiving apertures, using circularly polarized microstrip antennas with four elements. Input microstrip lines are in the same position.

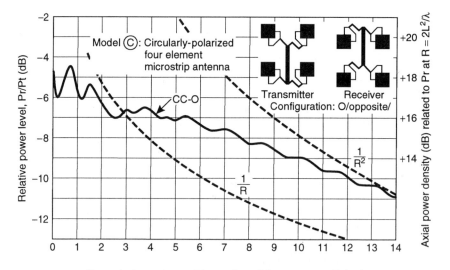

FIGURE 15.11. Received relative power level as a function of range between transmitting and receiving apertures, using circularly polarized microstrip antennas with four elements. Input microstrip lines are in the opposite position.

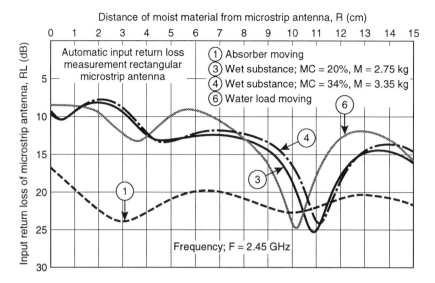

FIGURE 15.12. Measured input return loss of the transmitting rectangular microstrip antenna loading it with different moist materials at the distance of R [cm]: (1) absorber, ECCOSORB AN 79, (3) wet book, $21 \times 30 \times 5$ cm^3, (4) wet book, (6) water in $35 \times 32 \times 5.5$ cm^3 plastic container.

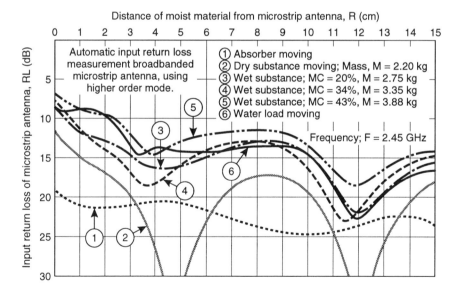

FIGURE 15.13. Measured input return loss of the transmitting broadband microstrip antenna as a function of distance of the moist material from it: (2) dry book, $21 \times 30 \times 5$ cm^3, (5) wet book; (1), (3), (4), and (6) are the same as in Figure 15.12.

which may be useful in certain cases (low-loss reflecting plate or the wall of the dielectric sample holder).

15.5. CONCLUSIONS

The development and widespread use of instrumentation for microwave moisture content measurement have matured. We now have to take the next step: cooperation between agricultural, microwave, and field engineers. One purpose of this paper is to show (experimentally) the possible reduction of calibration errors and improvement of equipment stability using integrated antennas and copper thick-film microwave circuits. It was found that in terms of propagation loss and near-field reflections, there is an optimum size (or element number) for the antenna, suitably a microstrip antenna with circular polarization.

REFERENCES

[1] F. Völgyi, "Versatile microwave moisture sensor," in *Conf. Rec. SBMO/89*, Sao Paulo, Brazil, 1989, Vol. II, pp. 456–462.

[2] F. Völgyi, L. Jachimovits, and I. Bozsóki, "Design of hybrid integrated microwave circuits on plastic substrate," in *Conf. Rec. IV. Nat. Conf. on Microwave Solid State Electronics*, Gdansk, Poland, 1977, pp. 45–53.

[3] A. W. Kraszewski, "Microwave aquametry—needs and perspectives," *IEEE Trans. MTT*, vol. 39, no. 5, pp. 828–835, May 1991.

[4] J. Mladek and Z. Beran, "Sample geometry, temperature and density factors in the microwave measurement of moisture," *J. Microwave Power*, vol. 15, no. 4, pp. 243–250, 1980.

[5] R. C. Hansen (Ed.), *Microwave Scanning Antennas*, New York: Academic Press, 1964, Vol. I, ch. 1, pp. 1–46.

[6] S. O. Nelson, "RF and microwave dielectric properties of agricultural products and their applications," in *Conf. Rec. SBMO/89*, Sao Paulo, Brazil, 1989, Vol. I, pp. 29–35.

[7] A. W. Kraszewski, "Microwave monitoring of moisture content in grain—further considerations," *J. Microwave Power Electromag. Energy*, vol. 23, no. 4, pp. 236–246, 1988.

[8] F. Völgyi, "Microstrip antenna array application for microwave heating," in *Conf. Rec. 23rd European Microwave Conf.*, Madrid, Spain, 1993, pp. 412–415.

16

Kaida b. Khalid
Zulkifly b. Abbas
Universiti Pertanian Malaysia, Serdang,
Selangor, Darul Ehsan, Malaysia

Development of Microstrip Sensor
for Oil Palm Fruits

Abstract.   This paper deals with the analysis and design of a microstrip sensor for quick and accurate determination of moisture content (MC) in fresh mesocarp of oil palm fruits. The moisture content in fresh mesocarp is higher at early stages of fruit development and decreases rapidly to about 30% to 40% in the ripe fruit. This period is almost coincident with the accumulation of oil in the mesocarp. The close relationship between moisture content and oil content in mesocarp gives a possibility of using %MC/fresh mesocarp as a parameter to gauge fruit ripeness. A functional relationship has been developed between insertion loss, $|S_{21}|$, of the sensor and moisture content in mesocarp, and close agreement has been found between computed and experimental results. The analysis of the electromagnetic wave propagation in the sensor structure is simply represented by a signal flow graph, and prediction of dielectric properties of mesocarp of oil palm fruits is given by a dielectric mixture model.

16.1. INTRODUCTION

Microwave aquametry or measurement of moisture content in liquids, semisolids, and solids by microwave techniques is known to be accurate and rapid. For a small sample size, it is conveniently measured by using a microstrip, since only a small part of the sample interacts with the line [1]. The application of this sensor for microwave aquametry in lossy liquids such as hevea rubber latex [2] and determination of fat/water in fish has been successful [3]. This paper describes the analysis and design of a microstrip sensor for oil palm fruits.

It was found that the amount of moisture content is higher at an early stage of fruit development and decreases rapidly to about 30% to 40% in the ripe fruit at 20 to 23 weeks after anthesis, and this period is almost coincident with the accumulation of oil in the mesocarp [4]. Therefore the close relationship between moisture content and oil content in mesocarp gives a possiblity of using

%MC/fresh mesocarp as a parameter to gauge fruit ripeness. A functional rela-
tionship between insertion loss, $|S_{21}|$, of the sensor and the moisture content in
the mesocarp is developed and then compared with the experimental results. The
performance and application of this sensor in setting up a harvesting system for
oil palm fruits has been discussed in detail by the authors [5].

16.2.   DESIGN AND ANALYSIS

The microstrip sensor and cross section of the sensing area are shown in Figure
16.1. The sensor consists of three parts: the coupling system (input/output) representing the transition between coaxial and stripline, the 50-Ω stripline section, and
the semi-infinite double-covered microstrip or sensing area. The substrate material
is RT-Duroid with a relative dielectric constant of 10.7 and thickness of 1.27 mm.
The microstrip section is protected by a thick layer of polymeric material about
0.4 mm thick.

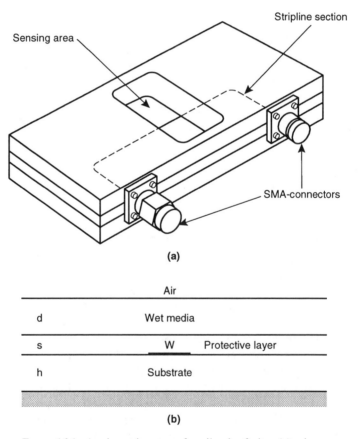

Figure 16.1. A microstrip sensor for oil palm fruits: (a) microstrip sensor, (b) cross section of microstrip sensing area.

The dimensions of the sensing area were selected for single fruit measurements, with a volume of about 31.5 mm × 17 mm × 7.27 mm. The impedance of the microstrip line (in air) is about 61 Ω. This impedance is approximately matched to a 50-Ω system at higher moisture content.

In this analysis, we consider the complex amplitude of an electromagnetic plane wave propagating through a medium which can be represented by the equation

$$E(d) = E(0)e^{-\gamma d}, \tag{16-1}$$

where $E(0)$ is the complex amplitude of the wave at some reference point, $E(d)$ is the complex amplitude at a distance d, and γ is the propagation constant of the medium.

The reflection and transmission phenomena in the sensor structure can be represented by a signal flow graph as shown in Figure 16.2. This structure consists of an input transition region from coaxial to stripline section a of the sensor, stripline section a, transition from stripline section to the sample front of reference plane 1, microstrip section, and similar transition portions from the sample back-reference plane 2. The signal flow graph in Figure 16.2 can be simplified as in Figure 16.3a and Figure 16.3b by using Mason's nontouching loop rules [6]. From Figure 16.3b the insertion loss $|S_{21}|$ is

$$|S_{21}| = \frac{S_{21}'' S_{21}' e^{-\gamma_m \ell_2}}{1 - S_{22}' S_{11}'' e^{-2\gamma_m \ell_2}}, \tag{16-2}$$

where elements of the scattering matrices S_{21}'', S_{21}', S_{22}', and S_{11}'' are

$$S_{21}'' = e^{-\gamma_s \ell_1}(1 - \Gamma_a)(1 - \Gamma_m)/S'', \tag{16-3i}$$

$$S_{21}'' = e^{-\gamma_s \ell_1}(1 - \Gamma_a)(1 - \Gamma_m)/S', \tag{16-3ii}$$

$$S_{22}' = \frac{-\Gamma_m(1 + \Gamma_a\Gamma_m e^{-2\gamma_s \ell_1}) - \Gamma_b(1 - \Gamma_m)(1 + \Gamma_m)e^{-2\gamma_s \ell_1}}{S'}, \tag{16-3iii}$$

$$S_{11}'' = \frac{-\Gamma_m(1 + \Gamma_b\Gamma_m e^{-2\gamma_s \ell_1}) - \Gamma_b(1 - \Gamma_m)(1 + \Gamma_m)e^{-2\gamma_w \ell_1}}{S''}, \tag{16-3iv}$$

where

$$S' = 1 + \Gamma_a\Gamma_m e^{-2\gamma_s \ell_1} \quad \text{and} \quad S'' = 1 + \Gamma_b\Gamma_m e^{-2\gamma_s \ell_1}.$$

The reflection coefficient at the stripline-microstrip transition, Γ_m, is

$$\Gamma_m = \left| \frac{Z_m - Z_0}{Z_m + Z_0} \right|, \tag{16-4}$$

where Z_m and Z_0 (= 50 Ω) are the characteristic impedance of the microstrip section and stripline section, respectively. The reflection coefficients at coaxial-stripline transitions Γ_a (input) and Γ_b (output) are determined by using a network analyzer. These values were found to be $\Gamma_a = 0.003$ and $\Gamma_b = 0.004$. The

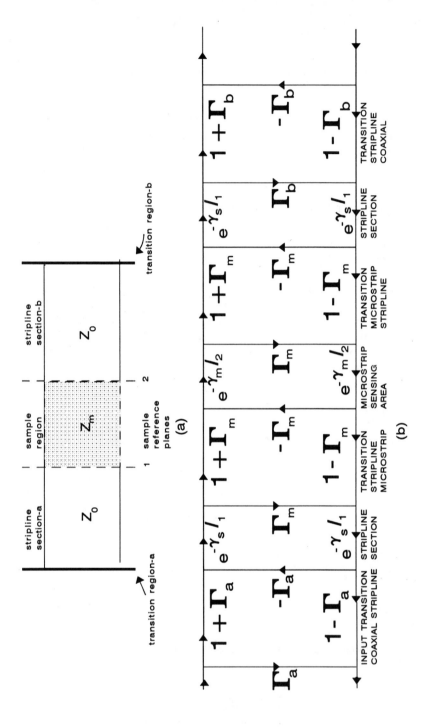

FIGURE 16.2. (a) Microstrip sensor with sample inserted; (b) equivalent two-port network.

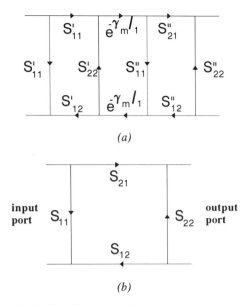

(a)

(b)

FIGURE 16.3. Simplified signal flow graph using Mason's nontouching loop rule: (a) simplified signal flow graph of Figure 16.2; (b) final form in terms of scattering parameters of the input and output ports.

terms γ_m and γ_s are the complex propagation constants for the microstrip section (sensing area) and stripline section, respectively, and are

$$\gamma_m = (\alpha_m + j\beta_m) \quad \text{and} \quad \gamma_s = (\alpha_s + j\beta_s), \tag{16-5}$$

where α is the attenuation constant, in this case due to the dielectric loss, and β is the phase constant. The calculation of Z_m, γ_m, and γ_s by TEM analysis is given in detail by Khalid et al. [7].

In order to calculate the above parameters, a relation between the moisture content of the mesocarp and its permittivity must first be established. To achieve this the mesocarp will be considered as a mixture of water, oil, and fiber. Using mixture theory the relative complex permittivity of the mesocarp can then be expressed as

$$\sqrt{\epsilon_m^*} = V_w\sqrt{\epsilon_w^*} + V_t\sqrt{\epsilon_t^*} + V_f\sqrt{\epsilon_f^*}, \tag{16-6}$$

where V_w, V_f, and V_t are the volume fractions of water, fiber, and oil, respectively, and ϵ_w^*, ϵ_t^*, and ϵ_f^* are the corresponding complex permittivities. The volume fraction for fiber V_f is considered to be a constant volume of 16% [8], whereas the water content and oil content vary with ripeness. Then V_t and V_w can be written as

$$V_t = 1 - V_w - V_f \tag{16-7}$$

and

$$V_w = \frac{M(\rho_f V_f + \rho_\iota - \rho_\iota V_f)}{\rho_w - M\rho_w + \rho_\iota M},$$ (16-8)

where M is the moisture content (wet basis), and ρ_f, ρ_ι, and ρ_w are the densities of fiber, oil, and water, respectively. The values of ρ_f, ρ_ι, and ρ_w are 0.92, 0.93, and 1.0, respectively.

The moisture dependence of the dielectric properties of the oil palm fruit at 9.7 GHz and 26°C is shown in Figure 16.4. The lines are the values predicted from mixture equation (16-6). From Figure 16.4 the experimental values almost fit with the prediction values of dielectric loss. Measurement of dielectric properties was carried out with a 4-mm open-ended coaxial-line probe coupled with an automatic network analyzer (HP 8720B) and computer. The input reflection coefficient of the sensor is related to the permittivity of the sample, and the accuracy of the measurement is found to be about ±5% for ϵ' and ±3% for ϵ''. The variations of characteristic impedance, effective dielectric constant, and dielectric loss of the loaded microstrip section with the moisture content are given in Figures 16.5, 16.6, and 16.7, respectively.

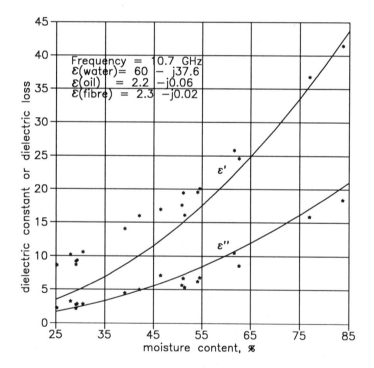

FIGURE 16.4. Dielectric constant ϵ' and dielectric loss ϵ'' in fresh mesocarp as functions of moisture content.

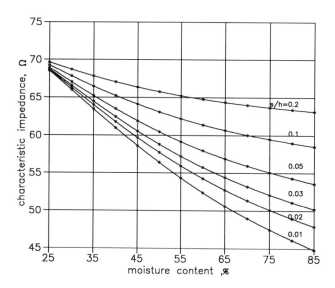

FIGURE 16.5. Characteristic impedance of loaded microstrip as functions of moisture content at various s/h ratios.

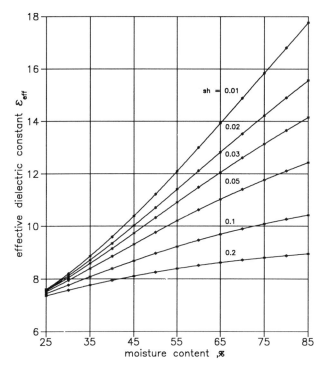

FIGURE 16.6. Effective dielectric constant of loaded microstrip against moisture content at various s/h ratios.

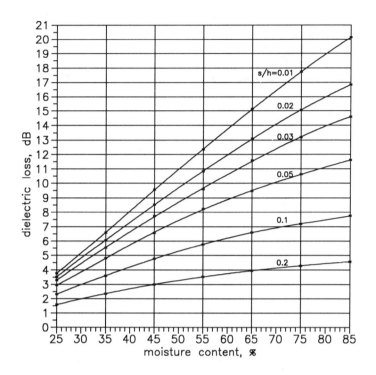

FIGURE 16.7. Dielectric loss of loaded microstrip against moisture content at various s/h ratios.

16.3. RESULTS

In this study, six bunches of Tenera variety from 11-year-old oil palm were selected. The samples were taken from the outer fruit around the equatorial of the bunch. The fruits were analyzed once a week from 12 weeks after anthesis until they were fully ripe and detached. The fresh mesocarp was separated from the nut and then cut into small pieces, and these pieces were crumbled until they become a uniform semisolid sample. The sample was pressed firmly into the sensor to ensure perfect contact. The magnitude of the insertion loss $|S_{21}|$ of the sensor was measured with an automatic swept frequency network analyzer. All measurements were made at 26°C. The actual moisture content of the sample was obtained by the Karl Fisher method with an accuracy of ±0.1%. Figure 16.8 shows the computed (eq. (16-2)) and experimental results of the insertion loss $|S_{21}|$ of the sensor as a function of moisture content at 10.7 GHz. The results show that the difference between computed and experimental results is approximately 4.3%. The sensitivity of this sensor is about 0.6 dB/%MC.

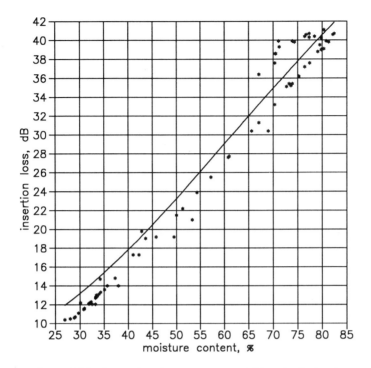

FIGURE 16.8. Comparison between measured (*) and calculated
(−) insertion loss as a function of moisture content.

16.4. CONCLUSIONS

In this paper the analysis and design of microstrip sensor for oil palm fruit have
been discussed. A close agreement has been obtained between computed and
experimental results of the insertion loss as a function of moisture content in fresh
mesocarp. This agreement suggests that a realistic optimization-based design
approach with respect to the effects of the geometrical and electrical parameters
of the sensing area can be developed. Therefore a sensor with highest sensitivity
can be designed which will better predict the ripeness time of the oil palm fruit.

ACKNOWLEDGMENTS

This work was supported by IRPA Research Grant (4-07-05-021) coordinated by
Ministry of Science, Technology and Environment of Malaysia. The authors wish
to thank Dr. Azis Ariffin of PORIM for helpful discussions during this work, the
management of the Farm Department, University Pertanian Malaysia, for their excellent cooperation, and the technical staff of Applied Electromagnetic Laboratory,
Physics Department, UPM, for their assistance in carrying out this project.

REFERENCES

[1] M. Kent, "The use of stripline configuration in microwave moisture measurement. II," *J. Microwave Power*, vol. 8, no. 2, pp. 189–194, 1973.

[2] K. B. Khalid, "The application of microstrip sensors for determination of moisture content in hevea rubber latex," *J. Microwave Power Electromag. Energy*, vol. 21, no. 1, pp. 45–52, 1988.

[3] M. Kent, "Hand-held instrument for fat/water determination in whole fish," *Food Control*, vol. 1, pp. 47–53, 1990.

[4] A. Ariffin, "Biochemical aspects of ripeness standards," in *Proc. Symp. on Impact of Pollination Weevil the Malaysian Oil Palm Industry*, Kuala Lumpur, PORIM, 1984.

[5] K. Khalid and Z. Abbas, "A microstrip sensor for determination of harvesting time for oil palm fruits (*Tenera: Elacis Guineensis*)," *J. Microwave Power Electromag. Energy*, vol. 27, no. 1, pp. 3–10, 1992.

[6] F. L. Warner, *Microwave Attenuation Measurement*, London: Peter Peregrinus, 1977, pp. 7–8.

[7] K. B. Khalid, T. S. M. Maclean, M. Razaz, and P. W. Webb, "Analysis and optimal design of microstrip sensors," *Proc. IEEE*, vol. 135, pt. H, no. 3, pp. 187–195, 1988.

[8] C. W. S. Hartley, *The Oil Palm*, London: Longman Group, 1977, pp. 222–223.

Yansheng Xu
Renato G. Bosisio
Ecole Polytechnique de Montréal, Montréal,
Québec, Canada

17

Calculation of Sensitivity of Various Coaxial Sensors Used in Microwave Permittivity Measurements

Abstract.   In this paper, a new method to enhance the measurement sensitivity of coaxial sensors (terminated by a circular waveguide or free space) at low frequencies for microwave permittivity measurements is presented. A study of some sensors based on this method shows that it is very flexible and especially effective in low-dielectric-constant measurements. Detailed calculations are performed, and the numerical data are presented, which are especially useful in the design of such sensors and selection of their parameters.

17.1.   INTRODUCTION

The measurement of microwave permittivity by using a coaxial line terminated by a circular waveguide or free space has been studied by many authors [1–13] and very wide band measurements have been achieved. However, the measurement sensitivity at low frequencies is often low and the accuracy becomes quite poor. In this paper we present a simple and effective method to improve the measurement sensitivity at low frequencies. Detailed numerical calculated data are presented which are very helpful in the selection of the sensor parameters.

17.2.   THEORY

A simple and effective method to enhance the measurement sensitivity at low frequencies is to fill a short section of the coaxial line inside the sensor by the test sample material as shown in Figure 17.1. The coaxial line in Figure 17.1a–c is terminated by a circular waveguide [14] and in Figure 17.1d it is terminated by free space. The length of section s may be arbitrary; however, in our case, we assume that the value of s equals one to three times the inner radius of the outer conductor of coaxial line a. It is clear that if s equals 0 then we have the ordinary sensors described in the literature [1–14]. The coaxial line and circular waveguide operate in the principal modes and do not support propagation of higher-order modes. It is also assumed that the outer radius of the inner conductor of coaxial line b is small and the amplitudes of higher-order modes attenuate toward zero when they travel through distance s. The calculation formula of the admittance at plane A for the cases shown in Figure 17.1a–c is derived from a full-wave analysis:

$$Y_a = j\frac{2k_0\, a}{\ln(a/b)}\left(Y_0 - \sum_{q=1}^{\infty} Y_q x_q\right),\qquad (17\text{-}1)$$

where Y_a is the input admittance (normalized) of the coaxial line at plane A and $k_0 = 2\pi/\lambda$, where

$$Y_0 = \sum_{i=1}^{\infty} \frac{A_i}{\beta_i a\lambda_i^2 a^2}\frac{J_0^2(\lambda_i b)}{J_1^2(\lambda_i a)},$$

$$Y_q = -\sum_{i=1}^{\infty} \frac{A_i}{\beta_i a(\lambda_i^2 a^2 - \zeta_q^2 a^2)}\frac{J_0^2(\lambda_i b)}{J_1^2(\lambda_i a)}.$$

$$(17\text{-}2)$$

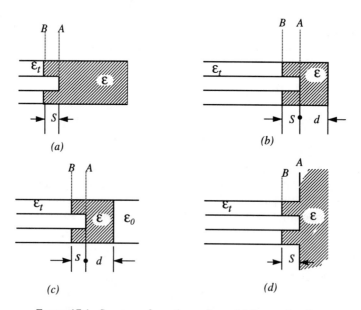

FIGURE 17.1. Some configurations of coaxial line probes for measuring the microwave permittivity.

The values of x_q can be shown to be the solution of the following equation, which is a simplified form of a more complex equation given in [2]:

$$\frac{x_n}{4\,\xi_n a}\left[\frac{J_0^2(\zeta_n b)}{J_0^2(\zeta_n a)}-1\right]=\sum_{i=1}^{\infty}\frac{-A_i}{\beta_i a(\lambda_i^2 a^2-\zeta_n^2 a^2)}\frac{J_0^2(\lambda_i b)}{J_1^2(\lambda_i a)} \tag{17-3}$$

$$-\sum_{i=1}^{\infty}\sum_{q=1}^{\infty}x_q\frac{\lambda_i^2 a^2 A_i}{\beta_i a(\lambda_i^2 a^2-\zeta_q^2 a^2)(\lambda_i^2 a^2-\zeta_n^2 a^2)}\frac{J_0^2(\lambda_i b)}{J_1^2(\lambda_i a)},$$

where $\lambda_i a$ ($i = 1, 2, \ldots$) are the ordered zeros of the Bessel function $J_0(\lambda_i a)$ and $\zeta_n a$ ($n = 1, 2, \ldots$) are the ordered zeros of the mixed Bessel function $J_0(\zeta_n a)Y_0(\zeta_n b)-J_0(\zeta_n b)Y_0(\zeta_n a)=0$,

$$\begin{aligned}
\beta_i a &= (\lambda_i^2 a^2-\epsilon k_0^2 a^2)^{1/2},\\
\xi_n a &= (\zeta_n^2 a^2-\epsilon k_0^2 a^2)^{1/2},
\end{aligned} \tag{17-4}$$

$$\begin{aligned}
A_i &= 1 &&\text{for case Figure 17.1a}\\
A_i &= \coth(\beta_i d) &&\text{for case Figure 17.1b}
\end{aligned} \tag{17-5}$$

and

$$A_i = \frac{1+\left(\dfrac{\epsilon_0-\epsilon}{\epsilon_0+\epsilon}\right)e^{-2\beta_i d}}{1-\left(\dfrac{\epsilon_0-\epsilon}{\epsilon_0+\epsilon}\right)e^{-2\beta_i d}} \qquad\text{for case Figure 17.1c.} \tag{17-6}$$

ϵ is the permittivity of the test material.

Note that λ_i and ζ_n are real constants for a given i and n.

The calculation formulas for static analysis are the same as eqs. (17-1)–(17-6) with the substitution $\beta_i \rightarrow \lambda_i$ and $\xi_n \rightarrow \zeta_n$. The calculation formula of the admittance at plane A for the cases shown in Figure 17.1d is derived to be

$$Y_a = \frac{jk_0\sqrt{\epsilon}}{\log(a/b)}\int_0^{\infty}[J_0(\tau a)-J_0(\tau b)]^2\frac{d\tau}{\tau\gamma}, \tag{17-7}$$

where $\gamma = \sqrt{\tau^2-k_0^2\epsilon}$. The admittance at plane B may be obtained by using the formula as follows:

$$Y = \sqrt{\frac{\epsilon}{\epsilon_t}}\frac{Y_a+j\tan(ks)}{1+jY_a\tan(ks)}, \tag{17-8}$$

where $k = k_0\sqrt{\epsilon}$. Here, Y is normalized to the coaxial line, filled by a material with dielectric constant ϵ_t.

17.3. NUMERICAL RESULTS

In our calculation the normalized quantity d/a and k_0a are used to make the calculated curves universal (that is, suitable for all cases with the same d/a and k_0a, and at the same time the values of a, d, and k_0 or the frequency point may be different). The horizontal coordinate is selected to be the dielectric constant of the test sample ϵ because the slope of the calculated curves characterizes the measurement sensitivity. For simplicity we put $a/b = 2.303$ and $\epsilon_t = 1$ in our calculation. The numerical data of $Y_T = Y/j2k_0a$ for the air-filled 50.0-Ω termination, $a/b = 2.303$, are shown in Figures 17.2–17.9 for cases (b) and (c) respectively (case (a) may be considered as a special case of (b) or (c) in which $d \to \infty$).

The data in Figures 17.2–17.5 are calculated for cases $k_0a = 0.001$ and may be considered as the static solutions of these cases. Among them the solid curves Figures 17.2 and 17.4 are obtained for $ks = k_0a$ and Figures 17.3 and 17.5 for $ks = 3k_0a$. Comparing these curves with the dotted ones which are plotted for the same cases with $s = 0$ for comparison, it is clear that the behavior of these curves is basically the same; i.e., the admittance of the sensor is seen to increase when the thickness of the test material d/a decreases (for case (b), see Figures 17.2 and 17.3) and the admittance of the sensor decreases when d/a decreases as seen in Figures 17.4 and 17.5 for case (c). It is seen for case (b) that Y_T (which is proportional to the measurement sensitivity at low frequencies

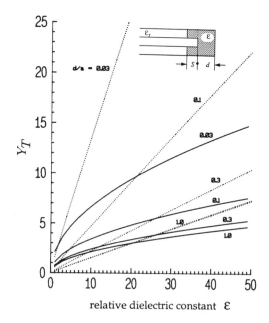

FIGURE 17.2. Dependence of Y_T on the dielectric constant of the test material ϵ for case (b) with $a/b = 2.303$: solid line (——) for $ks = k_0a$ and $k_0a = 0.001$ and dotted line (.....) for case $s = 0$.

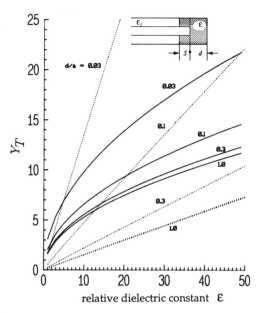

FIGURE 17.3. Dependence of Y_T on the dielectric constant of the test material ϵ for case (b) with $a/b = 2.303$: solid line (——) for $ks = 3k_0a$ and $k_0a = 0.001$, and dotted line (...) for case $s = 0$.

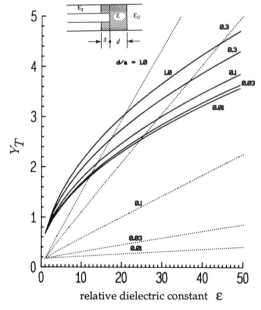

FIGURE 17.4. Dependence of Y_T on the dielectric constant of the test material ϵ for case (c) with $a/b = 2.303$: solid line (——) for $ks = k_0a$ and $k_0a = 0.001$, and dotted line (...) for case $s = 0$.

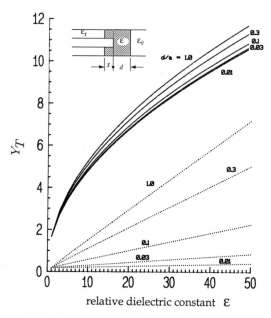

FIGURE 17.5. Dependence of Y_T on the dielectric constant of the test material ϵ for case (c) with $a/b = 2.303$: solid line (—) for $ks = 3k_0a$ and $k_0a = 0.001$, and dotted line (....) for case $s = 0$.

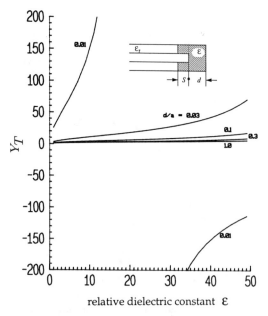

FIGURE 17.6. Dependence of Y_T on the dielectric constant of the test material ϵ for case (b) with $k_0a = 0.3$ and $ks = k_0a$, $a/b = 2.303$.

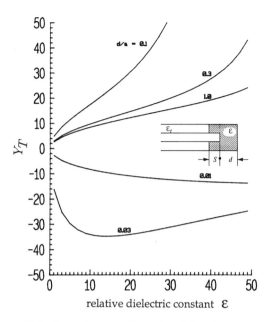

FIGURE 17.7. Dependence of Y_T on the dielectric constant of test material ϵ for case (b) with $k_0a = 0.3$ and $ks = 3k_0a$, $a/b = 2.303$.

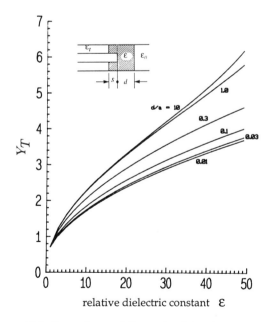

FIGURE 17.8. Dependence of Y_T on the dielectric constant of the test material ϵ for case (c) with $k_0a = 0.3$ and $ks = k_0a$, $a/b = 2.303$.

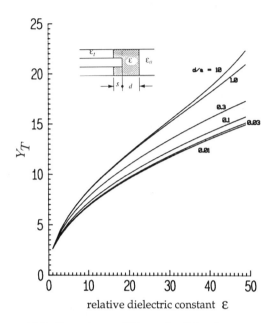

FIGURE 17.9. Dependence of Y_T on the dielectric constant of test material ϵ for case (c) with $k_0a = 0.3$ and $ks = 3k_0a$, $a/b = 2.303$.

since the admittances Y are small) increases after adding section s only when ϵ is small and d is thick (see Figures 17.2 and 17.3). Nevertheless, this should not be a problem since the measurement sensitivity for high ϵ is generally much higher than low-ϵ case and we may further increase Y_T for high ϵ case by using longer s if necessary. On the other hand, the resonance phenomena at high frequencies restrict the selection of small d and large s for case (b) (see Figures 17.6 and 17.7 for $k_0a = 0.3$).

It is clear from Figures 17.4 and 17.5 that for case (c) the improvements are present for almost all cases except very high dielectric constant ϵ and thick d. However, after using longer s (see Figure 17.5, where $ks = 3k_0a$) the improvements become much more significant (up to 4–10 times). Meanwhile, the resonance effect does not take place for case (c) at high frequencies (see Figures 17.8 and 17.9 for $k_0a = 0.3$). Therefore, it is of special interest to use this method of improvement for case (c).

The calculation results of case (d) (coaxial line terminated by a bulk of test material) are shown in Figure 17.10 where the admittance of the sensor Y is plotted against the dielectric constant of the test material ϵ. It is clear that the measurement sensitivity is improved for all ϵ values of the test material. The improvement is more significant for low-ϵ materials and can be enhanced further by using longer s.

In the cases when ϵ is complex, the admittance Y becomes complex also. When $\epsilon'' < \epsilon'$ (this condition is satisfied in most practical cases), Y may be

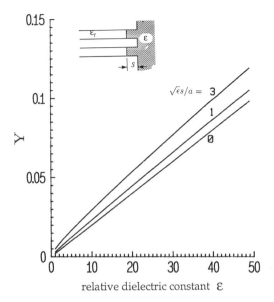

FIGURE 17.10. Dependence of Y on the dielectric constant of the test material ϵ for case (d) with $k_0 a = 0.001$ and $a/b = 2.303$.

expanded in a Taylor's series [14]:

$$Y = Y\Big|_{\substack{\epsilon''=0 \\ \epsilon=\epsilon'}} + \frac{\partial Y}{\partial \epsilon}\Big|_{\substack{\epsilon''=0 \\ \epsilon=\epsilon'}} (-j\epsilon'') + \cdots, \qquad (17\text{-}9)$$

where $\epsilon = \epsilon' - j\epsilon''$.

From Figures 17.2–17.5, 17.8, and 17.9 it is clear that Y is close to proportional to ϵ, and hence the higher-order derivatives of Y with respect to ϵ are not large. It is enough to retain only one or two higher-order terms in eq. (17-9), and the calculation of complex Y may be performed without difficulty. The above calculation is performed with $a/b = 2.303$, which corresponds to 50-Ω air-filled coaxial line. The results of calculation for the 50-Ω coaxial line filled with dielectrics (with other ratios a/b) show the same tendency of Y_T.

17.4. CONCLUSIONS

1. A simple and effective method of improvement of the test sensitivity of microwave permittivity at low frequencies for coaxial sensors is presented. The method is especially valuable for low-ϵ test materials where the test sensitivity is usually quite low at low frequencies.

2. Improvement of some coaxial sensors based on this method is demonstrated. Among them sensor (c) is most suitable for wide-bandwidth applications. Sensor (b) shows also significant improvement for thick test samples with low ϵ. For sensor (d) the improvement is obtained for all

ϵ values of the test material and is most significant also for low-microwave-permittivity case.

3. For lossy materials the calculation may be performed by using eq. (17-9) without difficulty.

REFERENCES

[1] N. E. Belhadj-Tahar and A. Fourrier-Lamer, "Broad-band analysis of a coaxial discontinuity used for dielectric measurements, *IEEE Trans. Microwave Theory Tech.*, vol. MTT-34, pp. 346–350, 1986.

[2] N. E. Belhadj-Tahar, A. Fourrier-Lamer, and H. Chanterac, "Broad-band simultaneous measurement of complex permittivity and permeability using a coaxial discontinuity," *IEEE Trans. Microwave Theory Tech.*, vol. MTT-38, pp. 1–7, 1990.

[3] M. A. Stuchly and S. S. Stuchly, "Coaxial line reflection methods for measuring dielectric properties of biological substances at radio and microwave frequencies—a review," *IEEE Trans. Instrum. Meas.*, vol. IM-29, pp. 176–183, 1980.

[4] B. Bianco, G. P. Drago, M. Marchesi, C. Martini, G. S. Mela, and S. Ridella, "Measurements of complex dielectric constant of human sera and erythrocytes," *IEEE Trans. Instrum. Meas.*, vol. IM-28, pp. 290–295, 1979.

[5] T. W. Athey, M. A. Stuchly, and S. S. Stuchly, "Measurements of radio frequency permittivity of biological tissues with an open-ended coaxial line: Part 1," *IEEE Trans. Microwave Theory Tech.*, vol. MTT-30, pp. 82–87, 1982.

[6] L. S. Anderson, G. B. Gajda, and S. S. Stuchly, "Analysis of an open-ended coaxial line sensor in layered dielectrics," *IEEE Trans. Instrum. Meas.*, vol. IM-35, pp. 13–18, 1986.

[7] A. Kraszewski, M. A. Stuchly, and S. S. Stuchly, "ANA calibration method for measurement of dielectric properties," *IEEE Trans. Instrum. Meas.*, vol. IM-32, pp. 385–387, 1983.

[8] A. Kraszewski, S. S. Stuchly, M. A. Stuchly, and S. A. Symonds, "On the measurement accuracy of the tissue permittivity in vivo," *IEEE Trans. Instrum. Meas.*, vol. IM-32, pp. 37–42, 1983.

[9] T. P. Marsland and S. Evans, "Dielectric measurements with an open-ended coaxial probe," *Proc. Inst. Elec. Eng.*, vol. 134, pp. 341–349, 1987.

[10] J. R. Mosig, J. C. E. Besson, M. Gex-Fabry, and F. E. Gardiol, "Reflection on an open-ended coaxial line and application to nondestructive measurement of Materials," *IEEE Trans. Instrum. Meas.*, vol. IM-30, pp. 46–51, 1981.

[11] D. K. Misra, "A quasi-static analysis of open-ended coaxial lines," *IEEE Trans. Microwave Theory Tech.*, vol. MTT-35, pp. 925–928, 1987.

[12] D. Misra, M. Chabbra, B. R. Epstein, M. Mirotznik, and K. R. Foster, "Noninvasive electrical characterization of materials at microwave frequencies using

an open-ended coaxial line: test of an improved calibration technique," *IEEE Trans. Microwave Theory Tech.*, vol. MTT-38, pp. 8–13, 1990.

[13] K. F. Staebell and D. Misra, "An experimental technique for in vivo permittivity measurement of materials at microwave frequencies," *IEEE Trans. Microwave Theory Tech.*, vol. MTT-38, pp. 337–339, 1990.

[14] Y. Xu and R. G. Bosisio, "Analysis of different coaxial discontinuities for microwave permittivity measurements," *IEEE Trans. Instrum. Meas.*, vol. IM-42, pp. 538–543, 1993.

G. Biffi Gentili
M. Leoncini
University of Florence, Florence, Italy

G. Salvetti
E. Tombari
IFAM-CNR, Pisa, Italy

18

Analysis of Electromagnetic Sensors for Dielectric Spectroscopy by Using the (FD)²TD Method

Abstract.   A general method based on the frequency-dependent finite-difference time-domain ((FD)²TD) solutions of Maxwell's equations is used to numerically simulate time-domain reflectometry (TDR) measurements of the dielectric properties of dispersive/nondispersive materials. A Gaussian pulse is launched in a coaxial line which is terminated by a sample cell containing the material under investigation. The reflected pulse is stored and the reflection coefficient determined. The numerical approach proposed allows one to model the electromagnetic (EM) behavior of a large class of sensors configurations. In order to validate the suggested (FD)²TD method and verify its flexibility, a classical cell configuration, which consists of a section of a coaxial line filled with the material under test, is analyzed. The case of an inhomogeneous material filling the sample cell has also been considered, for which a closed-form solution of the EM response is available only in the quasi-static approximation (very low frequencies). TDR measurements have been carried out with a calibration routine optimized by describing the cell response in terms of a numerically computed transfer function. The results obtained show the effectiveness of the proposed approach and indicate its usefulness in designing and testing of EM dielectric sensors.

18.1. INTRODUCTION

The estimation of physical parameters of moist substances and matrix systems through wideband microwave measurements is very stimulating in view of an extensive use of this approach in several fields as agriculture, building materials, artistic goods, industrial, biomedical, and scientific applications. For each application, a specific measuring circuit and a sensor are required, because of the different measuring environment and operating conditions.

The development of sensors for measuring physical properties of moist substances and matrix systems in the EM fields is conditioned by the availability of theoretical models of electromagnetic interaction between sensor and test media. Indeed, the optimization of the sensor structure and the implementation of a simple calibration procedure are more easily achieved with a combined theoretical analysis and experimental characterization.

The characteristic response of the sensor embedded in the measuring environment is an observable quantity such as the reflected or transmitted signal in the time or frequency domain. In the time domain, the measurable quantity is a real, transient waveform, contrary to the frequency domain where a complex steady-state waveform is measured which expresses the response to an alternating field at a given frequency. Measuring techniques have been developed for both domains, and the conversion from one to another is possible via Fourier transform. The time-domain measurement method, TDR, is very effective for performing high-frequency dielectric measurements, typically over 10^7–10^{10} Hz [1]. This broad band of frequencies is covered by a single measurement through the time-domain acquisition of a step-pulse reflected at the end of a coaxial line, terminated with a sample cell. The termination consists of the electromagnetic sensor loaded by the test material. The knowledge of the transfer function describing the electromagnetic characteristics of the sensor allows one to extract the dielectric properties of the material from the reflected measured pulse. In moist substances and porous media, water can be free and/or bound, exhibiting in each of the two phases a different dielectric response with frequency.

In this study the finite-difference time-domain (FDTD) method is used to simulate TDR measurements. The FDTD approach, first proposed by Yee in 1966 [2] for isotropic, nondispersive media, is based on the finite difference approximation of Maxwell's equations in the space and time domains. More recently, the formulation has been extended to include the frequency dependence of dispersive substances [3]. This last formulation, known as $(FD)^2TD$, appears to be very appropriate for the modeling of dielectric relaxation phenomena, provided the dielectric relaxation behavior can be described by an exponential decay in the time domain.

In the simulation experiment procedure, the input coaxial line and the sample cell, consisting of a section of an open-circuited coaxial line, are modeled according to their structures. A Gaussian electromagnetic pulse (superposition of TEM modes) is launched in the input line; with the dielectric properties of the material assigned as input data, the reflected waveform is stored and the reflection coefficient determined. This allows simulation of the sample cell response, thus giving us a deep insight into the sensor electromagnetic properties.

Two cases are considered: in the first the test material completely fills the cell; in the second a cylindrical insulating ring has been inserted between the material and the outer conductor in order to overcome some practical experimental problems (further details are reported in Section 18.3).

The following observations should be made regarding EM analysis:

- The transfer function of the open-circuited transmission line, filled with the homogeneous material, is well known [4] and this configuration can be used as validation test of the proposed method.

- When inhomogeneities are introduced in the dielectric which fills the cell, the EM response cannot be easily determined analytically. A numerical solution is needed in order to find the correct transfer function.

A series of TDR measurements on a few materials with a different dielectric behavior was performed on a Tektronix sampling oscilloscope. The values of dielectric permittivities have been obtained from the measured data using a calibration routine based on the results coming from the numerical simulations.

18.2. METHOD

A schematic description of the model of the EM sensor configuration is sketched in Figure 18.1. The sensor consists of a proper termination of a feed coaxial cable loaded with the sample material. A Gaussian electromagnetic pulse is launched at the input section (section B), while the time-varying voltage along the coaxial line is sampled in section C.

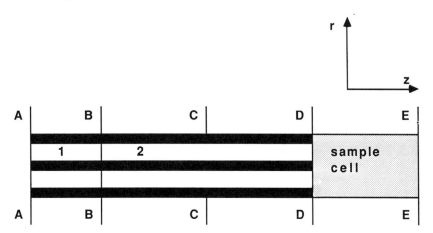

FIGURE 18.1. $(FD)^2TD$ mesh domains: region 1, reflected field domain; region 2, total field domain. A Gaussian pulse is launched at section B, the total field is sampled at section C, boundaries are A and E sections, while L is the reference section for load admittance computations.

The FDTD mesh domain is divided at the input section in two subdomains: (1) the total field domain (region 2); (2) the reflected field domain (region 1). In region 2, the total field is computed, while in region 1 only the reflected field is determined. This is achieved by subtracting the incident field from the total field at the input section using a technique similar to that proposed by Taflove and Umashankar for scattering problems [5]. The sampling section is far enough away from the cable termination so that higher-order evanescent modes, which are generated at the load discontinuity, are eliminated before reaching that section. This allows us to consider the reflected waveform due only to TEM propagating modes and then to correctly determine the reflection coefficient in terms of measurable quantities.

The numerical code is a straightforward formulation of Yee's approach [2], implemented in a 2D cylindrical coordinate system. Since most of the sensor configurations commonly used are axially symmetric, the 3D problem reduces to an exact 2D problem in the coordinate system selected. When dispersive materials are considered in the analysis, the same procedure proposed by Luebbers et al. [3] is used and the code becomes an $(FD)^2TD$ code; a further assumption is that the dielectric dispersion of the sample presents a relaxation characteristic describable with an exponential decay in the time domain (i.e., Debye's relaxation behavior).

Since the $(FD)^2TD$ is a discretization method, and then only a limited simulation space can be taken into account, appropriate boundary conditions have to be imposed to limit the solution domain. Mur's first-order absorbing boundary conditions [6] were used to absorb the reflected wave inside the line, while boundary conditions at section E depend on the EM characteristics of the sample cell under analysis. Details on the boundary conditions used for each case analyzed in this paper are given in the next section.

The numerical code is structured in two basic parts: a preprocessor and an $(FD)^2TD$ kernel code. The preprocessor is used to define the geometry of the feed coaxial cable and of the sample cell and to grid the entire spatial domain. The output of the preprocessor is a matrix which contains all the information about the mesh domain. This matrix is then read by the kernel code, which solves Maxwell's equations using their $(FD)^2TD$ approximation and stores the results in an external file. A feature of the model is that a variable grid step size can be employed anywhere in the spatial domain, provided the local rectangular mesh is unaffected. This allows one to define a finer grid where necessary. Particularly, this can be helpful in modeling sharp geometrical changes to improve the accuracy of the results and interfaces between materials with different dielectric characteristics.

The numerical code runs in about 2 hr on a workstation SUN SPARC 2, and usually a grid of 2–3×10^4 nodes is required to cover the entire domain.

18.3. SENSOR CONFIGURATION AND MODELING

Two basic configurations, depicted in Figures 18.2a and 18.2b, have been analyzed. The first one (Figure 18.2a) consists of a section of an open-circuited transmission line of length d_m completely filled with a homogeneous dielectric material. The EM behavior of this termination can be described in terms of an ideal section of a coaxial line of effective length d and characteristic admittance Y_L [4]. The fringing field which exists at the M section can be indeed described, within the frequency range of interest, in terms of an equivalent electrical length Δd. The characteristic admittance Y_L directly depends on the permittivity of the test material and can be written as

$$Y_L = Y_0\sqrt{\epsilon},$$

where Y_0 is the characteristic admittance of the coaxial line filled with air.

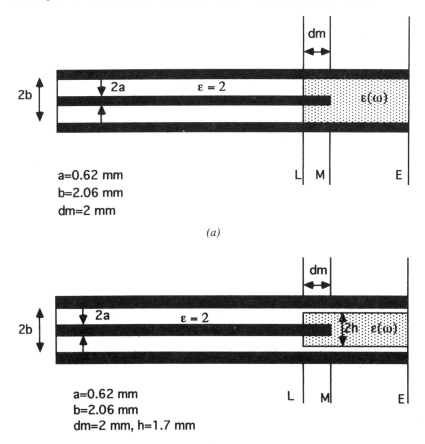

FIGURE 18.2. Schematic view of cell configurations: (a) homogeneous coaxial line filled with the sample material, (b) inhomogeneous coaxial line filled with the sample material, plus an insulating ring.

The sensor behaves as a coaxial termination load presenting the following admittance value at plane L-L (Figure 18.1):

$$Y = jY_L \tan(\beta d), \qquad (18\text{-}1)$$

with $\beta = \omega\sqrt{\epsilon}/c_0$ where ω is the angular frequency, c_0 is the velocity of light in vacuum, and d is the sensor electrical length.

All the quantities in eq. (18-1) are known except d which depends on the unknown fringing field equivalent length Δd. However, a good approximating value for Δd is $(b - a)/2$ where b and a are the outer and inner conductor radii, respectively [4]. A more accurate way to determine d can be carried out by using the method here proposed with the following procedure:

- Numerical simulations of the sensor response in presence of known dispersive/nondispersive materials are run and the reflection coefficient computed.
- The load admittance versus frequency is determined.
- Data are then fitted using the analytical function previously defined for the admittance and the d values extracted for each case.

As shown in Section 18.5, the values obtained for Δd are quite close to the approximate values of $\Delta d = (b-a)/2$ independently of the dielectric characteristics of the test material.

The second sensor (Figure 18.2b) consists of the same section of the coaxial line previously described, with a cylindrical insulating ring inserted between the test material and the outer conductor. The sample cell in Figure 18.2b can help to overcome some practical difficulties which could arise during the experiments. For instance, in low-temperature measurements, Teflon shrinking can create a leakage of liquid samples inside the coaxial cable. Furthermore, this sample cell is quite analogous to that obtained by coating the inner and the outer conductors in order to prevent damage of the metallic parts of the sensor when corrosive media are measured.

From the EM point of view, the insulating ring introduces inhomogeneities in the section of the coaxial line. This means that an analytical solution, describing the propagation in the line, can be found only at low frequencies by using a quasi-static approximation and then describing the propagation in terms of quasi-TEM modes. Indeed, in the high-frequency range, the quasi-TEM approach does not correctly describe the propagation phenomena and the transfer function of the sensor cannot be analytically determined. The $(FD)^2TD$ approach here proposed fills this gap since the transfer function can be numerically determined by a fitting procedure of the computed data.

In the low-frequency range the admittance at plane L-L obeys the equation applied for the sensor in Figure 18.1a, where Y_L and β depend on an effective dielectric constant ϵ_{eff} which can be determined within zero-order approximation as

$$\epsilon_{\text{eff}} = \frac{\epsilon \epsilon_{\text{in}} \ln(b/a)}{\epsilon_{\text{in}} \ln(h/a) + \epsilon \ln(b/h)}, \tag{18-2}$$

where ϵ and ϵ_{in} are the permittivities of the test material and of the insulating ring, respectively.

However, two points should be made:

1. The values of d and of the fringing field equivalent length Δd depend on the value of the equivalent dielectric constant (18-2).

2. The inhomogeneous coaxial line presents natural dispersion in the frequency domain, so the propagation constant β is no longer a linear function of ω.

Again, the procedure that has been followed uses the numerical simulation results to determine d and the dispersion characteristics of the line.

As depicted in Figures 18.2a and 18.2b, plane E-E corresponds to the boundary of the mesh domain. For both sensor configurations, a short-circuit boundary condition has been imposed there for the numerical analysis. Provided plane E-E is sufficiently far from plane M-M, the only field that can reach this boundary is that due to the cylindrical waveguide modes launched. Multiple reflections may then appear resulting in standing waves inside the line. The frequencies at which standing waves occur can be determined and the upper limit of the measuring frequency of the loaded sensor can be calculated.

18.4. TDR MEASUREMENTS

TDR measurements have been carried out by using a dual-channel apparatus based on a Tektronix 7854 waveform-processing oscilloscope configured as a sampling oscilloscope. The first channel, consisting of a delay line, was used for time referencing, while a second channel was configured (in terms of coaxial line length between signal source–sampling head–cell) in order to obtain the separation of the incident and reflected step pulses at the sampling section and a time window free of spurious reflections [4]. The measuring time window has been selected to be 2 ns, and the waveforms were sampled and digitized in 1000 points. The equivalent frequency range covers 200 MHz to 10 GHz.

The first sensor (Figure 18.2a) was made by assembling the shell and the pin of a male SMA connector on a female/female SMA adapter. The second sensor (Figure 18.2b) was realized by adding an insulating Teflon ring ($\epsilon = 2$) inside the connector housing. The dimensions of the sensors are reported in Figure 18.2.

In order to extract the permittivity values from TDR measurements a calibration procedure is needed to correct for spurious reflections due to the connectors and other line mismatches [4]. The calibration routine relates the measured admittance $\bar{y}$ (which is affected by connectors and line mismatches) to its theoretical value y by the equation

$$\bar{y} = (a + by)/(1 + cy) \qquad (18\text{-}3)$$

where a, b, and c are complex frequency-dependent parameters determined through the calibration routine by using air and two reference media of known permittivity [4]. The improvement introduced is that the transfer functions, used as a theoretical representation of the load admittance y, are obtained by fitting the numerical simulation results obtained from the $(FD)^2TD$ analysis.

All measurements reported here have been performed at a temperature of $T = 25.0 \pm 0.1°$ C.

18.5. RESULTS

18.5.1 Numerical Simulations

First simulation runs for the open-circuited transmission line of Figure 18.2a filled with nondispersive/dispersive materials were taken. The admittance Y values obtained for various nondispersive materials with ϵ ranging from 1 to 21 are shown in Figure 18.3, while in Figures 18.4a and 18.4b the values obtained for a dispersive material with $\epsilon_s = 10.5$, $\epsilon_\infty = 2.5$, and $\tau = 2$, 20, and 180 ps, respectively, are reported. In all these figures, the continuous lines correspond to the theoretical results obtained by using the analytical function of eq. (18-1) with $d = 2.70 \pm 0.05$ mm as determined by the fitting of the numerical data. The agreement between numerical and theoretical results is quite satisfactory and confirms the validity of the proposed approach. As expected, the electrical length d does not depend on the loading material permittivity values, and the equivalent fringing field contribution Δd is quite close to $(b - a)/2$ (see Section 18.4).

The dispersive propagation behavior in the inhomogeneous coaxial line included between planes L-L and M-M (see Figure 18.2b) was also analyzed. The length d_m was increased to 10 cm, the voltages at two distinct sections 1 cm apart were sampled, and the phase velocity was computed. In Figure 18.5, β/β_c, where β_c is the propagation constant for the homogeneous coaxial line filled with Teflon, is shown for $\epsilon = 4.86$, 10.5, and 21, at frequencies for which propagation is quasi-

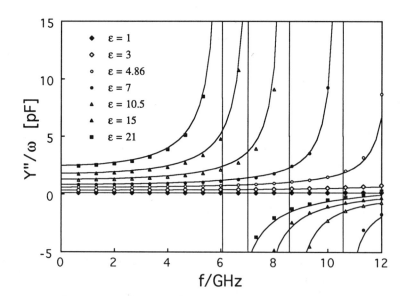

FIGURE 18.3. Imaginary part vs. frequency of the load admittance for the sample cell in Figure 18.2a as from $(FD)^2TD$ simulations for various nondispersive materials; continuous lines are obtained from eq. (18-1) with $d = 2.7$ mm.

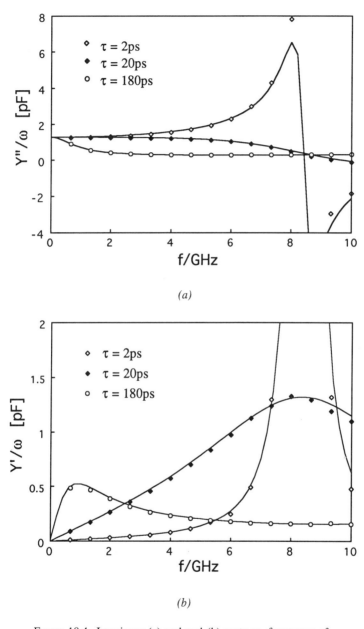

FIGURE 18.4. Imaginary (a) and real (b) parts vs. frequency of the load admittance for the sample cell in Figure 18.2a as from $(FD)^2TD$ simulations. Data are shown for dispersive materials with $\epsilon_s = 10.5$, $\epsilon_\infty = 2.5$, and $\tau = 2$, 20, and 180 ps, respectively; continuous lines are obtained from eq. (18-1) with $d = 2.7$ mm.

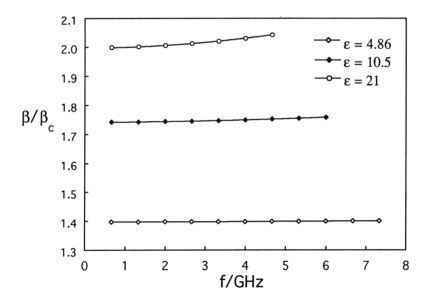

FIGURE 18.5. Propagation constant vs. frequency for the quasi-TEM mode in an inhomogeneous coaxial line as that depicted in Figure 18.2b. Simulations were run for three different nondispersive materials and data normalized to the propagation constant of the feed coaxial cable.

TEM. The simulation results indicate a weak dispersive behavior of the line that can be analytically described in terms of a $k\omega^2$ correction with k dependent on the permittivity of the dielectric material.

The numerical results obtained for the cell in Figure 18.2b are reported in Figures 18.6 and 18.7, for nondispersive and dispersive test materials, respectively. The permittivity values for each case are indicated in the above figures. For all simulations the transfer functions of the sensors have been analytically retrieved by fitting procedures. As could be expected, eq. (18-1), with the electrical length d now depending on ϵ_{eff}, can be successfully used to fit the data in the low-frequency range. On the contrary, in the high-frequency range, it has been verified that the empirical equation

$$Y = jY_L \tan \left(\omega^\alpha d \sqrt{\epsilon_{eff}}/c_0 \right) \tag{18-4}$$

describes the behavior of the numerical data. The fitting parameters d and α appearing in eq. (18-4) are listed in Table 18.1 together with ϵ and ϵ_{eff}.

18.5.2 TDR Experimental Results

In Figures 18.8a and 18.8b, the admittance of the first sensor configuration, measured for two liquids of known permittivities, chloroform and methylene chloride [7], are reported together with their theoretical values obtained using the transfer function of eq. (18-1) with the parameter $d = 2.7$ mm as determined by fitting

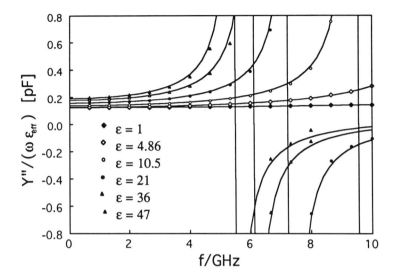

FIGURE 18.6. Imaginary part vs. frequency of the load admittance for the sample cell of Figure 18.2b as from $(FD)^2TD$ simulations for various nondispersive materials; continuous lines correspond to the theoretical curves of eq. (18-4) where d and α values are those reported in Table 18.1.

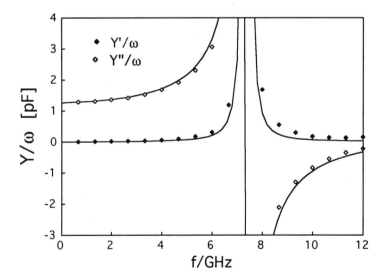

FIGURE 18.7. Imaginary (Y'') and real (Y') parts vs. frequency of the load admittance for the sample cell of Figure 18.2b as from $(FD)^2TD$ simulation for a medium with $\epsilon_s = 21, \epsilon_\infty = 2$, and $\tau = 2$ ps; continuous lines correspond to the theoretical curves of eq. (18-4) where $d = 3.54$ mm and $\alpha = 1.0036$.

TABLE 18.1. Fitting Parameters Obtained for Various Nondispersive Loading Materials

ϵ	ϵ_{eff}	d (mm)	α
1	1.092	2.78 ± 0.05	1.005 ± 0.003
4.86	3.916	2.85 ± 0.05	1.0069 ± 0.0004
10.5	6.12	3.12 ± 0.05	1.0078 ± 0.0002
21	8.07	3.54 ± 0.05	1.0036 ± 0.0004
36	9.31	4.00 ± 0.05	1.0030 ± 0.0007
47	9.84	4.31 ± 0.05	1.003 ± 0.002

ϵ_{eff} was calculated from eq. (18-2), d from eq. (18-1) applied to low-frequency range data, and α from eq. (18-4) applied in the whole range of frequencies.

the numerical data. The deviations between experimental and theoretical curves, particularly evident in the high-frequency range, are due to spurious reflections and connector mismatches.

In Figure 18.9, the admittance measured for an unknown liquid (1-chlorobutane) is reported both as measured and corrected using the calibration procedure.

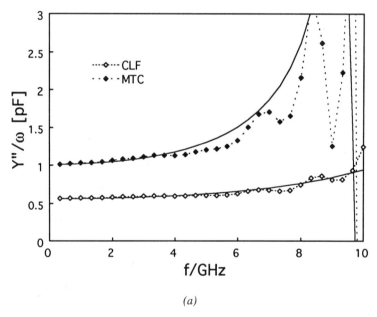

(a)

FIGURE 18.8. Imaginary (a) and real (b) parts vs. frequency of the load admittance for the sample cell of Figure 18.2a as from experimental TDR measurements for two reference media, chloroform ($\epsilon_s = 4.86$, $\epsilon_\infty = 2.38$, and $\tau = 6.5$ ps) and methylene chloride ($\epsilon_s = 8.81$, $\epsilon_\infty = 2$, and $\tau = 1.8$ ps); continuous lines correspond to the theoretical curves of eq. (18-1).

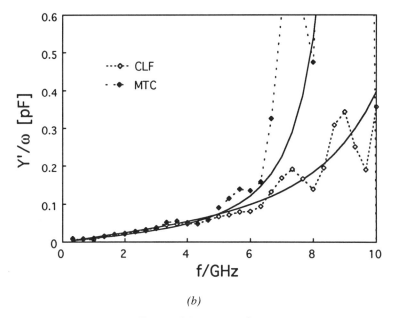

(b)

FIGURE 18.8. *continued*

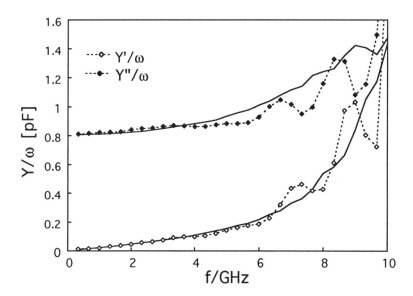

FIGURE 18.9. Imaginary (Y'') and real (Y') parts vs. frequency of the load admittance for the sample cell of Figure 18.2a as from experimental TDR measurements on 1-chlorobutane; continuous lines correspond to the corrected admittance values with the calibration performed as described in Section 18.4.

The calibration was carried out using the data shown in Figures 18.8a and 18.8b as references and following the procedure described in Section 18.4. It is worthwhile to observe how the corrected data are smoothed out. The dielectric permittivity values computed for the 1-chlorobutane by numerically solving eq. (18-1) in terms of the calibrated admittance data are shown in Figure 18.10. The experimental results obtained for 1-chlorobutane indicate Debye's dielectric relaxation behavior with $\epsilon_s = 7.05$, $\epsilon_\infty = 2.5$, and $\tau = 7$ ps.

The second set of measurements concerns the sensor configuration shown in Figure 18.2b. Again, admittance measurements have been carried out for the two reference media (chloroform and methylene chloride) and for the unknown 1-chlorobutane material. However, since this sensor configuration was used for the first time, a measurement on a third known material, 1,2-dichloroethane [7], has been carried out to validate the use of the empirical transfer function (see eq. (18-4)) both in the calibration routine and for the ϵ determination. As shown in Figure 18.11, the permittivity values obtained for the of 1,2-dichloroethane are in accordance with the expected Debye behavior of the medium. The results obtained for 1-chlorobutane are in Figure 18.12 and confirm what was obtained using the first type sensor (see Figure 18.9).

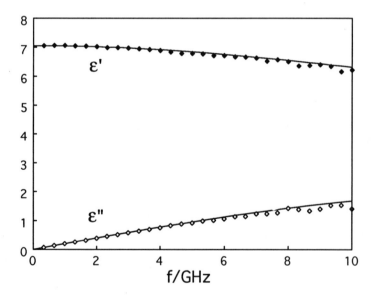

FIGURE 18.10. Imaginary (ϵ'') and real (ϵ') parts of permittivity vs. frequency for 1-chlorobutane as extracted from the transfer function of eq. (18-1) by using the corrected admittance data shown in Figure 18.9. The continuous lines represent the characteristic Debye relaxation behavior for $\epsilon_s = 7.05$, $\epsilon_\infty = 2.5$, and $\tau = 7$ ps.

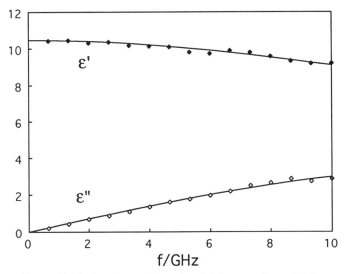

FIGURE 18.11. Imaginary (ϵ'') and real (ϵ') parts of permittivity vs. frequency for the 1,2-dichloroethane as extracted from the transfer function of eq. (18-4) by using the TDR data measured for the sample cell in Figure 18.2b; continuous lines correspond to the Debye relaxation behavior of this material with $\epsilon_s = 10.43$, $\epsilon_\infty = 2.56$, and $\tau = 7.2$ ps [7].

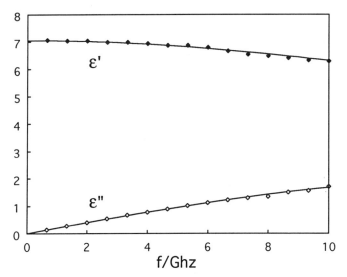

FIGURE 18.12. Imaginary (ϵ'') and real (ϵ') parts of permittivity vs. frequency for 1-chlorobutane as extracted from the transfer function of eq. (18-4) by using the TDR data measured for the sample cell in Figure 18.2b. The continuous lines represent the characteristic Debye relaxation behavior for $\epsilon_s = 7.05$, $\epsilon_\infty = 2.5$, and $\tau = 7$ ps.

18.6. CONCLUSIONS

The $(FD)^2TD$ method was applied to simulate TDR permittivity measurements and to determine the EM response of a sample cell filled with an homogeneous and a nonhomogeneous material, respectively. Moreover, TDR experimental measurements have been carried out by considering a few materials with various dielectric permittivity values.

The results of this study show that

- The $(FD)^2TD$ method allows simulation of permittivity measurements without artefacts due to connectors and line mismatches, always present in any experimental setup. These effects appear as a large oscillation in the EM response of the sample cell and do not allow the determination of the transfer function of the cell if only the measured data are used.

- The transfer function can be determined through the fitting of the numerical results. Then, proper calibration routines can be defined and used to extract the value of permittivity from measured data.

Basically, the approach proposed allows a detailed analysis of the EM response for any kind of sensor configurations provided they are based on a termination of a coaxial line and are rotationally symmetric. In particular, it is worthwhile to stress that sensors with more complex structures or operating in different conditions (like unbounded media) can be studied with a detailed theoretical analysis of their EM behavior. The design and optimization of new kinds of sensors fulfilling a wide range of experimental requirements is then possible. This would cover, for example, the study of water content in porous materials, the in vivo characterization of biological media, and measurements in high- or low-temperature conditions for which sensors with specific mechanical properties are necessary.

REFERENCES

[1] R. H. Cole, "Time domain reflectometry," *Ann. Rev. Phys. Chem.*, vol. 28, pp. 283–300, 1977.

[2] K. S. Yee, "Numerical solution of initial boundary value problems involving Maxwell's equations in isotropic media," *IEEE Trans. Ant. Propagat.*, vol. AP-14, pp. 302–307, May 1966.

[3] R. Luebbers, F. P. Hunsberger, K. S. Kunz, R. B. Standler, and M. Schneider, "A frequency-dependent finite-difference time-domain formulation for dispersive materials," *IEEE Trans. Electromag. Compat.*, vol. 32, pp. 222–227, August 1990.

[4] D. Bertolini, M. Cassettari, G. Salvetti, E. Tombari, and S. Veronesi, "Time domain reflectometry to study the dielectric properties of liquids: some problems and solutions," *Rev. Sci. Instrum.*, vol. 61, pp. 450–456, Dec. 1990.

[5] A. Taflove and K. R. Umashankar, "The finite-difference time-domain (FDTD) method for electromagnetic scattering and interaction problems," *J. Electromag. Waves Appl.*, vol. 1, pp. 243–267, 1987.

[6] G. Mur, "Absorbing boundary conditions for the finite-difference approximation of the time-domain electromagnetic-field equations," *IEEE Trans. Electromag. Compat.*, vol. EMC-23, pp. 377–382, Nov. 1981.

[7] R. H. Cole, J. G. Berberian, S. Mashimo, G. Chryssikos, A. Burns, and E. Tombari, "Time domain reflection methods for dielectric measurements to 10 GHz," *J. Appl. Phys.*, vol. 66, pp. 793–802, July 1989.

R. N. Clarke
A. P. Gregory
T. E. Hodgetts
G. T. Symm
National Physical Laboratory,
Teddington, Middlesex, UK

19

Improvements in Coaxial Sensor Dielectric Measurement: Relevance to Aqueous Dielectrics and Biological Tissue

Abstract.   Recent developments in traceable coaxial sensor metrology have led to improved accuracy in the determination of dielectric properties of aqueous materials and biological tissues. In measurements from 10 MHz to 5 GHz, it has been demonstrated that homogeneous or stratified dielectrics can often be characterized more accurately in a cylindrical cell which is fitted coaxially to the sensor than in a "probe" or "infinite half-space" geometry. Three computer programs for relating complex permittivities of multilayer structures to measured reflection coefficients have been written, compared, and shown to agree—residual discrepancies between their numerical outputs being significantly smaller than expected measurement uncertainties. The programs cover both "infinite half-space" and cylindrical cavity geometries. The findings of this work have been brought to bear upon measurements of water, aqueous solutions, and laminar and tissue-equivalent dielectrics. The implications for the measurement of stratified biological tissues are discussed.

19.1.   INTRODUCTION

This work originated some years ago in studies of the characterization of the dielectric properties of human tissues in vivo for the detection and treatment of tumors [1–3]. Aqueous dielectric and biological tissue measurements have been an ongoing concern in this program of work and have largely determined the sensor size and frequency range employed: 14-mm-diameter sensors have been used from 10 MHz to 3 GHz [4–8] (since extended to 5 GHz). There have been three other major targets for this work: (1) the general improvement of accuracy in coaxial sensor dielectric metrology; (2) the extension of the technique to more complex geometries [7,9]; (3) the requirement that all measurements be ultimately traceable to UK national standards. The latter requirement gave rise to further work on the characterization of a number of reference liquids commonly used in the calibration of coaxial sensors [10] (see also [17]). Traceability also requires that the computer programs used to relate specimen complex permittivity, $\epsilon^* = \epsilon' - j\epsilon''$, to measured reflection coefficient, $\Gamma = |\Gamma| \times \exp(j\theta)$, must be *shown* to be functioning correctly. This has been accomplished by comparing computer programs [6] and comparing their numerical output with actual measurements in calculable geometries [1,7,9]. Our main aim is to show how software and hardware generated in this broader program of work can improve the quality of measurements on water, aqueous solutions, biological tissues, and moisture-holding dielectrics in general. The significant developments are concerned with treatment of practical specimen sizes and dielectric inhomogeneity—specifically with stratified dielectrics.

Computer treatments of coaxial sensors model the fringing electromagnetic fields which emerge from the open end of the sensor and pass through the dielectric. Γ depends on ϵ^* because the field boundary conditions at the sensor-to-dielectric interface demand continuity of certain components of the electromagnetic field [12–16]. Most earlier computer models tacitly or explicitly adopted an infinite half-space (IHS) approximation for the dielectric specimen. They also assumed complete homogeneity for the dielectric [1]. More recent work in this laboratory [7,9] and elsewhere has specifically sought to overcome limitations of this sort of model. Computer programs which handle laminar, stratified, finite-thickness, and finite-diameter specimens also have the advantage that they allow the practical measurement errors of the earlier, simpler models to be quantified.

Among the geometries studied previously has been the homogeneous dielectric lamina backed by a conducting plane. Jenkins et al. reported agreement in measurement/computer program intercomparisons in this geometry for the case where the "lamina" is a reference liquid [7]. Nishikata and Shimizu [17,18] and Fan and Misra [19] have been concerned with a similar geometry. Li et al. [20] were specifically concerned with measuring thin moisture layers. Measurement of multiple parallel layers each of which has a different thickness, t_i, and complex permittivity, ϵ_i^* (Figure 19.1), has been treated recently elsewhere [21,22], in this laboratory (NPL) [9] and by Moschuring and Wolff [23], who studied media suitable for microwave integrated circuit substrates. Xu et al. [24,25] have analyzed similar geometries, and their experimental work has included measurement of surface resistivity of thin films. Nishikata and Shimizu [18] have analyzed the "sandwich" case of a solid laminar specimen with a gap on either side and a backshort plane.

It was pointed out by Jenkins et al. [7] that edge effects can limit the accuracy of coaxial sensor measurements. That is, when a sensor is being used as a "probe" (as in Figure 19.1), reflections or field distortions at the edge of coaxial sensor flanges or the edge of specimens can produce measurement errors which cannot be fully quantified with IHS or similar theories. The latter effect can be much larger than the flange reflections. Measurements upon low-loss specimens are

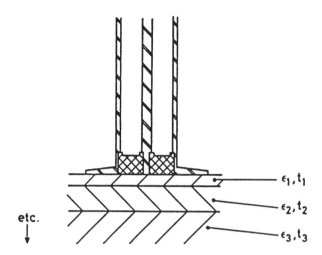

FIGURE 19.1. An NPL type B 14-mm coaxial sensor of the probe type, shown measuring a stratified dielectric. ϵ_1, t_1, etc., are respectively the complex permittivity and thickness of each layer.

particularly prone to this type of error, which can typically exceed 10% in ϵ^* with practically sized specimens and can even reach 100%. If field perturbations from the perimeters of specimens cannot actually be avoided, then it is good practice to ensure that they are at least calculable. By using disk-shaped specimens and enclosing them in a cylindrical cavity (Figure 19.2), calculable conditions can be preserved. Two of the three "dielectric-multilayer" computer programs developed at NPL recently were primarily intended to handle this geometry. The third dealt directly with the "multilayer infinite half-space" (multilayer-IHS) case. The latter is best suited to high-loss measurements, while measurements in cylindrical cells can be performed on both high- and low-loss specimens. Saed [26], Saed et al. [27], and Otto and Chew [28] have also provided numerical analyses of similar cavities fitted to coaxial sensors. Belhadj-Tahar et al. [29,30] have analyzed one- and two-port cells which have the same diameter as the coaxial outer conductor. A related geometry is that studied by Macphie et al. [31], who analyzed the input impedance of a coaxial line opening into a larger circular waveguide section, where the inner conductor is extended some distance into the waveguide. This paper covers models and measurements with both "probe" (Figure 19.1) and a number of "cell" geometries, one of which is shown in Figure 19.2.

The computer programs which were developed for this work can be used with all types of flat-faced open-ended coaxial sensors across the full microwave frequency range. For the purposes of testing the measurement techniques empiri- cally, however, two types of coaxial sensor based upon standard 14-mm precision airline and GR-900-compatible connectors were employed up to 5 GHz. One was an NPL type B 14-mm sensor probe with a 50-mm-diameter flange (Figure 19.1)

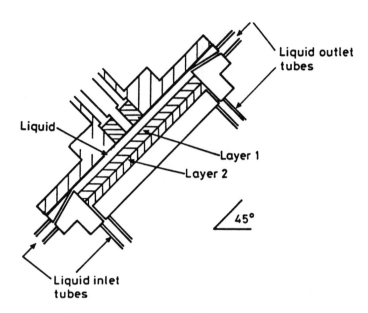

FIGURE 19.2. A coaxial sensor modified so as to fit onto a
cylindrical cell. The cell has inlet and outlet tubes for liquid
measurements and liquid-immersion measurements upon solids.
It is tilted to remove bubbles from the sensor face.

[9]. The second was a new type C sensor which has a thicker 100-mm diameter flange, specifically designed for the fitting of a cylindrical cell (Figure 19.2). These sensors are usually referred to as 14-mm sensors because they are constructed from 14-mm precision airline, but the diameter of the sensor aperture is actually machined to 15.09 mm in both cases. They have polymer beads (permittivity, ϵ_b', taken to be 2.538) to support the inner conductor. The inner and outer conductor radii of the coaxial line are changed in the bead so as to retain a 50-Ω characteristic impedance in both bead and airline sections. In addition to these two geometries, some of the measurements detailed below have been performed in a 14.29-mm-diameter cylindrical *unmatched* cell in which the inner and outer conductor diameters do not change along the length of the sensor (i.e., they have the same diameter in bead and cell). These relatively large sensor/cell diameters are ideal for empirically checking computer programs and sensor theory because measurement errors caused by manufacturing dimensional tolerances are not as large as with smaller sensors. They also provide optimal sensitivity for biological tissue and many other aqueous dielectric measurements in the RF region of the electromagnetic spectrum.

This paper covers the computer models used and a numerical comparison of program outputs against measurements upon water and stratified dielectrics. It concludes with a discussion of the consequences of this work for improving measurements upon aqueous dielectrics and biological tissue.

19.2.   THE COMPUTER PROGRAMS

The programs have been described elsewhere [9], so only a summary is given here, together with a demonstration of the level of agreement that can be achieved between independent analytic treatments. All of the programs resolve the electromagnetic field into a sum of TM_{0n} modes in the coaxial and cylindrical cell sections, with the addition of forward and reflected TEM waves in the coaxial line. Γ is equal to the ratio of these TEM waves only, since the TM waves are evanescent in the coaxial line. As in previous work, it was found advisable to use a large number of TM modes in the analyses: 50 or over in the coaxial line, scaled in proportion to radius in the cell. More conveniently, in order to speed up computations, a quadratic extrapolation from two, four, and eight modes to an effectively "infinite" number [1,6,9] also provides sufficient accuracy. Apart from these similarities, the programs differ in detail.

One program, MIC2, was developed by G. T. Symm from the proposals of Warham [15]. It handles the multilayer-IHS geometry of Figure 19.1 where the half-space fields are treated in terms of antenna theory similar to the treatment of Mosig et al. [14]. The probe flange diameter is assumed to be infinite in extent, however, potentially giving rise to the "edge effect" errors discussed above in practical measurements. The computer programs TEH and RCAV were written to cover the finite-diameter cell geometry. The former was written by Hodgetts and is based on a variational approach of his own [12,13]. The latter is based on the theory of Saed et al. [27], extended by Symm. In the tables which follow, versions of these programs are variously called TEH2, TEH3, RCAV3, etc., to denote the maximum number of layers allowed for. Versions of the programs which handle more than three layers have not been written. Though this would be a trivial extension in the case of TEH and RCAV, practical measurements on more than three layers have not been performed since they would meet with diminishing returns in terms of accuracy and usefulness. A fourth program, COAX, was purchased [32] for intercomparison purposes. It is a general-purpose cylindrical-geometry analysis program. It proved to be very helpful in the task of estimating uncertainties caused by geometrical imperfections, but it was less convenient to use for actual dielectric measurements than the purpose-written programs.

The programs described above are based upon "forward" algorithms which compute Γ from known permittivities, ϵ_i^*, and layer thicknesses, t_i, where i is the index number of each dielectric layer (Figure 19.1). For dielectric *measurements* an "inverted" algorithm is required which, for a three-layer dielectric, computes any one ϵ_i^* ($i = 1, 2, 3$) from the known frequency, f, layer thicknesses, t_i, radius, r, measured reflection coefficient, Γ, and known ϵ_i^* of the *other* two layers (plus the coaxial sensor parameters). For a homogeneous (nonlayered) material the subscripts can be dropped so that its permittivity, ϵ^*, is computed from its thickness and radius, t and r. Inversion of the algorithm is achieved by placing the forward routine in an iterative loop (see, for example, [1]), starting with an estimate of ϵ_i^* or ϵ^* and iterating its value until the computed value of Γ equals its measured value.

Confidence in the above programs was gained by direct numerical comparison between them and by comparisons with earlier coaxial-sensor programs that deal only with homogeneous dielectrics and IHS or "infinite lamina" geometries. These earlier programs were regarded as being already tried and tested [6], so for appropriate geometries they could be treated as benchmarks against which the new programs could be compared. They are designated TEH1 [12], GTS [1], and AGPW [15]. In all cases the numerical comparisons produced satisfactory results. They were all performed for approximately 14-mm-diameter sensors, but there is every reason to believe that similar agreement will be obtained for other sensor sizes. In particular, if all linear dimensions (of the sensor and specimen) were scaled *down* by the factor, say, α, then—everything else being equal, for example, ϵ^*, etc.—exactly the same numerical computation would be performed at a frequency which is scaled *up* by α. Thus the software should perform just as well for any sensor that is optimized in size for its intended frequency range.

19.2.1 Numerical Intercomparisons

We present examples of computer output for purely *hypothetical* measurements and calibrations. Successful comparisons of *actual* measurements with computer predictions are presented in the next section. Computations of this sort are useful for three purposes: (1) to compare programs in order to generate confidence in their correct operation; (2) to provide Γ values for *calibration* of coaxial sensors (by use of known reference liquids, for example); and (3) to compute uncertainties for proposed measurements. In some cases such purely theoretical analyses may show, for example, that (for a given probe, dielectric, and frequency range) a measurement is barely worth performing because the uncertainties would be too high, or they may indicate that a change in experimental geometry would improve measurement uncertainties significantly.

Automatic network analyzer (ANA) reflectometry requires at least three known standards to be attached to the measurement port for calibration. With coaxial sensors, it is convenient to use (1) a short-circuit at the end of the sensor, (2) an open circuit into air, and (3) a reference liquid of appreciable permittivity and loss (e.g., methanol and ethanol). For all of these calibration "standards," Γ can be calculated because ϵ^* is assumed to be known [9,17]. To reduce measurement errors, it is often advisable to use a least-squares calibration algorithm with more than three standards [9]. Whichever calibration procedure is followed, however, the open circuit remains one of the most important and commonly used of calibration geometries. On computing Γ-values up to 3 GHz for a 14-mm sensor in air, one finds that radiative power loss into an IHS of air is virtually negligible. Therefore $|\Gamma_{air}| \approx 1.0$ and the calibration procedure requires only the theoretical value of $\arg(\Gamma_{air}) = \theta$ to be effective. Computed values of θ are compared in Table 19.1. The finite-cell-diameter programs TEH2, RCAV, and COAX can only be made to approximate the IHS geometry by computing θ for large cell radii, r. As shown, this gives rise to discrepancies as the frequency approaches the TM_{01}-mode cutoff frequency, which depends on r (4.41 GHz for $r = 26$ mm; 3.28 GHz for

TABLE 19.1. Numerical Intercomparison of Computed Reflection Coefficient Phases, θ in degrees ($|\Gamma| \approx 1.0$), for a Coaxial Sensor Opening into Air.

Freq. (GHz)	Coax $r = 26$ mm	Coax $r = 48$ mm	TEH2 $r = 35$ mm	RCAV $r = 35$ mm	MIC2 $r = \infty$	AGPW $r = \infty$
0.300	−1.15	−1.15	−1.15	−1.15		−1.15
0.500	−1.92	−1.92	−1.92	−1.92		−1.91
1.000	−3.86	−3.85	−3.85	−3.85	−3.83	−3.84
1.500	−5.81	−5.80	−5.80	−5.80		−5.78
2.000	−7.77	−7.80	−7.78	−7.78	−7.73	−7.75
2.100	−8.17	−8.21	−8.18	−8.19		−8.15
2.200	−8.60	−8.64	−8.59	−8.59		−8.54
2.300	−8.98	−9.13	−9.00	−9.00		−8.94
2.400	−9.41	−9.28	−9.41	−9.41		−9.35
2.500	−9.79	−9.66	−9.83	−9.83		−9.75
2.600	−10.23	−10.06	−10.25	−10.26		−10.16
2.700	−10.62	−10.46	−10.69	−10.69		−10.56
2.800	−11.04	−10.86	−11.14	−11.14		−10.97
2.900	−11.46	−11.27	−11.60	−11.62		−11.38
3.000	−11.88	−11.68	−12.12	−12.13	−11.76	−11.79

The figures are for a 14-mm coaxial sensor with inner conductor radius $a = 2.333$ mm, outer conductor radius $b = 7.549$ mm, and bead permittivity $\epsilon_b = 2.15 - 0.0j$. The column headings correspond to the computer programs described in the text. The assumed cell radius, r, is given for each column.

$r = 35$ mm; 2.39 GHz for $r = 48$ mm). It will nevertheless be seen that excellent agreement between the different algorithms is obtained for θ at all frequencies more than 300 MHz below such cutoffs. Discrepancies are less than 0.1° in $|\theta|$, while *measurement* uncertainties are typically about ±0.3°. The final two columns in the table are computed for infinite half-spaces and do not exhibit such cutoff phenomena. They assume that the coaxial aperture lies in an infinite conducting plane. The value of $|\theta|$ for, say, a 26-mm cell is slightly higher than that of the IHS because the fringing capacitance is higher (this is most noticeable at the higher frequencies).

By varying cell radii, as in Table 19.1, one can estimate the magnitude of edge effects from finite-radius specimens. Thus, Figure 19.3 shows the effect of changing the cylindrical-cell radius, r, upon the value of Γ at 3 GHz. Both parts of the figure are computed for deionized water when the cell is 3.0 mm thick. The dashed line (Figure 19.3a) and cross (+, Figure 19.3b) show the extrapolated value for $r = \infty$. Clearly there is little direct practical interest in a "variable-diameter" metal cell, but the computations do have a practical relevance: they put an upper limit upon likely discrepancies in finite-specimen-diameter measurements with a probe. The variations in Figure 19.3 are commensurate with those estimated by Jenkins et al. [7] from actual probe measurements on water when $r = 25$ mm.

One of the most important decisions to be made when using a coaxial sensor as a probe is just how large the specimen needs to be to give reliable values for ϵ^*.

Clearly for deionized water in this short cell even a 35-mm radius is not large enough to be treated as effectively "infinite." With lossier liquids (e.g., methanol) or lossy solids such as muscle tissue, similar evidence to that of Figure 19.3 [9] shows that specimen or sensor-flange radii of 25 mm are probably just sufficiently large with a 14-mm sensor up to 3 GHz to be treated as effectively infinite without introducing errors into ϵ^* of more than a few percent.

Computations such as those in Figure 19.3 can be obtained from any of the finite-diameter programs (RCAV, TEH, or COAX), discrepancies in Γ again being smaller than estimated measurement uncertainties. Table 19.2 demonstrates this for hypothetical dielectrics of permittivity $\epsilon^* = 100 - j100$. The figures in this table are broadly consistent with the assertion that specimen radii greater than $r = 25$ mm can be expected to appear effectively infinite when the dielectric material is very lossy, as it is in this case with $\tan\delta = 1$.

As a final example of a program comparison, Table 19.3 shows computations for a two-layer dielectric "specimen" in a cylindrical metal cell 35 mm in radius

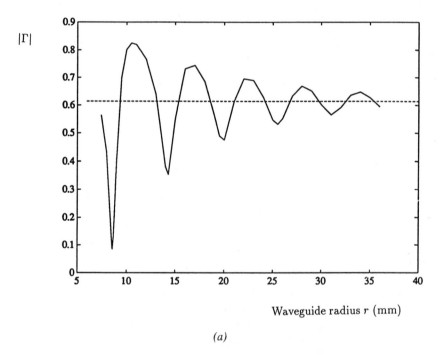

Waveguide radius r (mm)

(a)

Figure 19.3. Reflection coefficient, Γ, at 3 GHz computed for deionized water in a perfectly conducting cylindrical cell 3.0 mm thick, as the radius r is varied: (a) variation of $|\Gamma|$ with r; (b) full variation in complex Γ on a polar chart. The dashed line and the cross are the limiting values for $r = \infty$ as obtained from program AGPW. The computations are for the same 14-mm sensor as Table 19.1.

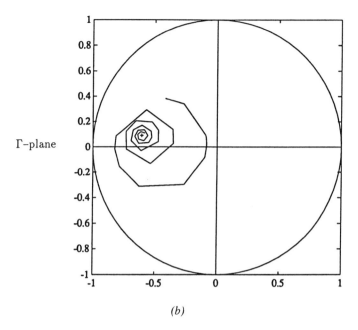

(b)

FIGURE 19.3. *continued*

TABLE 19.2. Intercomparison of Computed Values of Reflection Coefficient, Γ, for a Lamina 1 mm Thick for Hypothetical Dielectrics of Permittivity $\epsilon = 100 - j100$.

Freq. (GHz)	Coax $r = 26$ mm	TEH2 $r = 35$ mm	RCAV $r = 35$ mm	MIC2 $r = \infty$	AGPW $r = \infty$
0.100	0.4186 $-82.51°$	0.4176 $-83.34°$	0.4170 $-83.63°$	0.4159 $-83.39°$	0.4177 $-82.69°$
1.000	0.8380 $-175.22°$	0.8382 $-175.39°$	0.8390 $-175.33°$	0.8385 $-175.33°$	0.8391 $-175.36°$
2.000	0.8945 $+176.68°$	0.8941 $+176.60°$	0.8945 $+176.59°$	0.8942 $+176.59°$	0.8942 $+176.58°$
3.000	0.8911 $+171.19°$	0.8908 $+171.09°$	0.8911 $+171.09°$	0.8909 $+171.08°$	0.8908 $+171.09°$

The dielectric is contained in a perfectly conducting cylindrical cell of radius r and thickness 1 mm which is placed coaxially at the end of the sensor. The details of the 14-mm sensor are as for Table 19.1. The table entries are $|\Gamma|$ and $\theta = \arg(\Gamma)$.

and approximately 1.5 mm thick overall. Similar comparisons between computer programs are published elsewhere [9]. In the whole program of comparisons, the level of agreement was typically better than ±0.001 in $|\Gamma|$ and $\pm0.1°$ for θ at all frequencies not close to high Q-factor cell-resonances [ibid.].

TABLE 19.3. Values of Γ from RCAV3 and TEH2 for a Two-Layer Dielectric System in a 35-mm-Radius Cell Attached to a 14-mm Sensor.

FREQUENCY r	Γ FROM RCAV3		Γ FROM TEH2		ϵ^*-ETHANOL	
(GHz)	$\|\Gamma\|$	θ DEG	$\|\Gamma\|$	θ DEG	ϵ'	ϵ''
0.3	0.9707	−12.50	0.9707	−12.48	22.85	6.69
0.5	0.9332	−19.58	0.9332	−19.55	19.69	9.23
0.8	0.8790	−28.50	0.8790	−28.46	15.19	10.41
1.0	0.8493	−33.91	0.8493	−33.85	12.89	10.23
1.5	0.7589	−49.85	0.7590	−49.76	9.29	8.81
2.0	0.7711	−49.69	0.7711	−49.58	7.48	7.35
2.5	0.7628	−63.38	0.7628	−63.27	6.49	6.21
3.0	0.7483	−76.02	0.7482	−75.89	5.91	5.33

The sensor details are as follows: inner conductor radius $a = 1.994$ mm, outer conductor radius $b = 7.545$ mm, bead permittivity $\epsilon_b = 2.358 - j0.0$. Layer 1 (nearest the sensor): thickness 0.514 mm, is ethanol, for which ϵ^* is given in the table. Layer 2: thickness 1.003 mm, is notionally the polymer Delrin, $\epsilon^* = 2.96 - j0.0$ at all frequencies.

19.3. MEASUREMENTS

Studies of numerical output from programs have their uses, as described above, but the most convincing demonstration of the effectiveness of computer programs is a comparison of their output with actual measurements on well-characterized materials. Measurements were therefore performed upon reference solids and liquids [9]. The dielectric properties of most of these materials had been previously measured at NPL by a number of other techniques [10]. Values for the complex permittivity used for n-alcohol reference liquids are those obtained in a recent measurement program at NPL, rather than figures in the existing scientific literature. The complex permittivity of water has been taken from Kaatze and Uhlendorf [33,34].

The level of agreement that can be obtained in a *calculable* geometry is demonstrated by Figure 19.4. This is a polar chart of the complex Γ-plane which shows measured and computed values of Γ for deionized water and for four organic reference liquids in a 25-mm-radius cylindrical cell at the end of a 14-mm sensor. The ANA/sensor *calibration* for this measurement used the same cell, the "standards" being (1) a short-circuit, (2) an air-filled cell, and (3) an ethanol-filled cell. The agreement between the computed and measured values for water, for example, is therefore obtained using standards which present Γ-values very different from those of water at all frequencies. The agreement is nevertheless very close, reproducing all of the electromagnetic resonances in the body of the water as loops in the traces. Discrepancies are magnified on the resonances, as expected, and in most cases can be accounted for by small geometrical imperfections and metal wall or contact losses in the cell which are not modeled in the theory. In consequence, the Q-factors of the resonances are slightly lower in the measurements than in the theory and the resonance frequencies are also slightly different, but typically by no more than 10% and 1%, respectively, at most.

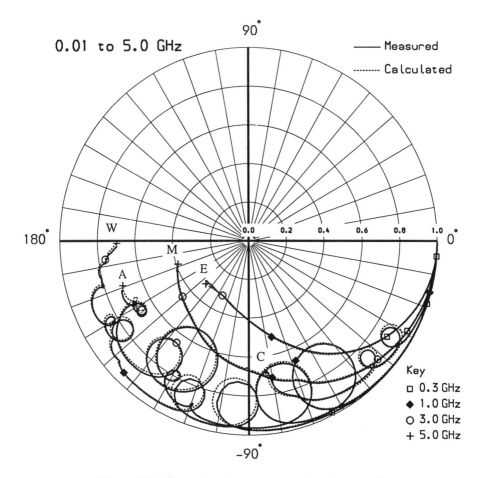

FIGURE 19.4. Comparison between measured and computed values of Γ, 100 MHz–5 GHz, for deionized water and four organic liquids in a 25-mm-radius cell, 35.6 mm thick. The computations are for the 14-mm sensor of Table 19.3. The traces are marked as follows: A, acetone; C, chloroform; E, ethanediol; M, methanol; W, water. Ethanol was used as a calibration reference liquid.

In a "perfect" cylindrical cavity (i.e., one having a *small* external coupling factor), the frequencies of the resonances would be readily calculable from first principles. TM_{0n}-mode resonances, for example, should occur at frequencies, f_n, such that $f_n \approx \Re e(j_{0n} \times c/2\pi r \sqrt{\epsilon^*})$, where the j_{0n} are the zeros of the Bessel function J_0 (2.4048, 5.5201, 8.6537, . . .). This gives $f_1 \approx 520$ MHz, $f_2 \approx$ 1.2 GHz, etc., for water. In practice, the coupling of the cell "cavity" to the coaxial line is by no means "small," and some perturbation to the resonant frequencies must therefore be expected. The other interposed resonances are of the form TM_{0nm}, but their frequencies are still further perturbed through electromagnetic interactions with the coaxial line aperture. Methanol and ethanediol are sufficiently lossy to damp resonances completely.

Figure 19.4 represents a significant advance over the situation reported by Grant and Clarke in 1989 [1] (cf. Figure 15 of that paper), in which resonances in a glass beaker of water were observed and regarded as noncalculable. Considerable effort was at that time required to produce a geometry in which deionized water could meaningfully be measured *at all* with a 14-mm coaxial sensor. This gives yet another reminder of the failings of simple IHS models for coaxial sensor fringing fields. However, because of the large number of resonances shown in Figure 19.4, it is not to be recommended that a cell of such large radius (25 mm) actually be used in serious measurements of aqueous dielectrics. Figure 19.5 shows measured and computed traces of Γ for a more suitable cell whose radius is the *same* as that of the coaxial line (7.144 mm), where the response is much more well behaved because the TM_{0n}-mode resonances are shifted above the measurement frequency range.

Figures 19.4 and 19.5 demonstrate that good theory/measurement agreement can be obtained for water even when a reference liquid with a very different permittivity is used for calibration: discrepancies in the complex Γ of water are typically less than 0.01 in magnitude at frequencies away from resonances. In both figures ethanol has been used as a calibration standard. Its reference Γ-values, as used in the calibration, are also shown in Figure 19.5. Nevertheless, the measured values for ϵ'' of water are undoubtedly too high for the data in Figure 19.5, as can be seen from Table 19.4. This table compares the measured permittivity, ϵ_m^*, computed by TEH2 from the Γ-values of Figure 19.5, against literature permittivity values ϵ_l^* derived from the work of Kaatze and Uhlendorf [33]. The discrepancy shows up as a small difference in $|\Gamma|$, which is just discernible in the figure. The most significant causes of this discrepancy are (1) small errors in the attributed permittivity of ethanol, (2) residual calibration errors with the short- and open-circuit geometries, (3) the fact that metal and contact losses in the cell are not taken into account, (4) small temperature differences from 20.0°C which change the permittivity of both the standard (ethanol) and the water test sample. But many of these problems can easily be overcome: in measurements on water and aqueous dielectrics the effects of (1)–(3) can be greatly reduced, and sometimes rendered negligible, by using a substitution technique. To implement this it is good practice to calibrate with pure deionized water itself as a reference liquid rather than ethanol. This procedure has been used at NPL for measurements on dilute aqueous solutions where small changes of the permittivity from that of pure water are of interest and are readily discernible by substitution. Provided the temperature is stable to within 0.1°C, repeatability for either Cartesian component of Γ is typically found to be better than ±0.001 in 14-mm cells. This can correspond to an uncertainty of under ±0.08% at the 95% confidence level for measuring ethanediol concentration in water, for example.

Figures 19.4 and 19.5 show good agreement between measured and computed values for Γ up to 5 GHz, but it is not recommended that 14-mm sensors be used

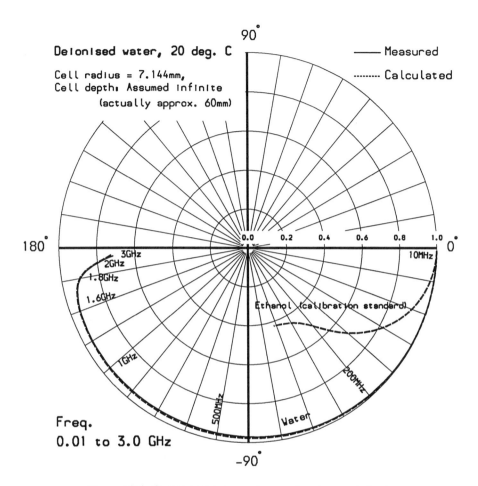

FIGURE 19.5. Comparison between measured and computed values of Γ, 10 MHz–3 GHz, for deionized water in a 60-mm-deep,
7.144-mm-radius cell, treated as being infinite in length. The
measurements were performed in a 50-Ω precision air line (inner
conductor radius $a = 3.102$ mm, outer conductor radius $b =$
7.144 mm). The bead material has permittivity $\epsilon_b = 2.538 - j0.0$.
Data from these measurements are given in Table 19.5. The
dashed trace is the reflection coefficient computed for ethanol,
which was used as a reference liquid to calibrate the sensor in
this measurement.

for measurements on aqueous dielectrics much above 2 GHz. The sensitivity of
the technique ($\partial \Gamma / \partial \epsilon^*$) falls quite rapidly above that frequency for $\epsilon' \approx 70$–80, so
large errors in measuring ϵ^* can result. Maximum sensitivity for 14-mm sensors
for aqueous dielectrics is in the range 50 MHz–1 GHz, the frequency range for
which they were designed. Smaller sensors are more suitable at higher frequencies. Provided sufficiently large specimens are used, as noted in the last section,

TABLE 19.4. Comparison of ϵ^* Measured at 20.0 $\pm 0.4°$C for Deionized Water with Literature Data [33] Using Ethanol as a Calibration Liquid.

Frequency f	Reflection Coefficient		Measured ϵ^*		ϵ^* from [33]	
(GHz)	$|\Gamma|$	θ deg	ϵ'_m	ϵ''_m	ϵ'_l	ϵ''_l
0.01	0.9997	−2.67	80.38	0.50	80.21	0.04
0.2	0.9890	−50.45	80.63	1.16	80.20	0.88
0.5	0.9643	−100.72	80.34	2.89	80.15	2.19
1.0	0.9541	−139.03	79.32	4.89	79.95	4.37
1.6	0.9414	−162.62	79.02	7.27	79.56	6.96
1.8	0.9008	−170.19	79.22	8.25	79.38	7.81
2.0	0.8267	−173.56	79.06	9.55	79.19	8.66

The measurements are shown in Figure 19.5 where the cell details are also given.

14-mm coaxial sensors with a 25-mm-radius flange can nevertheless profitably be used with lower-permittivity, higher-loss dielectrics up to 3 GHz or even higher. This has been demonstrated for methanol and other alcohols [9] and for tissue equivalent materials [35]. A comprehensive uncertainty analysis has been carried out for these sensors [9]. In measurements upon three-layer systems, 1 to 3 mm thick, agreement in ϵ^* to better than $\pm 3\%$ has been obtained for the layer nearest the sensor, and better than $\pm 10\%$ for the second and even the third layers (ibid.). These findings have relevance to measurements upon stratified biological tissues, and these are discussed in the next section.

19.4. THE RELEVENCE TO MEASUREMENTS ON AQUEOUS DIELECTRICS

The aim of this paper is to draw conclusions of relevance to the characterization of aqueous and biological dielectrics. It has long been recognized that the open-ended coaxial sensor is a convenient and powerful tool for measuring malleable materials, such as biological tissues, and for other lossy dielectrics with a significant water content. But it has been shown [9] that it can only be used meaningfully in conjunction with theoretical treatments which are appropriate to the measurement geometry used. Dielectric inhomogeneities and specimen size limitations can cause errors of up to 100% in derived permittivities if a homogeneous-IHS model is used for analysis, as it has been in much published work. As demonstrated above, these difficulties can now be avoided.

The IHS model leaves much to be desired in the case of tissues such as skin (especially when measured in vivo) where outer layers are dry and inner layers are moist. Attribution of a single "effective permittivity," ϵ^*_{eff} to such a stratified tissue can only be an approximation (see, e.g., [2]), and one should not be surprised if the behavior of ϵ^*_{eff} runs counter to what is expected for a *truly* homogeneous, structureless dielectric. Structured dielectrics can always produce values for Γ

which are incompatible with an assumption of homogeneity. Figure 19.4 illustrates this point. It shows responses containing multiple resonances which originate from very strong "inhomogeneities," specifically from the abrupt changes in permittivity at the liquid/metal interfaces on the cylindrical cell walls. Such responses could not be obtained at all from a homogeneous IHS specimen. If values for ϵ^*_{eff} of water were to be computed from the appropriate trace in Figure 19.4 using an IHS algorithm, they would not correspond to the permittivity of water and would not comply with any accepted dielectric theory. While all of this is clear, the consequences of such considerations for the measurement of tissues can benefit from closer examination.

Tables 19.5a and b illustrate the problem for a purely hypothetical stratified dielectric. This may notionally be taken as a model of three-layer tissue (e.g., dry

TABLE 19.5a. Details of a Hypothetical Three-Layer Dielectric Specimen Used for the Computations in Table 19.5b.

Layer i	t_i (MM)	ϵ_s	ϵ_∞	f_r
1	0.3	6.5	4.0	2.0 GHz
2	0.3	3.5	2.5	0.5 GHz
3	40.0	50.0	40.0	10.0 GHz

Each layer is a dielectric which follows a Debye relaxation law, characterized by the parameters in the table (see text). The thickness of the ith layer is t_i, layer 1 being closest to the coaxial sensor.

TABLE 19.5b. Computations of Γ by the Program TEH3 for the Composite Dielectric Described in Table 19.5a.

Frequency f	Γ from TEH3		ϵ^*_{eff}	
(GHz)	$|\Gamma|$	θ DEG	ϵ'_{eff}	ϵ''_{eff}
0.5	0.9529	−16.89	11.05	1.92
1.0	0.9331	−35.76	12.16	1.20
1.5	0.8951	−52.28	11.93	1.37
2.0	0.7673	−68.76	11.49	2.73
2.5	0.6569	−72.58	9.69	3.29
3.0	0.6784	−81.03	9.35	2.58
3.5	0.6853	−92.69	9.18	2.08
4.0	0.6635	−103.78	9.18	1.48
4.5	0.6285	−112.36	9.65	1.74
5.0	0.6018	−118.72	9.10	1.98

The columns headed ϵ^*_{eff} give the effective permittivity computed from Γ, assuming the dielectric to be homogeneous. The trends of permittivity with frequency as shown in the table would not be physically possible for a homogeneous medium.

skin and fat over muscle). However, the permittivity differences between the layers have been exaggerated to show the effects that can occur. Table 19.5a gives the assumed properties of each layer. The analysis is for specimen radius of 35 mm and a total depth of 40.6 mm. These dimensions are, however, sufficiently large to prevent the remote perimeters of the (lossy) specimen from affecting the computations significantly. Each dielectric layer is assumed to exhibit a pure Debye relaxation: $\epsilon^* = \epsilon_\infty + (\epsilon_s - \epsilon_\infty)/(1 + jf/f_r)$, where ϵ_∞, ϵ_s, and f_r are, respectively, the high-frequency permittivity, "static" permittivity, and relaxation-frequency parameters which characterize the Debye relaxation, and f is the frequency. Thus, the permittivity of each layer exhibits a single loss peak (at f_r), and its real permittivity, ϵ', falls monotonically with frequency from ϵ_s to ϵ_∞. Using the program TEH3 the reflection coefficient, Γ, of this structure has been computed and is tabulated in Table 19.5b. From these Γ-figures, the value of ϵ_{eff} for the same specimen is computed *assuming the specimen to be homogeneous*. The behavior of ϵ_{eff} obtained, rapid rises and falls of ϵ' and ϵ'', is counter to normal physical behavior of (and would be impossible to obtain from) a homogeneous dielectric. Evidence for such anomalous behavior in values for ϵ_{eff} computed from IHS-programs is found in Grant et al. [1] and Jenkins [4], as acknowledged in these references.

The lessons to be drawn from these considerations will depend upon the use to which the tissue measurements are being put. In cases where the basic structure of a dielectric is known (e.g., layer thicknesses, etc.) it would obviously be better to use a multilayer program in analyzing measurements, but this is not always possible in tissue measurements. For example, Grant et al. [2] and Jenkins [4] were characterizing tissues in vivo in research programs aimed at (1) *detecting* tumorous tissue and (2) its *treatment* by means of RF hyperthermia. It may be argued that in the case of detection, *any* change in ϵ_{eff}, whether caused by true permittivity changes or structural inhomogeneity, could be effective. Similar considerations apply to use of coaxial sensors in food processing, for example, in checking the freshness of foods. In case (2), though, (treatment) the ϵ_{eff} which is relevant for a given coaxial sensor would not be expected to coincide with that which is appropriate for a hyperthermia applicator. Precision in such cases strictly requires a better structural knowledge.

19.5. CONCLUDING COMMENTS

This paper has drawn attention to a number of ways in which coaxial sensor dielectric measurements can and have been improved. It has reviewed work which has shown that use of cylindrical cells, where practical, can make coaxial sensors into semiprecision tools in which measurements can be performed with absolute uncertainties of $\pm 1\%$ or better. Alternatively, resolutions of better than $\pm 0.1\%$ by weight can be obtained for measuring concentrations of organic material in water. Multilayer theory both highlights and provides possible solutions to the difficulties of measuring stratified tissues. Although the work reported here is for

14-mm sensors at frequencies up to 5 GHz, the conclusions can be applied across the whole RF and microwave range if the correct scaling parameters are applied.

ACKNOWLEDGMENTS

We wish to acknowledge the valuable exchange of ideas we have had with Dr. Soe Min Tun of SMT Consultants Ltd. We would also like to thank Dr. A. W. Kraszewski, of the United States Department of Agriculture, for his encouragement to write this paper.

REFERENCES

[1] J. P. Grant, R. N. Clarke, G. T. Symm, and N. M. Spyrou, "A critical study of the open-ended coaxial line sensor technique for RF and microwave complex permittivity measurement," *J. Phys. E: Sci. Instrum.*, vol. 22, pp. 757–770, 1989.

[2] J. P. Grant, R. N. Clarke, G. T. Symm, and N. M. Spyrou, "*In vivo* dielectric properties of human skin from 50 MHz to 2.0 GHz," *Phys. Med. Biol.*, vol. 33, pp. 607–612, 1988.

[3] J. P. Grant and N. Spyrou, "Complex permittivity differences between normal and pathological tissues: mechanisms and medical significance," *J. Bioelectricity*, vol. 4, pp. 419–458, 1985.

[4] S. Jenkins, "Measurement of the complex permittivity of dielectric reference liquids and human tissues," Ph.D. thesis, Bristol University, Faculty of Medicine, 1991.

[5] S. Jenkins, R. N. Clarke, M. Horrocks, and A. W. Preece, "Dielectric measurements on human tissues between 100 MHz and 3 GHz," in *Proc. European Assoc. of Thermology Microwave Group Meeting on Advances in Medical Microwave Imaging, Lille, France*, 1989.

[6] S. Jenkins, T. E. Hodgetts, G. T. Symm, A. G. P. Warham, R. N. Clarke, and A. W. Preece, "Comparison of three numerical treatments for the open-ended coaxial line sensor," *Electron. Lett.*, vol. 26, no. 4, pp. 234–236, 1990.

[7] S. Jenkins, A. G. P. Warham, and R. N. Clarke, "Use of an open-ended coaxial line sensor with a laminar or liquid dielectric backed by a conducting plane," *Proc. IEE*, Pt. H, vol. 139, pp. 179–182, 1992.

[8] R. H. Johnson, J. L. Green, M. P. Robinson, A. W. Preece, and R. N. Clarke, "A resonant open-ended coaxial line sensor for measuring complex permittivity," *Proc. IEE*, Pt. A, vol. 139, pp. 261–264, 1992.

[9] A. P. Gregory, R. N. Clarke, T. E. Hodgetts, and G. T. Symm, "RF and microwave dielectric measurements upon layered materials using a reflectometric coaxial sensor," NPL Report DES 125, NPL, Teddington, Middlesex, UK, 1993.

[10] S. Jenkins, T. E. Hodgetts, R. N. Clarke, and A. W. Preece, "Dielectric measurements on reference liquids using automatic network analysers and calculable geometries," *Measure. Sci. Technol.*, vol. 1, pp. 691–702, 1990.

[11] A. Nyshadham, C. L. Sibbald, and S. S. Stuchly, "Permittivity measurements using open-ended sensors and reference liquid calibration—An uncertainty analysis," *Trans. IEEE Microwave Theory Tech.*, vol. 40, pp. 305–314, 1992.

[12] T. E. Hodgetts, "The calculation of the equivalent circuits of coaxial-line step discontinuities," *R. Signals Radar Establish. Memor. No. 3422*, 1981.

[13] T. E. Hodgetts, "The open-ended coaxial line: a rigorous variational treatment," *R. Signals Radar Establish. Memor. No. 4331*, 1989.

[14] J. R. Mosig, J. C. E. Besson, M. Gex-Fabry, and F. E. Gardiol, "Reflection of an open-ended coxial line and application to non-destructive measurement of materials," *IEEE Trans. Instrum. Meas.*, vol. IM-30, pp. 46–51, 1981.

[15] A. G. P. Warham, "Annular slot antenna radiating into lossy material," *NPL Report DITC 152/89*, 1989.

[16] R. D. Nevels, C. M. Butler, and W. Yablon, "The annular slot antenna in a lossy biological medium," *Trans. IEEE Microwave Theory Tech.*, vol. MTT-30, pp. 314–319, 1985.

[17] A. Nishikata and Y. Shimizu, "Analysis for reflection from coaxial end attached to lossy sheet and its application to nondestructive measurement," *Electron. Commun. Jpn. 2*, vol. 71, pp. 95–101, 1988.

[18] A. Nishikata and Y. Shimizu, "Effectiveness of gaps for lossy dielectric sheet measurements by coaxial electrode," *Electron. Commun. Jpn. 2*, vol. 73, pp. 14–22, 1990.

[19] S. Fan and D. Misra, "A study on the metal-flanged open-ended coaxial line terminating in a conductor-backed dielectric layer," *Proceedings of the Conference on Microwave Theory and Techniques, IEEE Proceedings CH2735-9/90*, pp. 43–46, 1990.

[20] L. L. Li, N. H. Ismail, L. S. Taylor, and C. C. Davis, "Flanged coaxial probes for measuring thin moisture layers," *Trans. IEEE, Biomed. Eng.*, vol. 39, pp. 49–57, 1992.

[21] L. S. Anderson, G. B. Gajda, and S. S. Stuchly, "Analysis of an open-ended coaxial line sensor in layered dielectrics," *Trans. IEEE Instrum. Meas.*, vol. IM-35, pp. 13–18, 1986.

[22] S. Fan, K. Staeball, and D. Misra, "Static analysis of an open-ended line terminated by layered media," *Trans. IEEE Instrum. Meas.*, vol. IM-39, pp. 435–437, 1990.

[23] H. Moschuring and I. Wolff, "The measurement of inhomogeneities and of the permittivity distribution in MIC-substrates," *IEEE Conference on Instrumentation and Measurement, Boston, IEEE conference proceedings IMTC/87, Cat. no. 87CH2405-9*, 1987.

[24] Y. Xu, R. G. Bosisio, and T. K. Bose, "Some calculation methods and universal diagrams for measurement of dielectric constants using open-ended coaxial probes," *Proc. IEE Pt. H*, vol. 138, pp. 356–360, 1991.

[25] Y. Xu and R. G. Bosisio, "On the measurement of resistive thin films and dielectric slabs using an open-ended coaxial line," *J. Electromagn. Waves and Appl.*, vol. 6, pp. 1247–1258, 1992.

[26] M. A. Saed, "A method of moments solution of a cylindrical cavity placed between two coaxial transmission lines," *Trans. IEEE Microwave Theory Tech.*, vol. 39, pp. 1712–1717, 1991.

[27] M. A. Saed, S. M. Riad, and W. A. Davis, "Wide-band dielectric characterization using a dielectric filled cavity adapted to the end of a transmission line," *Trans. IEEE Instrum. Measure.*, vol. 39, pp. 485–491, 1990.

[28] G. P. Otto and W. C. Chew, "Improved calibration of a large open-ended coaxial probe for dielectric measurements," *Trans. IEEE Instrum. Measure.*, vol. IM-40, pp. 742–746, 1991.

[29] N.-E. Belhadj-Tahar, A. Fourrier-Lamer, and H. De Chanterac, "Broad-band simultaneous measurement of complex permittivity and permeability using a coaxial discontinuity," *Trans. IEEE Microwave Theory Tech.*, vol. MTT-38, pp. 1–7, 1990.

[30] N.-E. Belhadj-Tahar and A. Fourrier-Lamer, "Broad-band simultaneous measurement of complex permittivity and permeability for uniaxial or isotropic materials using a coaxial discontinuity," *J. Electromagn. Waves Appl.*, vol. 6, pp. 1225–1245, 1992.

[31] R. H. Macphie, M. Opie, and C. R. Ries, "Input impedance of a coaxial line probe feeding a circular waveguide in the TM_{01} mode," *IEEE Trans. Microwave Theory Tech.*, vol. MTT-38, pp. 334–337, 1990.

[32] S. M. Tun, Manual for 'COAX,' Version 1.01, January 1992, Antenna Software Ltd., Gt. Malvern, UK.

[33] U. Kaatze and V. Uhlendorf, "The dielectric properties of water at microwave frequencies," *Z. Phys., Neue Folge*, vol. 126, pp. 151–165, 1981.

[34] U. Kaatze, "Complex permittivity of water as a function of frequency and temperature," *J. Chem. Eng. Data*, vol. 34, pp. 37–40, 1989.

[35] M. P. Robinson, M. J. Richardson, J. L. Green, and A. W. Preece, "New materials for dielectric simulation of tissues," *Phys. Med. Biol.*, vol. 36, pp. 1565–1571, 1991.

Technical Applications

Measuring Methods and Equipment

20

Charles W. E. Walker
Pacific Automation Instruments Ltd.,
Richmond, B.C., Canada

Accurate Percent Water Determination by Microwave Interaction Alone: 1954–Present

Abstract. The paper includes an outline of early work on the interaction of microwaves with water molecules in their rotational energy spectrum, much of it prior to the substantial wartime development of microwave techniques. The remainder of the paper describes the problems encountered in applying this technology to the measurement of water in solid and liquid industrial materials and to means of overcoming these problems. It is emphasized that the problems encountered differ with each product measured and with each industrial process and often with each individual plant, so that these differences must be taken into account in designing moisture meters for specific applications.

20.1. INTRODUCTION

20.1.1 Historical Background

Methods of detecting the presence of a particular chemical substance and of measuring the quantity or proportion of that substance which is present in any material frequently depend on some form of spectroscopy. For molecular substances, such as water, this essentially involves quantum transitions between molecular energy levels. Schiff [1] points out that these may be classified as "electronic," "vibrational," and "rotational" types. He estimates that the electronic and vibrational energy levels are all likely to be in the ultraviolet, visible, or near-infrared regions of the electromagnetic spectrum, with only the rotational energy levels falling in the far-infrared region.

For the measurement of simple molecules, such as water, the rotational energy spectrum seems particularly appropriate because it involves the rotation of the molecule as a whole rather than electron rotation energies or bond vibration energies which may be common to many different molecules. Work by Randall et

301

al. [2], prior to World War II, had shown a water vapor absorption in the microwave region at around 23 GHz, but no use appears to have been made of this means for moisture measurement until after the end of the war, perhaps principally because it was not until then that microwave techniques were fully developed, particularly at the higher frequencies. It was the need for greater resolution and accuracy in target location that pushed the radiolocation (or RADAR) technology to higher and higher microwave frequencies, and this culminated in what was called K-band, extending from 18 to 26 GHz, which gave a somewhat dramatic demonstration of the moisture interaction when an intervening cloud effectively obliterated the RADAR signal.

Becker and Autler [3] investigated the matter in the Columbia Radiation Laboratory at Columbia University in 1946, following on the work of Randall et al., and, with improved accuracy, showed the peak moisture absorption to lie between 22.15 and 22.45 GHz. They also showed that the absorption line is broadened as the water vapor density is increased.

About this time or shortly after, two books were published in this field. The one by Herzberg [4] seems to be directed primarily toward an understanding of the structure of polyatomic molecules, but it covers rotational spectra in the infrared and microwave regions. The other, by Townes and Schawlow [5], is clearly and primarily directed to the application of microwave techniques. It was published in 1955 and discussed the use of the newly developed electronic techniques to obtain more precise measurements on atomic and molecular spectra in the microwave region, which could, in turn, give precise information on the structure and behavior of the absorbing substances. Their book is therefore particularly concerned with gases, especially at low pressures where the spectral lines are narrow and the resonant frequencies precisely defined. In addition to other data, they give the microwave absorption frequency of H_2O vapor as 22,235 MHz. They include much data, both theoretical and experimental on shapes and widths of spectral lines, which particularly concern "pressure broadening."

From the theoretical point of view, pressure broadening arises from closer proximity of neighboring molecules, with shorter times between collisions. More generally the broadening is due to increased interaction between molecules and what may be called cooperative phenomena. Clearly in a condensed state—in liquids and amorphous solids—these effects can be expected to be still stronger, but neither Herzberg nor Townes and Schawlow give any information on this, and there appears to have been little published experimental work in this area up to the mid-1950s.

Townes and Schawlow also discuss the effect of centrifugal distortion in producing frequency shift, which can be significant for asymmetric rotors, and they point out that, in the gaseous state, absorption intensities are inversely proportional to $T^{-2.5}$ due to the Boltzmann distribution, where T is the temperature on the Kelvin scale.

They provide essentially no information on millimeter-wave transitions, since suitable millimeter-wave electronic techniques were not fully developed by 1955.

20.2.   PRINCIPLES OF APPLICATION

There were only two issued U.S. patents in this field cited by the U.S. Patent Office in reference to my first patent application on moisture measurement [6] filed in January 1958. One was to Wild et al. [7], which was filed in 1945 and issued in 1956, and the other was to Breazeale [8], issued in 1953. In response to the Wild patent application, the Patent Office stated that the basic concept of using microwaves for moisture measurement was covered by spectroscopy patents that had been issued in the previous century.

To successfully apply microwaves for measurement of percent water in industrial materials, important factors must be considered, factors which may be particular to the material being measured and may change with both the chemical and physical conditions of the industrial operation in which it is involved. In addition to the general factors of temperature dependence and the form in which the material is available for measurement, there is the effect of bonding between the water and the material, either by physical forces (e.g., surface tension) or by chemical combination; there is also the possible presence of other molecules exhibiting microwave absorption in the same frequency range (e.g., ethyl alcohol) and the effect of direct conductivity by metal atoms or graphite (carbon atoms) or by ionic conductivity. The effect of density of the material being measured must also be considered, because microwave absorption by water is directly proportional to the mass of free water molecules in the *volume* that is being sensed, whereas percent water (moisture content) is proportional to the mass of water in the *mass* of material that is being sensed. Yet another physical factor which can be important is the effect of particle or lump size; if the material being measured, such as coal or wood chips for example, contains pieces comparable in size to a resonant half-wavelength at the microwave frequency used for the measurement, there will be a strong effect on the microwave signal. More generally, particle size directly affects surface area and so directly affects surface bonding. In this connection, porosity is also important, particularly for large lumps, because it can greatly increase the surface area available for bonding. This effect is encountered, for example, with some lower-grade coals.

Yet another factor to be considered is the range of moisture to be encountered or, more specifically, the range of total free water to be measured. This can be as little as a few parts per million, as in some oil products, to as much as 75% or more, wet basis, as in wood chips or bark, for example. Furthermore, the measurement may be on a thin sheet, as on paper-making machines, or may be of material on a conveyor belt or across the base of a storage silo or a chute where the microwaves must sense through many inches of material.

The low end of this whole range clearly requires the maximum sensitivity and calls for use of the peak absorption frequency. But for the high end with high total water content, the microwave signal at the peak frequency would be so heavily attenuated that accurate measurement would not be economically viable. The choice of the best operating microwave frequency is therefore of great importance.

All these factors will now be considered individually and in detail, starting with the choice of frequency.

20.2.1 Frequency

In work on materials involving high water content, measurements were made at two frequencies. For these measurements, two instruments were used which had been developed in England at about that time by Associated Electrical Industries Ltd. called their S-band meter, using 2.45 GHz, and their X-band meter, using 10.7 GHz.[1] These measurements confirmed the further fall-off in sensitivity as the frequency is lowered, and they followed the expected curve of a broad relaxation or absorption peak.

For measurement of concrete sand, in which the moisture may be as high as 15% or 20%, and the measurement is required at the base of the sand storage hopper which is about 16 inches (40 cm) across, a frequency of 2.45 GHz has proved to be effective, and in iron foundries, where the molding sand rarely exceeds 4% and measurement is to be made on a conveyor belt with only about 3- or 4-inch (8–10 cm) sand depth, a frequency of 10.7 GHz has given good results.

For measurement of wood chips, where the moisture content can be as high as 75% wet basis (in the northern rain forests of British Columbia) and measurement is required on conveyor belts with belt loading of several inches, a low frequency is mandated, but this can then run into the problem of some chips acting like resonant half-wavelengths ($\lambda/2$) antennas. (It is the wavelength of the microwaves in the chips with their dielectric constant which is pertinent, not the wavelength in air.) Investigation of this problem has shown that the attenuation-versus-particle-size relationship at any given microwave frequency follows the form in Figure 20.1. For small particles (e.g., sawdust) there is no change in sensitivity with particle size up to some point beyond which, due to random variation, there are a small number of $\lambda/2$ particles. Again, for larger particles there is an upper plateau of constant sensitivity which results from a saturation condition where there is a high density of $\lambda/2$ particles. Terman [10] shows that in resonant antenna arrays the transmitted or received signal intensity increases with the number of antennas per unit length of the array (in other words, with the antenna density) but that the increase per added antenna falls off toward a sensitivity plateau with no further increase per added antenna beyond this. The antenna spacing at which this plateau becomes effective is directly related to the wavelength with which the antennas resonate. Correspondingly, the start of the upper plateau in Figure 20.1 is a function of the microwave frequency and of the mean size of the wood chips being measured and, therefore, of the number of resonant $\lambda/2$ chips per unit volume. Between the upper and lower plateaux there is a range of mean particle sizes for which the sensitivity is uncertain and measurement is therefore inaccurate, but, since the start of the

---

1. Associated Electrical Industries Ltd. engineers told me during a visit to their plant in England that the S-band meter had been developed initially in the 1950s to measure moisture in brick walls, caused by ground water seeping up—by capillary and chemical action—along the mortar between the bricks, a phenomenon popularly called "rising damp."

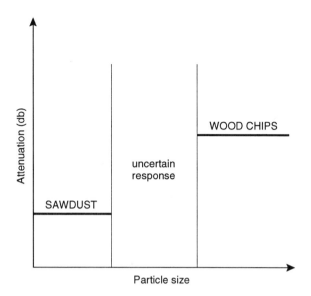

FIGURE 20.1. Effect of particle size on attenuation of microwaves in wood chips, using a fixed microwave frequency.

upper plateau is a function of frequency, the location of this uncertain range is also a function of frequency. Thus, for any given mean particle size there is a "forbidden" frequency range within which measurement is likely to be inaccurate. The result for wood chip measurement is that a microwave frequency of 2.45 GHz or higher provides accurate measurement for chips made in a $\frac{7}{8}$-inch (2.2 cm) chipper but not for $\frac{5}{8}$-inch (1.6 cm) chips, for which the frequency must be increased to at least 3.4 GHz. Likewise, an early attempt to measure wood chips at Potlatch with microwaves using a frequency of about 1 GHz proved to be inaccurate.

With wood chips there is a further problem due to the fact that material transported on a conveyor belt can have a variable cross section, being sometimes like the shape of a convex lens and sometimes more like the shape of a prism, with corresponding effects on the microwave signal. With most granular materials, this can be controlled or at least held constant with a plow but wood chips do not slide over one another and any form of plow proves disastrous. A driven roller can be used to make a flat surface, but this only packs the chips to increase their density where the surface was higher and, as seen by the microwaves, there is little or no change. There appear to be only two possible solutions to this problem with wood chips: the one is for the mill personnel to maintain a close watch on the chip profile and to maintain or adjust the flow through transfer points so as to maintain a constant profile past the moisture sensor; the other is to use a different type of microwave sensor.

For measurement at extremely low moisture content, in the parts per million range, the much higher frequency millimeter-wave peak in the water rotational spectrum, located at 183 GHz, appears interesting, since theory indicates higher sensitivity at higher-frequency peaks. Furthermore, at this frequency the wavelength in oil, for example, will be about a millimeter, making it possible to send a well-collimated "pencil" beam down a long tube or pipe containing the material to be measured. In this way, measurement can be made through several feet of the material and the sensitivity could be increased several-thousand-fold over what can be obtained with the 22-GHz peak.

20.2.2 Temperature

For most industrial materials more than one factor causes the microwave response to change with temperature and these do not all act in the same direction. As stated, Townes and Schawlow [5] show a theoretical decrease of attenuation with increasing temperature due to the basic quantum absorption, whereas both surface bonds and chemical bonds are likely to be broken as the Boltzmann energy increases with temperature, leading to increased attenuation as the temperature rises. Temperature effects arising from ionic conductivity, if present, are dealt with in the following section on direct conductivity.

The decreased attenuation in gases, following a 2.5 times power of the absolute temperature would, for example, mean a decrease in attenuation by 0.36 dB from 4.47 to 4.11 dB due to a 10°C rise in temperature from 20 to 30°C. This is exactly what was found in measurements (unpublished) on silica sand containing 8.2% water, as shown in Figure 20.2. Other measurements made on the same grade of silica sand showed almost no bonding.

Kaye and Laby [11] give the Boltzmann energy as a little under 0.04 eV/molecule at normal room temperature, rising to near 0.05 eV/molecule at the boiling point of water. With bond energies around 0.04 eV/molecule, as calculated from

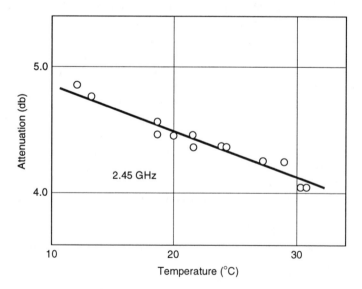

FIGURE 20.2. Effect of temperature on attenuation of microwaves by 8.2% water in silica sand.

data given by Kay and Laby for surface tension of water in air, an increase of microwave attenuation with increased temperature will be expected, due to bonds having been broken by the increased temperature. I do not know of any hard data confirming this effect of temperature, but in experiments on pieces of moist lumber, where bonding is significant, the effects of temperature on attenuation at 10.7 GHz and at 2.45 GHz are plotted in Figure 20.3.

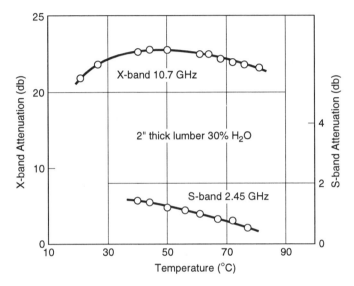

FIGURE 20.3. Effect of temperature on attenuation of microwaves in lumber.

20.2.3 Bonding

Surface bonding is probably the most widely encountered complicating factor, because there are relatively few solid substances on which water does not "wet" the surface—i.e., substances on which the water surface tension is negative. Since the energy of surface tension of water in air is about 0.04 eV/molecule, as calculated from data listed by Kaye and Laby [11], a water molecule held by surface tension forces will not be as readily rotated by a microwave field. Therefore, it will not interact so strongly with the microwaves, and will be more difficult to measure.

The same is true of water which is hydrogen bonded to cellulose or starch molecules, as in wood products and grains, for which the bond energy is of the same order.

Experimentally I have found (unpublished) that in all cases of simple surface bonding, the graph of microwave absorption follows the form shown for coal in Figure 20.4, in which the slope of microwave attenuation per percent water has exactly half its full value up to some breakpoint beyond which the full slope is effective; the change of breakpoint with particle size, as shown in Figure 20.4, is due to the change of surface area per unit mass of the material being measured.

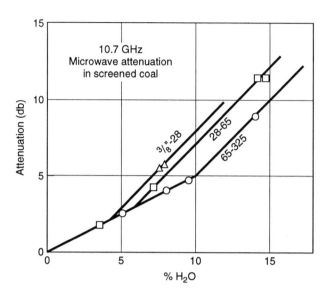

Figure 20.4. Attenuation of microwaves by water in coal, showing effect of particle size.

In the case of hydrogen bonding to cellulose, my experiments (unpublished) showed an exact square-law relationship between microwave attenuation and moisture content, dry basis, up to the fiber saturation point ($33\frac{1}{3}\%$ dry basis, 25% wet basis) which corresponds to two water molecules per cellulose ring ($C_6H_{10}O_5$). Beyond this point the relationship is linear, corresponding to full attenuation. More recently I have confirmed the square-law relationship in experiments on oats, barley, and rye, as shown in Figure 20.5, in which the solid-line curve follows the square-law according to the formula on the graph. It would seem that this square-law dependence may arise from the fact that the hydrogen bonding is a reversible chemical reaction between the water and the cellulose or starch molecules and that two water molecules react with each cellulose ring so that

$$[H_2O]^2 \text{ [cellulose ring]} = K[H_2O\text{–cellulose–}H_2O]$$

where the brackets stand for concentration.

20.2.4 Direct Conductivity

If the measured material contains any uncombined metal atoms or carbon atoms, these will provide direct conductivity and will attenuate the microwave signal through direct ohmic resistivity even if present in only a few parts per million, because the microwaves act on an essentially atomic scale.

Uncombined metal atoms, other than noble metals (gold, platinum, etc.), are seldom present, except when produced by chemical action in an industrial process, and this is true of most ores coming directly from a mine.

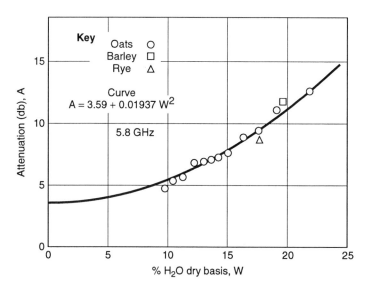

FIGURE 20.5. Attenuation of microwaves by moisture in grains, showing square-law relationship.

Free carbon atoms are found in some coal but generally are not present. In foundry molding sand, however, to which is added a few percent powdered coal, contact with hot iron (or other molten metal) results in a coking action which sets free carbon atoms, and the sand which is returned for reuse contains an appreciable amount of atomic carbon.

Measurement of ground water, which may be desired for agriculture or for orchards and perhaps in connection with irrigation, will be affected by any ionic conductivity. This can be due to chemical fertilizers or can come from pollution such as from salt sprayed on highways for ice and snow removal.

In all these cases the effect of direct conductivity on the microwave attenuation can be eliminated and/or measured by making the measurement simultaneously at two microwave frequencies, located on the rising slope of the water absorption peak [12]. Attenuation due to direct conductivity is not frequency dependent, as seen from Maxwell's basic electromagnetic equations [13]. By taking the difference between the attenuations at the two frequencies, the contributions from direct conductivity are eliminated, leaving only the difference between the attenuations due to water molecule rotation, which is stronger at the higher frequency, closer to the absorption peak.

If a measure of the direct conductivity is desired—either for gold assay, pollution control, or some other reason—it can be obtained by subtracting the contribution due to the now-determined water content from the total attenuation of the signal at one of the two frequencies. Where ionic conductivity is involved, temperature effects will need to be considered when it is the ionic content that is required to be measured, as for pollution control. The measurement of moisture is not affected, because the whole ionic conductivity effect is canceled between the two microwave frequencies. Temperature, however, does affect the interpretation of ionic conductivity in terms of ionic content because the ionic mobility increases with temperature, thereby increasing the ionic conductivity per ion. The effect of temperature on ionic conductivity is illustrated in Figure 20.6, which shows the added attenuation due to sodium and chloride ions in water over the temperature range from 10 to 70°C at salt concentrations of 0.5%, 0.625%, and 0.75%. The attenuations due to the water were separately measured on pure water and subtracted from the total attenuations due to salt and water over this same temperature range.

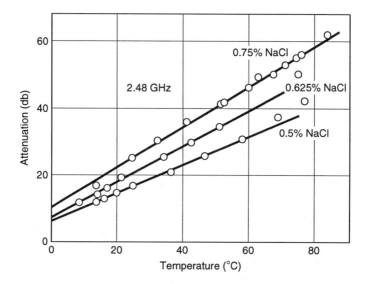

FIGURE 20.6. Attenuation of microwaves by ionic conductivity due to sodium chloride in water, showing increase with temperature due to increased ionic mobility.

20.2.5 Density

The quantity required in conjunction with the measure of microwave absorption by water is seldom the true density of the material; it is, as pointed out, the mass of material in the volume sensed by the microwaves; this is the density as measured and involves the packing fraction as well as the true density of the solid material. For most dry granular materials the packing fraction is in the region of 0.5 but can be much less, particularly under moist conditions.

The measurement of what may be termed "effective density" can be made by a gamma density gauge [14]. This has the advantage that both sensors use electromagnetic radiation and therefore operate under similar conditions, but their frequencies are very different, which means that their collimation is very different so that they do not sense identically the same volume of material. Another problem is that their operating time constants may differ, the gamma gauge being dependent on the statistics of its radioactive gamma ray source. Under steady-state conditions, this does not matter, but their responses to changes in the measured material are time-dependent and can temporarily produce wide fluctuations.

The readings of a weightometer, either strain gauge or other, can be interpreted as effective density, but only if the volume weighed is known and remains precisely constant, and this can only be relied on under controlled laboratory conditions and, for a manually operated instrument, requires use by trained laboratory personnel.

The ideal solution appears to be to use the same microwave signal for measuring the effective density as is used for measuring the water content. This can be done by sensing the change in phase of the signal in passing through the material, simultaneously with sensing the attenuation [12]. The reason for this is that the phase change is a function of the dielectric constant of the material through which the signal passes, and this in turn is a function of the effective density. The only uncertainty is that there is no completely satisfactory theoretical relation which has so far been developed in solid-state physics, which gives the dielectric constant of a mixture in terms of the known constants of its constituents. There are a number of empirical relations, and one which has on many cases proved useful for granular and other materials that are about half air is

$$\log \epsilon_m = \sum f_i \log \epsilon_i,$$

where
ϵ_m = dielectric constant of the mixture
ϵ_i = dielectric constant of constituent i
f_i = volume fraction of constituent i.

20.3. CONCLUSIONS

The primary conclusion from this work is that microwaves provide an accurate and versatile means for measuring the moisture content of solid and liquid materials. For these measurements, the microwave frequencies used must be those which interact with the rotational energy of the water molecule. This includes all frequencies which lie within the resonant peaks. There are, however, two essential requirements:

1. The material to be measured must be analyzed chemically and physically to determine its interactions with water and its interactions with microwaves due to constituents other than water. This analysis must include minor constituents, some of which may not always be present.

2. The available and possible physical conditions of measurement must be known; in some cases, a choice must be made, e.g., conveyor belt, storage silo, transfer chute, auger, etc. The range of moisture content to be measured and, where appropriate, the minimum and maximum depth of material to be sensed must all be agreed in writing, together with any other pertinent facts such as the possible ranges of known interfering substances.

REFERENCES

[1] L. I. Schiff, *Quantum Mechanics*, New York: McGraw-Hill, 1955, ch. 11, pp. 298–299.

[2] H. M. Randall, D. M. Dennison, N. Ginsberg, and L. R. Webber, "Far infrared spectrum of water vapor," *Phys. Rev.*, vol. 52, pp. 160 et seq. 1937.

[3] G. E. Becker and S. H. Autler, "Water vapor absorption of electromagnetic radiation in the centimeter wave-length range," *Phys. Rev.*, vol. 70, pp. 300–307, Sept. 1946.

[4] G. Herzberg, *Molecular Spectra and Molecular Structure*, vol. II, New York: Van Nostrand, 1945.

[5] C. H. Townes and A. L. Schawlow, *Microwave Spectroscopy*, New York: McGraw-Hill, 1955.

[6] C. W. E. Walker, "Apparatus and method for measurement of moisture content," U.S. Patent 3,079,551 filed Jan. 1958, issued Feb. 1963.

[7] R. Wild et al., U.S. Patent 2,729,786 filed 1945, issued Jan. 1956.

[8] W. Breazeale, U.S. Patent 2,659,860 issued Nov. 1953.

[9] C. W. E. Walker, "Microwave moisture measuring instrument," presented at the ISA Analysis Instrumentation Symposium, Houston, TX, April 29–May 1, 1963.

[10] F. E. Terman, *Radio Engineering*, New York: McGraw-Hill, 1937, ch. 15, sect. 132, pp. 663–665, fig. 383.

[11] G. W. C. Kaye and T. H. Laby, *Tables of Physical and Chemical Constants*, 11th ed., Longmans Green, 1956, pp. 39–41.

[12] C. W. E. Walker, "Microwave moisture measurement using two microwave signals of different frequency and phase shift determination," U.S. Patent 4,727,311 filed Mar. 1986, issued Feb. 1988.

[13] J. D. Jackson, *Classical Electrodynamics*, New York: Wiley, 1962, ch. 6, eqs. (6.28), p. 178.

[14] C. W. E. Walker, "Apparatus for measuring percent moisture content of particulate material using microwaves and penetrating radiation," U.S. Patent 3,693,079 filed Apr. 1970, issued Sept. 1972.

21

K. Kupfer
School of Architecture and Building,
Weimar, Germany

Possibilities and Limitations of Density-Independent Moisture Measurement with Microwaves

Abstract. One of the most important characteristic values of aggregates for the production of concrete is moisture. Its determination within a wide range with small measuring errors is of great importance for the proportioning of water if a definite mix formula has to be strictly kept to. The requirements for a rapid and contactless measurement not influenced by the salt content of material as well as for a high degree of accuracy are met by the microwave moisture measuring method. For high accuracy it is necesary to know and to reduce influences of measuring errors caused by density differences, grain size, salt content, and thermal effects.

21.1. INTRODUCTION

Disturbances due to changes of bulk density are crucial factors in dielectric moisture measurements. The resulting measuring errors are bigger than those caused by influence of salt content at frequencies below 1 GHz and systematic errors of the measuring device. An increase in bulk density can simulate higher moisture in the material. The bulk density depends on the grain size distribution, the grain forms, and the capillary water. The relative thickness of the layer of adhesion water in the material and the water content increase with decreasing grain size. Without compression the bulk cargo has a stochastic distribution.

Using microwaves to measure moisture, the grain sizes of the material influence absorption and scattering of the electromagnetic field. Both attenuation and phase shift increase with higher density and larger grain size [1]. In order to eliminate density interferences or to keep them constant, many measuring systems have material stream-forming systems or the density is determined with a γ-radiation method [2]. When measuring a loose bulk cargo, the grain size of the material has a very high influence on the measurement accuracy. Any additional compression results in a homogenization of the mixture and its grain sizes, grain forms, and grain distributions. The influence of grain sizes of an aggregate mixture increases with higher measuring frequency. In the microwave region the particle diameter of a coarser mixture has near-wavelength dimensions. Transmitting electromagnetic waves through a water-solid mixture results in scattering at inhomogeneities. The amount of scattering increases with size and number of inhomogeneities and measurement frequency. In order to achieve high accuracy with the present measuring method, the influence of bulk density and layer thickness has to be compensated, and the systematic errors due to material grain size, salt content, and temperature should be minimized [3,4].

21.2. A MODEL FOR A DENSITY-INDEPENDENT MOISTURE MEASUREMENT

According to the literature [5–10] a density compensation is possible for some materials within acceptable given limits, for instance, by forming the quotients of moisture and density-dependent values of attenuation and phase shift. The complexity of water binding values, however, does not permit these conditions to be generalized. The range of validity has to be checked anew for a density-independent measurement in each case and thus also in the case of concrete aggregates.

The real part ϵ_r' and imaginary part ϵ_r'' of the complex permittivity are intrinsic parameters which can be measured by the attenuation A and phase shift Φ of an electromagnetic wave passing through the dielectric. Due to the proportionality of permittivity to moisture content of a solid, it is possible to relate attenuation and phase shift or their quotient to the moisture value (percentage dry weight). An increase of bulk density and layer thickness at constant moisture causes greater attenuation and phase shift (Figure 21.1a,b).

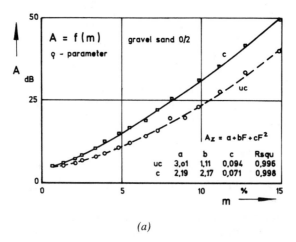

(a)

FIGURE 21.1. Influence of moisture-dependent bulk-density-differences on the calibration curves of gravel sand (c-compressed; uc-uncompressed) on (a) attenuation, (b) phase shift, (c) quotient.

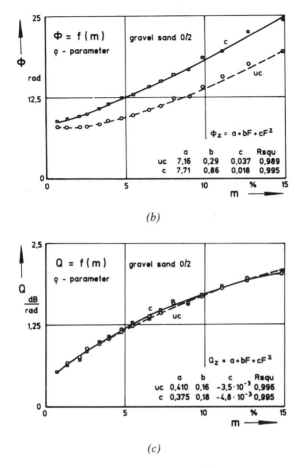

FIGURE 21.1. *continued*

The calibration curves of the quotient $Q = A/\Phi$ do not show these changes of material density (Figure 21.1c). By plotting the attenuation of the uncompressed (uc) and compressed (c) material versus phase shift, linearity is demonstrated.

Constructing straight lines through equal points of moisture on both calibrating curves (uc) and (c), one can read the gradient angle ψ of these lines (Figure 21.2). Note that tan ψ is independent of bulk density and layer thickness, but depends on the moisture of the material. It is defined by the quotient of attenuation and phase shift:

$$\tan \psi = A(\text{uc})/\Phi(\text{uc}) = A(\text{c})/\Phi(\text{c}). \tag{21-1}$$

Equation (21-1) shows that the quotients of attenuation and phase shift for both uncompressed and compressed material are the same. Figure 21.2 shows the A-Φ diagram for gravel sand 0/2; the corresponding calibration curves are plotted in Figure 21.1. In this case it was possible to carry out density-independent measurements in the moisture range 1–15% with a moisture error $\leq \pm 0.5\%$. In the

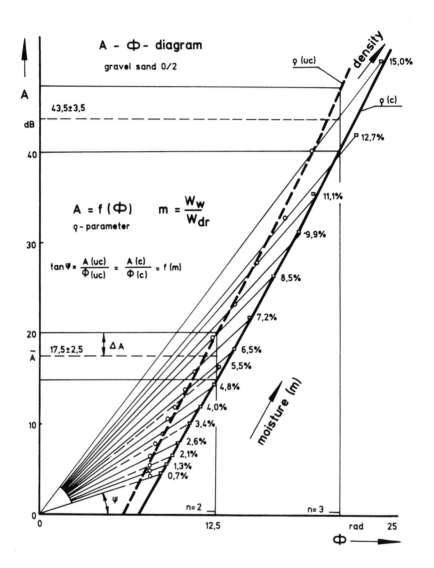

FIGURE 21.2. Moisture-dependent calibration curves with different densities in A-Φ diagram (the angle ψ shows independence of density).

moisture measurement process construction of calibration curves for specific materials is time consuming. In order to realize highly reproducible measurements for specific materials, it is necessary to store the values of attenuation A or phase shift Φ as a function of moisture, to compare the measurements and to calculate the mean resulting calibration curve. The scattering range of measurement values is decisive for the confidence interval of the calibration curve and defines the moisture measurement accuracy [3,4].

From the A-Φ diagram a mathematical model of calibration should be derived to avoid these time-consuming measurements. The A-Φ diagram is shown as a function of increasing bulk density in Figure 21.3a,b. All values of density lie on

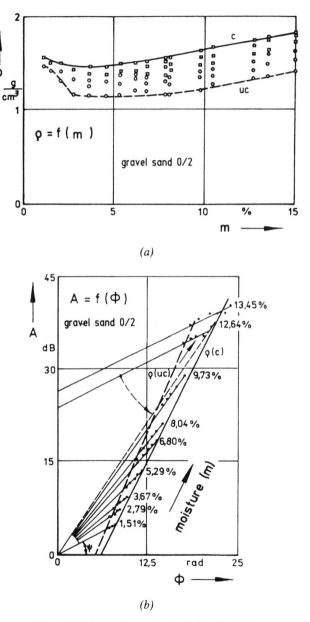

(a)

(b)

FIGURE 21.3. (a) Moisture-dependent bulk-density-differences as a function of moisture content, (b) Variation of phasor or tan ψ in A-Φ diagram as a function of density and moisture.

a phasor between the limiting curves ρ(uc) and ρ(c) (Figure 21.3b). The angle ψ of the phasor is independent of density and thickness and is only a function of moisture.

Randomly scattered measurement values are intensified in the moisture saturation range. That is the reason why the calibration curves of the quotient for the uncompressed and compressed materials in the most cases are divergent (Figure 21.1c). The density-independent measurement is incorrect in this range. The sand grains are surrounded by a water film and, under the influence of gravity and at low density ρ(uc), this water diffuses through the material to the base of the sample container. In repeated measurements the probability of getting reproducible values is low. At very high moisture in sand or for coarse grains, the phasor is displaced along the attenuation coordinate (Figure 21.3b). The possibility for a density-independent moisture measurement at high moisture is drawn in this figure with the phasor $M = 12.64\%$. Plotting all values of density the phasor is displaced to higher A values. Taking only the highest-density values into account, it is possible to get the phasor through the origin of the coordinate system. That is why a precompression must be carried out for materials with high moisture content to enable a density-independent moisture measurement in the saturation range. For reasons of reproductibility a calibration curve must be constructed, serving as a basic curve for further measurements at a medium or high compaction with higher accuracy than found for uncompressed material.

21.3. INTERFERENCES OF GRAIN SIZE ON A DENSITY-INDEPENDENT MEASUREMENT

The influence of grain size for gravel sand 0/2, 0/4, and gravel 0/8 on the gradient of the calibration curves is shown in Figure 21.4. The values of attenuation,

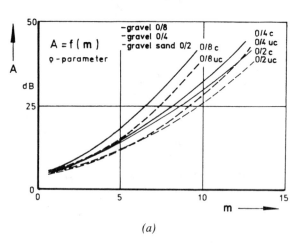

(a)

FIGURE 21.4. Influence of moisture-dependent bulk-density differences with various grain size distributions on (a) attenuation, (b) phase shift, (c) quotient.

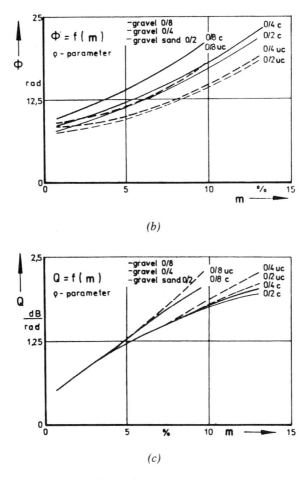

(b)

(c)

FIGURE 21.4. *continued*

phase shift, and their quotient increase with coarser grain. While the high gradient of the attenuation curve is caused by scattering of microwaves, the small change of phase shift results from the surface influence of the test material [10,11]. In Figure 21.4 it is shown that the moisture range for a density-independent moisture measurement decreases as grain size increases. Because of a smaller surface area the saturation range of a coarser-grain material is fixed at a lower moisture content. In this range the calibration curves of the quotient are divergent (Figure 21.4c). The changes of attenuation are bigger than those of the phase shift, and the gradient of the calibration curve is increased. Figure 21.4c shows that the moisture range for a density-independent measurement is reduced with increasing grain diameters, and the measurement errors are increased as shown by the comparison of calibration curves Q(uc) and Q(c). A coarse material with random grain shapes, sizes, and distributions causes reflection and scattering of the electromagnetic field. If moisture measurement of aggregates is utilized for concrete production, these values influence measurement accuracy with a high proportion of unknown systematic and random errors. Scattering of microwaves and influence of the specific surface together cause measurement errors at X-band. Compensation is achieved for coarse-grain materials with high fine-grain portions by homogenizing the material. Summation of error is possible in the moisture saturation range.

The A-Φ diagram with a successive compression from gravel 0/8 is the same as shown in Figure 21.3. It can be seen that the possibility exists for a density-independent moisture measurement at a coarser-grain size and a high moisture. In this case also a precompression has to be carried out, to enable a density-independent measurement in the moisture saturation range.

For gravel sand 0/2 it was possible to perform density-independent measurement in the moisture range from 1–10, 6% with errors $\leq \pm 0.25\%$ moisture and for gravel sand 0/8 in the moisture range from 1.5–9.2% (Figure 21.5).

Test materials with coarse-grain size are homogenized by precompression, and the influence of microwave scattering is reduced (Figure 21.5b). The A-Φ diagram verifies density-independent moisture measurement. In addition, it is possible to extend the application of a density-independent measurement to the moisture saturation range and to increase the accuracy of the measurement.

The influence of scatterings on microwave measurements has been shown before [11–15].

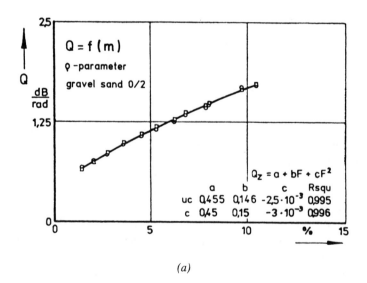

(a)

FIGURE 21.5. Extension of moisture-range by precompression: (a) gravel sand 0/2, (b) gravel 0/8.

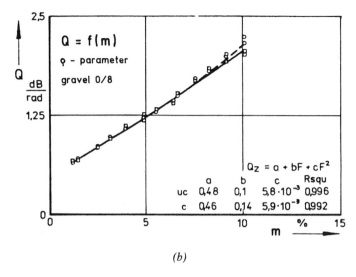

(b)

FIGURE 21.5. *continued*

21.4. INFLUENCE OF ION CONDUCTIVITY
ON A DENSITY-INDEPENDENT MOISTURE
MEASUREMENT

In aggregates, salts are found in solution. This salt content is variable. Further-more, it is possible that the salt content results from sewage in which salts are concentrated by reprocessing. Analyses of water showed that most dissolved salts were sodium chloride. In the microwave region polarization losses dominate com-pared to conductivity losses in influencing attenuation [1]. Figure 21.6 presents the influence of salt content on the measurement of attenuation, phase shift, and their quotient during a density-independent moisture measurement. By measure-ments in X-band of a compressed gravel sand 0/2, it was shown that a maximum allowable salt content of 0.04% does not influence attenuation, phase shift, and their quotient. The calibration curves with and without salt content are identical. If the salt content exceeds an allowable limit of 0.04%, it represents a noncom-pensating disturbance. A significant difference is seen between calibration curves (Figure 21.6a) with a salt content of 0.04% and 2%. Dielectric losses caused by salt content strongly influence the attenuation but affect the phase shift less. Be-cause of high attenuation the calibration curve of the quotient is displaced to higher absolute values. In this case a material-specific calibration curve has to be used. It is also possible to measure the conductivity by some other method and to integrate the systematic error into the calculation. Salt content in higher concentrations in aggregates shows the same effects as a rise in temperature. In both cases, the relaxation time becomes smaller and the maximum of ϵ_r'' is displaced to higher fre-quencies. The phasor in the A-Φ diagram is shifted to higher A-values. Because of the small influence on phase shift (Figure 21.6b), a phase-measuring method is applicable for those materials. It is necessary to fix the density at constant levels.

For calcium silicate brick production purposes moisture and ion conductivity were analyzed at 3 GHz [16]. The influence of ion conductivity on a microwave moisture measurement has been published in [17–22].

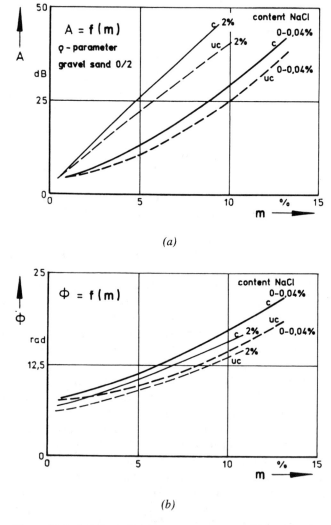

(a)

(b)

FIGURE 21.6. Influence of salt content of a fine-grained gravel sand on the calibration curves on (a) attenuation, (b) phase shift, (c) quotient at a middle compression (mc).

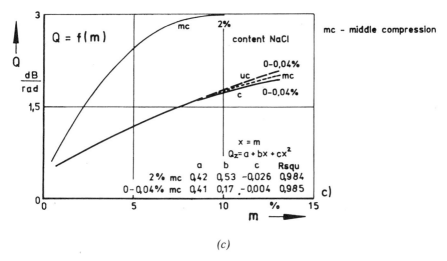

(c)

FIGURE 21.6. *continued*

21.5. THERMAL EFFECTS

Temperature as a systematic error influences both the accuracy of the measuring system and the measurement of aggregates. Depending on frequency, moist materials have different temperature coefficients. In a certain temperature and moisture range, it is possible to minimize the temperature dependence by using a suitable measuring frequency [18,23]. By measuring the moisture, the temperature variation is specific for each material mixture. For compensation it is necessary to know the temperature coefficient.

Temperature investigations of gravel sand served to specify the dependence of attenuation and phase shift on the temperature of the material. An increase of temperature displaces the maximum of ϵ_r'' to higher frequencies. The result of that fact is that the imaginary part of the complex permittivity and the proportional value of attenuation have a negative temperature coefficient for frequencies of measurement below the relaxation frequency. The temperature coefficient (TC) of the quotient results as the sum of the TCs of attenuation and phase shift. To find the influence of water on the bulk material, the attenuation and phase-shift values were compared with a 1-cm water layer.

The measured curves of attenuation, phase shift and their quotient for gravel sand 0/2 as a function of temperature with moisture content as a parameter are shown in Figure 21.7.

The negative TC of attenuation significantly influences the result of quotient A/Φ as the phase-shift curves are relatively constant in the temperature interval. With increasing moisture of the material, the attenuation and phase-shift curves approach those of water. For a relatively dry bulk material (0.6% moisture by gravel sand 0/2) the temperature influence is very small and the temperature coefficient is negligible. The following example shows that compensation for temperature influence is necessary. For gravel sand 0/2 with moisture 5.8%, temperature increases from 17.5 to 26.3°C ($\Delta T = 8.8$ K); the quotient changes by $\Delta Q = 0.21$ dB/rad and the resulting moisture error is $\Delta M = 1.6\%$.

The value of the phase shift is relatively independent of salt content and temperature variation [18,19]. In those cases there is an opportunity to apply a phase moisture measuring method.

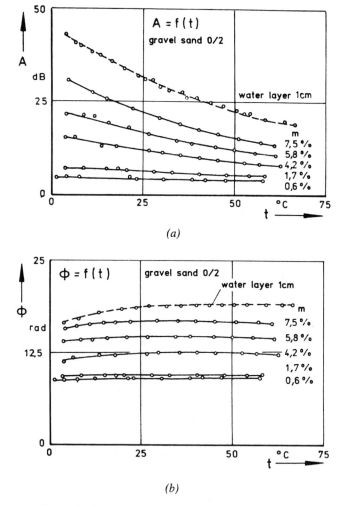

(a)

(b)

FIGURE 21.7. Influence of temperature dependence with increasing moisture on (a) attenuation, (b) phase shift, (c) quotient.

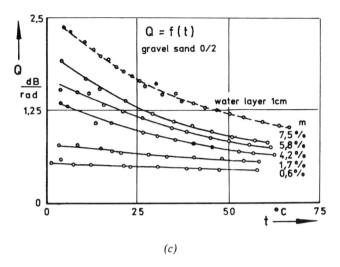

(c)

Figure 21.7. *continued*

Influences of temperature on the relaxation time of water during microwave measurements were given in [24–28].

21.6. CONCLUSION

Bulk density differences in concrete aggregates cause errors during microwave moisture measurements. According to the literature a compensation is given for various materials with different functions of attenuation and phase shift.

In conformity with practical tests an attenuation-phase-shift diagram could be used as a mathematical model for a theoretically founded proof for independence of the density. With the diagram it was possible to extend the application range and the accuracy of a density-independent measurement. Disturbances such as grain size, ion conductivity, and thermal affect influence attenuation values, in particular, causing measurement errors and limiting density-independent measurement. During moisture measurements of aggregates in the X-band, grain sizes must be smaller than $\lambda_\epsilon/2$ (8 mm). Influence of salt content with 0.04% is negligible. Thermal effects with differences smaller than 10 K have to be corrected during data processing with a linear function. Phase-shift values are relatively independent of these interferences. Therefore there is an opportunity to apply a phase moisture measuring method.

The microwave moisture measuring technique is not applicable to materials used by the construction industry. Moisture measurements of many raw materials, food, and agricultural products may contribute to guarantee their quality and avoid losses in trade and industry. A new generation of moisture measuring units must come up to the specific requirements of their users. The units should permit easy application, be reliable in operation, and be moderate in price as well as require low costs for installation and maintenance [3,29,30].

REFERENCES

[1] A. Kraszewski, "Microwave aquametrie—a review," *J. Microwave Power*, vol. 15, no. 4, pp. 209–220, 1980.

[2] "Experiences with the microwave moisture meter micro-moist," *Mineral Process.*, vol. 30, no. 9, pp. 549–557, 1989.

[3] K. Kupfer, "Determination of moisture in aggregates for the concrete production by means of the microwave measuring technique," Hochschule für Architektur und Bauwesen Weimar, Diss. A (Ph.D.), 1990.

[4] K. Kupfer, "Disturbances by the application of microwave moisture measurements," *Techn. Messen*, vol. 60, S. 19–28, 1993.

[5] A. Kraszewski and S. Kulinski, "An improved microwave method of moisture content measurement and control," *IEEE Trans. Ind. Electr. Contr. Instr.*, vol. IECI-23, no. 4, pp. 364–370, 1976.

[6] W. Meyer and W. Schilz, "A microwave method for density independent determination of the moisture content of solids," *J. Phys. D: Appl. Phys.*, vol. 13, pp. 1823–1830, 1980.

[7] W. Meyer and W. Schilz, "High frequency dielectric data on selected moist materials," *J. Microwave Power*, vol. 17, pp. 67–77, 1982.

[8] S. Nelson, "Observations on the density dependence of dielectric properties of particulate materials," *J. Microwave Power*, vol. 17, pp. 143–152, 1982.

[9] M. Kent, "Complex permittivity of fish meal: a general discussion of temperature, density and moisture dependence," *J. Microwave Power*, vol. 13, pp. 275–281, 1978.

[10] M. Kent and E. Kress-Rogers, "Microwave moisture and density measurements in particulate solids," *Trans. Instr.*, vol. MC 8, pp. 161–168, 1986.

[11] H. Chaloupka, O. Ostwald, and B. Schiek, "Structure independent moisture measurements," *J. Microwave Power*, vol. 15, pp. 221–231, 1980.

[12] J. D. Dyson, "Measurement of near fields of antennas and scatterers," *IEEE Trans. Antennas Propagation*, vol. IAP-21, pp. 446–460, 1973.

[13] J. D. Richmond, "Scattering by a dielectric cylinder of arbitrary cross section shape," *IEEE Trans. Antennas Propagation*, vol. IAP-13, pp. 334–341, 1965.

[14] A. Göller, "Mikrowellen-Feuchtemessung unter Einbeziehung wichtiger Material-parameter-ein verbessertes Konzept der Signalauswertung," Technische Universität Ilmenau, Diss. A, 1992.

[15] S. Nelson, G. Fanslow, and D. Bluhm, "Frequency dependence of the dielectric properties of coal. I, II," *J. Microwave Power*, vol. 15, pp. 277–282, 1980; vol. 16, pp. 319–326, 1980.

[16] K. Kupfer and A. Klein, "Experiments on the suitability of microwave measuring techniques for moisture measurement in calcium silicate brick production," *Miner. Process.*, vol. 33, pp. 213–221, 1992.

[17] M. Najim, Zyoute, and T. Okazzani, "Complex permittivity of Maroccan phosphate as function of moisture content," in *Conf. Proc. EuMC Kopenhagen*, pp. 261–266, 1977.

[18] A. Klein, "Untersuchungen der dielektrischen Eigenschaften feuchter Steinkohle im Hinblick auf die Anwendbarkeit des Mikrowellenfeuchtemeßverfahrens zur Wassergehaltsbestimmung," Universität Aachen, Fakultät für Bergbau u. Hüttenkunde, Diss., 1978.

[19] A. Klein, "Microwave determination of moisture in coal—comparison of attenuation and phase measurement," *J. Microwave Power*, vol. 16, pp. 289–304, 1981.

[20] G. Fuchs, "Mikrowellentechnische Sensoren für die Prozeßmeßtechnik u.—automatisierung unter besonderer Berücksichtigung der Mikrowellen-Feuchtemessung," Technische Hochschule Ilmenau, Diss. B, 1986.

[21] A. Kraszewski, S. Kulinski, and K. Checinski, "Measurement of moisture content in granular ammonium phosphate by a microwave method," *J. Microwave Power*, vol. 9, pp. 361–372, 1974.

[22] J. B. Hasted, T. H. Buchanan, and M. Hagis, "The dielectric properties of water in solutions," *J. Chem. Phys.*, vol. 20, pp. 1452–1465, 1952.

[23] S. Aggarwal and R. Johnston, "The effect of temperature on the accuracy of microwave moisture measurements on sandstone cores," *IEEE Trans. Instr. Measurement*, vol. IM-34, pp. 21–25, 1985.

[24] J. Kalinski, "Einige Probleme der industriellen Feuchtemessung mit Mikrowellen," *TIZ (Fachberichte)*, vol. 103, pp. 145–153, 1979.

[25] H. Zaghloul and H. A. Buckmaster, "The complex permittivity of water at 9,356 GHz from 10 to 40°C," *J. Phys. D: Appl. Phys.*, vol. 18, pp. 2109–2118, 1985.

[26] G. P. De Loor, "Dielectric properties of heterogenous mixtures containing water," *J. Microwave Power*, vol. 3, pp. 67–73, 1968.

[27] J. Kalinski and J. Rakowski, "Some possibilities of technological quality control on the basis of the laboratory measurements of materials," in *Proc. Int. Symp. Metrology for Quality Control*, Tokyo, pp. 31–34, 1984.

[28] J. Mladek and Z. Beran, "Sample geometry temperature and density-factors in the microwave measurement of moisture," *J. Microwave Power*, vol. 15, pp. 243–240, 1980.

[29] A. Kraszewski, "Microwave aquametry—needs and perspectives," *IEEE Trans. Microwave Theory Techniques*, vol. MTT-39, pp. 828–835, 1991.

[30] K. Kupfer, "Moisture determination of solid and liquid materials using microwave measuring methods—Bibliography 1961–1991," *Wiss. Z. Hochschule Arch. Bauwesen-B-Weimar*, vol. 38, S.273–282, 1992.

22

John R. Kendra
Fawwaz T. Ulaby
Kamal Sarabandi
University of Michigan, Ann Arbor, MI

Snow Probe for in Situ Determination of Wetness and Density

Abstract.   The amount of water present in liquid form in a snowpack exercises a strong influence on the radar and radiometric responses of snow. Conventional techniques for measuring the liquid water content m_v suffer from various shortcomings, which include poor accuracy, long analysis time, poor spatial resolution, and/or cumbersome and inconvenient procedures. This report describes the development of a hand-held electromagnetic sensor for quick and easy determination of snow liquid water content and density. A novel design of this probe affords several important advantages over existing similar sensors. Among these are improved spatial resolution and accuracy and reduced sensitivity to interference by objects or media outside the sample volume of the sensor. The sensor actually measures the complex dielectric constant of the snow medium, and the water content and density must be obtained through the use of empirical or semiempirical relations. To test the suitability of existing models and allow the development of new models, the snow probe was tested against the freezing calorimeter and gravimetric density determinations. From these comparisons, valid models were selected or developed. Based on the use of these models, the following specifications were established for the snow probe: (1) liquid water content measurement accuracy $= \pm 0.66\%$ in the wetness range from 0% to 10% by volume, and (2) wet snow density measurement accuracy $= \pm 0.05$ g/cm^3 in the density range from 0.1 to 0.6 g/cm^3.

22.1. INTRODUCTION

In the study of microwave remote sensing of snow, it is necessary to consider the presence of liquid water in the snowpack. The dielectric constant of water is large (e.g., $\epsilon_w = 88 - j9.8$ at 1 GHz [1]) relative to that of ice ($\epsilon_i \approx 3.15 - j0.001$ [2]), and therefore even a very small amount of water will cause a substantial change in the overall dielectric properties of the snow medium, particularly with respect to the imaginary part. These changes will, in turn, influence the radar backscatter and microwave emission responses of the snowpack.

Among instruments available for measuring the volumetric liquid water content of snow, m_v, under field conditions, the freezing calorimeter [3–5] offers the best accuracy ($\approx 1\%$) and has been one of the most widely used in support of quantitative snow-research investigations. In practice, however, the freezing calorimeter technique suffers from a number of drawbacks. First, the time required to perform an individual measurement of m_v is about 30 min. Improving the temporal resolution to a shorter interval would require the use of multiple instruments, thereby increasing the cost and necessary manpower. Second, the technique is rather involved, requiring the use of a freezing agent and the careful execution of several steps. Third, the freezing calorimeter actually measures the mass fraction of liquid water in the snow sample, W, not the volumetric water content m_v. To convert W to m_v, a separate measurement of snow density is required. Fourth, because a relatively large snow sample (about 250 cm^3) is needed to achieve acceptable measurement accuracy, it is difficult to obtain the sample from a thin horizontal layer, thereby rendering the technique impractical for profiling the variation of m_v with depth. Yet, the depth profile of m_v, which can exhibit rapid spatial and temporal variations [6,7], is one of the most important parameters of a snowpack, both in terms of the snowpack hydrology and the effect that m_v has on the microwave emission and scattering behavior of the snow layer.

In experimental investigations of the radar response of snow-covered ground, it is essential to measure the depth profile of m_v with good spatial resolution (2–3 cm) and adequate temporal resolution (a few minutes), particularly during the rapid melting and freezing intervals of the diurnal cycle. There have in recent years appeared a host of instruments [8] which retrieve snow parameters quickly and nondestructively, by measuring the dielectric constant of snow and relating it to the physical parameters. Of these techniques, the most attractive candidate has been the "Snow Fork," a microwave instrument developed in Finland [1]. The strengths of this technique are the simplicity of the equipment and speed of the measurement, high spatial resolution, and the ability to measure both the real and imaginary parts of the dielectric constant of snow, allowing for more powerful algorithms that enable determination of snow wetness and density with a single measurement. In the process of examining the Snow Fork approach, we decided to modify the basic design to improve the sensitivity of the instrument to m_v and reduce the effective sampled volume of the snow medium, thereby improving the spatial resolution of the sensor. Our modified design, which we shall refer to as the "Snow Probe," is described in Section 22.2. The snow probe measures the real and imaginary parts of the relative dielectric constant of the snow medium, from which the liquid water content m_v and the snow density ρ_s are calculated through the use of empirical or semiempirical relations. The degree to which such relations are valid is established through a comparison with direct techniques. Therefore, in the process of developing a snow probe algorithm, it was necessary to perform independent measurements of ρ_s and m_v. Density measurements were performed with a standard tube of known volume, whose weight is measured both empty and full of snow. For a direct technique for measuring m_v we evaluated two candidates: (1) the freezing calorimeter and (2) the dilatometer [9], which measures the change in volume that occurs as a sample melts completely. The dilatometer approach was rejected because of poor measurement accuracy and long measurement time (about 1 hr). The form of the relations which were ultimately established as a result of these comparison studies is described in Section 22.3.

22.2.   SNOW DIELECTRIC PROBE

22.2.1   Snow Probe Measurement System

Figures 22.1 and 22.2 show the snow probe measurement system and a schematic. The sweep oscillator, under computer control, sweeps (in discrete 10-MHz steps) over a relatively large frequency range. This serves to determine, within ± 5 MHz, the frequency at which the detected voltage is a maximum, corresponding to the resonant frequency of the probe. The RF power transmitted through the snow probe is converted to video by the crystal detector, measured by the voltmeter, which in turn sends the voltage values to the computer. The frequency spectrum is generated in real time on the monitor of the computer. In the second pass, a much narrower frequency range is centered around the peak location and swept with a finer step size (≈ 1 MHz). The center frequency and the 3-dB bandwidth

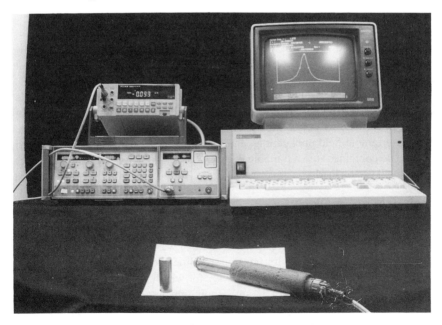

FIGURE 22.1. Snow probe system.

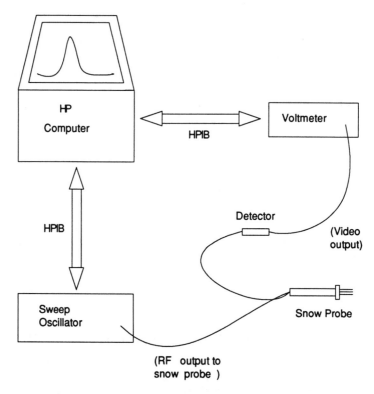

FIGURE 22.2. Schematic of snow probe system.

around it are found, and from these, first the dielectric constant and then the snow parameters m_v and ρ_s are determined according to procedures described in detail in Section 22.3.

As depicted in these figures, the snow probe is connected to a coaxial cable approximately 15 m long. This arrangement is suitable, for example, for cases in which measurements are required in an area fairly local to truck-mounted radars. For more remote field applications, the functions of the hardware shown would need to be combined into a portable unit. The technology for building a compact unit is well established.

22.2.2 Sensor Design

The snow probe is essentially a transmission-type electromagnetic resonator. The resonant structure used in the original design [1] was a twin-pronged fork. This structure behaves as a two-wire transmission line shorted on one end and open on the other. It is resonant at the frequency for which the length of the resonant structure is equivalent to $\lambda/4$ in the surrounding medium. The RF power is fed in and out of the structure using coupling loops.

For our design, we used a coaxial-type resonator, as illustrated in Figure 22.3. The skeleton of the outer conductor is achieved using four prongs. The principle is

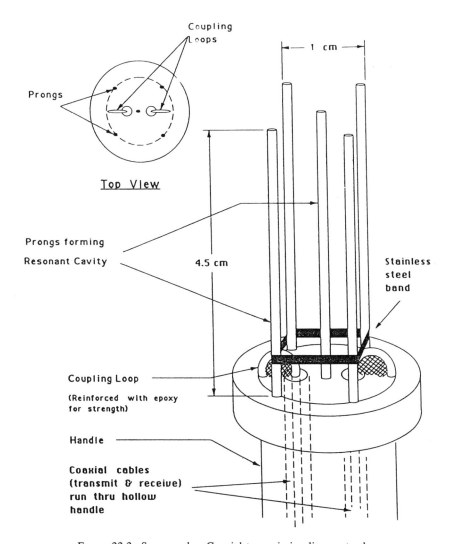

FIGURE 22.3. Snow probe. Coaxial transmission lines extend through handle. At the face of the snow probe, the center conductors of the coaxial lines extend beyond and curl over to form coupling loops.

basically the same: a quarter-wavelength cavity, open on one end, shorted on the other, with power delivered in and out through coupling loops. The coaxial design was chosen for purposes of spatial resolution. Being a shielded design, the electric field is confined to the volume contained within the resonant cavity, as opposed to the original design, which used only two prongs. The coaxial design also had a much higher quality factor ($\approx$110 vs. 40–70 for the original design) which, as discussed below, allows for more accurate determination of the complex dielectric constant. The snow probe is shown in Figure 22.4. The stainless-steel band

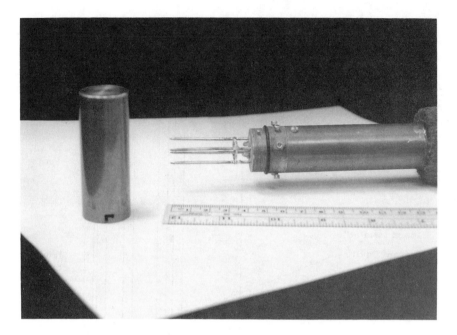

FIGURE 22.4. Snow probe with cap.

encircling the resonant structure near the bottom is necessary to defeat competing two-wire resonances which are otherwise excited between any given pair of the outer prongs.

The real part of the dielectric constant is determined by the resonant frequency of the transmission spectrum or, equivalently, the frequency at which maximum transmission occurs. As mentioned, this corresponds to the frequency for which the wavelength *in the medium* is equal to four times the length of the resonator. If the measured resonant frequency is f_a in air and f_s in snow, then the real part of the dielectric constant is

$$\epsilon_s' = \left(\frac{f_a}{f_s}\right)^2. \tag{22-1}$$

The imaginary part of ϵ_s is determined from the change in Q_m, the measured quality factor of the resonator. The quality factor may be determined by measuring Δf, the half-power bandwidth [10]:

$$Q_m = \frac{f_r}{\Delta f}, \tag{22-2}$$

where f_r is the resonant frequency (f_a or f_s, depending on whether the medium is air or snow). For the snow probe, power losses exist because of radiation, finite conductivity of the conductors, coupling mechanisms (i.e., coupling loops), and dissipation in a lossy dielectric. Thus the measured Q is [10]

$$\frac{1}{Q_m} = \frac{1}{Q_R} + \frac{\epsilon''}{\epsilon'}, \tag{22-3}$$

where Q_m is the measured Q when the probe is inserted in the snow medium, and Q_R is the quality factor describing collectively radiation losses, losses due to the finite conductivity of the conductors, and the power losses due to the external coupling mechanisms for the dielectric-filled snow probe. To calculate ϵ'', from eq. (22-3), one must not only measure Q_m and know ϵ', but the value of Q_R should be known also. As long as $\tan \delta = \epsilon''/\epsilon'$ is very small, which is the case for snow, it is reasonable to assume that Q_R is a function of ϵ' only. This assumption was verified experimentally by measuring Q_R for each of five materials with known dielectric properties (Table 22.1). For each material, Q_m was calculated by eq. (22-2) on the basis of measurements of Δf and f_r and then it was used in (22-3) to compute Q_R. The values of ϵ' and ϵ'' of the test materials given in Table 22.1 were measured with an L-band cavity resonator. This process not only validated the assumption that Q_R is dependent on ϵ' only, but it also produced an expression for computing it:

$$\frac{1}{Q_R} = \frac{(pf_r + b) \times 10^{-3}}{f_r}, \tag{22-4}$$

where p and b are constants and f_r is the resonant frequency associated with the material under test (which is related to ϵ' by $\epsilon' = (f_a/f_r)^2$). For the probe used in this study, $p = 8.381$ and $b = 0.7426$ when f_r is in gigahertz. Combining eqs. (22-2)–(22-4) and specializing the notation to snow (by adding a subscript s to ϵ'' and replacing the subscript r by s in f_r), we obtain the expression

$$\epsilon'' = \left(\frac{f_a}{f_s}\right)^2 \left[\frac{\Delta f_s}{f_s} - (p + \frac{b}{f_s})\right]. \tag{22-5}$$

TABLE 22.1. Summary of Test Material Properties and Measurements

MATERIAL	$\bar{\epsilon}$	f_r (GHz)	Q_m
Air	$1.0 - j0.0$	1.715776	125.2
Sand	$2.779 - j3.7e^{-2}$	1.036	51.7
Sugar	$1.984 - j7.778e^{-3}$	1.22947	89.3
Coffee	$1.497 - j3.32e^{-2}$	1.43125	30.4
Wax	$2.26 - j2.9e^{-4}$	1.150308	137.0

Equations (22-1) and (22-5) constitute the basic relations used for determining ϵ'_s and ϵ''_s from measurements of f_a, the resonant frequency when the probe is in air, and f_s and Δf_s, the resonant frequency and associated 3-dB bandwidth measured when the probe is inserted in the snow sample.

22.2.3   Spatial Resolution/Outside Interference

As mentioned, the partially shielded design of this sensor reduces its sensitivity to permittivity variations outside the sample volume. By sample volume, we refer to the volume inside the cylinder described by the four outside prongs (Figure 22.3). The coaxial design will tend to produce greater field confinement relative to a twin-prong structure.

The effective sample volume was tested in the following way: a cardboard box (30 cm × 30 cm) was filled with sugar to a depth of ≈16 cm. The snow probe was inserted into the sugar at a position in the center of the top surface, and then the dielectric constant was measured. Next, a thin metal plate (square, ≈25 cm on a side) was inserted into the sugar, parallel to and resting against one side of the box. The dielectric constant was remeasured. The metal plate was incrementally moved closer to the sensor position, with dielectric measurements recorded at each sensor-to-plate distance. The results of the experiment are shown in Figure 22.5, in which ϵ'' is plotted as a function of the sensor-to-plate separation.

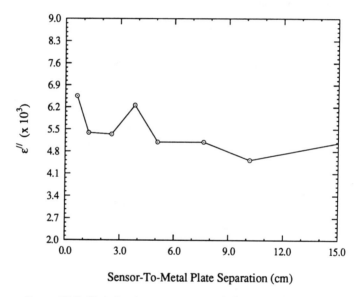

Sensor-To-Metal Plate Separation (cm)

FIGURE 22.5. Variation in measurement of ϵ'' of sugar as a function of sensor proximity to metal plate. (Real part ϵ' stayed in the range 2.00–2.01.)

The plate appears to have a weak influence on the measurement, even at a distance of only 0.6 cm. To put this variation into perspective, had the material been snow, and using the relations given in Section 22.3.1, the fluctuation in the estimate of liquid water would have ranged from $m_v = 0.6\%$ to $m_v = 0.8\%$. The real part of the dielectric constant (not shown in Figure 22.5) stayed within the range 2.00–2.01 during the experiment. The results of this experiment, which essentially confirm the expectation that the electric field is confined to the volume enclosed by the four prongs, translate into a vertical resolution of about 2 cm when the snow probe is inserted into the snowpack horizontally (the snow probe cross section is 1 cm × 1 cm).

There is necessarily a compromise between the ability to make high-spatial-resolution measurements and maximum ruggedness of design. The overall small size of the probe requires the use of small-diameter prongs as well to ensure that the snow volume which is being sampled is not compressed to the point of compromising the measurement. Though we used stainless steel for the prongs to afford maximum strength, it is still possible—for snow samples which are especially dense, coarse, or icy—to have some bending of the prongs occur. For some extreme cases the probe might not be a practical option for a measuring device. In these cases, it may also be the case that the simple relationships (described in Section 22.3) between dielectric constant and snow parameters no longer hold.

22.3. RETRIEVAL OF SNOW DENSITY AND LIQUID WATER CONTENT

The preceding section described the design and operation of the snow probe and the procedure used for measuring ϵ_s' and ϵ_s'' of the snow medium. The next step is to use these measurements to determine the density ρ_s and liquid water content m_v. This is accomplished by using a set of empirical or semiempirical relationships relating the dielectric constant of wet snow to its density and liquid water content. These relationships express the dielectric constant of wet snow ϵ_{ws} in terms of ϵ_{ds}, the dielectric constant of the snow in the absence of liquid water, plus additional terms that account for the increase in ϵ' and ϵ'' due to the presence of liquid water:

$$\epsilon_{ws}' = \epsilon_{ds}' + \Delta', \qquad (22\text{-}6)$$

$$\epsilon_{ws}'' = \epsilon_{ds}'' + \Delta'', \qquad (22\text{-}7)$$

where Δ' and Δ'' represent the incremental increases due to m_v. The particular expressions for these quantities which we adopt for evaluation are based on the dispersion behavior of liquid water [2]:

$$\Delta' = 0.02 m_v^{1.015} + \frac{0.073 m_v^{1.31}}{1 + (f_s/f_w)^2}, \qquad (22\text{-}8)$$

$$\Delta'' = \frac{0.073(f_s/f_w) m_v^{1.31}}{1 + (f/f_w)^2}, \qquad (22\text{-}9)$$

where f_s is the resonant frequency at which ϵ_{ws}' and ϵ_{ws}'' are measured by the probe, $f_w = 9.07$ GHz is the relaxation frequency of water at 0°C, and m_v is expressed in percent. Thus, the quantities measured by the snow probe are ϵ_{ws}', ϵ_{ws}'', and f_s, and the quantities we wish to retrieve are m_v and ρ_{ws}, the latter being the density of the wet snow medium.

22.3.1 Liquid Water Content

In the frequency range around 1 GHz, which is the operational frequency range of the probe, the dielectric loss factor of dry snow ϵ_{ds}'' is less than 4×10^{-4} (for a snow density ρ_{ds} less than 0.5 g/cm^3). For $m_v = 1\%$, the increment Δ'' given by eq. (22-9) is equal to 7.5×10^{-3}, which is approximately 20 times larger than the first term. Hence, ϵ''_{ds} may be ignored in eq. (22-7) and the equation can be solved to express m_v in terms of ϵ''_{ws}:

$$m_v = \left\{ \frac{\epsilon''_{ws}\left[1 + (f_s/f_w)^2\right]}{0.073(f_s/f_w)} \right\}^{1/1.31}. \qquad (22\text{-}10)$$

The applicability of this retrieval procedure was evaluated by comparing the results obtained using eq. (22-10) on the basis of the snow-probe measurements with those measured with a freezing calorimeter. The freezing calorimeter measures the liquid water mass fraction W, from which m_v was calculated from the relationship

$$m_v = 100\rho_{ws}W, \qquad (22\text{-}11)$$

where ρ_{ws} is the density of the wet snow sample, which was measured gravimetrically.

The results for the liquid water content comparison are shown in Figure 22.6. The error bars associated with the freezing calorimeter data points show the range of results obtained from typically two separate (and usually simultaneous) determinations. (Data points with no error bars indicate only a single measurement or that only the mean value of a set was available.) The freezing calorimeter has generally excellent precision.

The values for m_v obtained from the snow-probe dielectric measurements are computed by eq. (22-10). The data points and error bars shown for the snow probe are based on an average of 12 separate measurements made for each snow sample, and the uncertainty of the estimate of the mean value as represented by the error bars was computed as $\pm\sigma/\sqrt{N}$, where σ is the standard deviation of the set of measurements and N is the number of measurements in that set. From the figure, the agreement between the two techniques is generally very good, and, with the exception of an outlier at the 6% level, the use of the snow probe and eq. (22-9) gives results within $\pm0.5\%$ of the freezing calorimeter results. This result strongly supports the validity of eq. (22-9).

22.3.2 Snow Density

With m_v known, through the retrieval procedure described in the preceding section, we now turn our attention to using eqs. (22-6) and (22-8) in order to retrieve the wet snow density ρ_{ws} from ϵ'_{ws}, the dielectric constant of the wet snow medium measured by the snow probe. To do so, we first express ρ_{ws} in terms of ρ_{ds}, the density of the snow had the liquid water been removed:

$$\rho_{ws} = \rho_{ds} + m_v/100. \qquad (22\text{-}12)$$

Next, we use the expression [11]

$$\epsilon'_{ds} = 1 + 1.7\rho_{ds} + 0.7\rho_{ds}^2 \qquad (22\text{-}13)$$

and combine it with eq. (22-6) to obtain

$$\epsilon'_{ws} = 1 + 1.7\rho_{ds} + 0.7\rho_{ds}^2 + \Delta', \qquad (22\text{-}14)$$

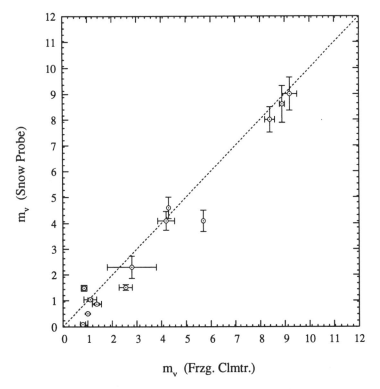

FIGURE 22.6. Comparison of snow wetness results obtained
via snow probe and freezing calorimetry respectively. Snow
probe data points are based on an average of 12
separate measurements.

where Δ' is given in eq. (22-8). Upon combining eqs. (22-12) and (22-14) and
solving for ρ_{ws}, we obtain

$$\rho_{ws} = m_v/100 - 1.214 + \sqrt{1.474 - 1.428(1 - \epsilon'_{ws} + \Delta')}, \qquad (22\text{-}15)$$

in which only the positive root is considered. To compute ρ_{ws} from eq. (22-15),
we use the value of m_v determined in the previous section through eq. (22-10),
the value of ϵ'_{ws} measured by the snow probe, and the value of Δ' calculated from
eq. (22-8). The values of ρ_s (for both wet and dry snow) determined through this
procedure are compared with gravimetric measurements of ρ_s in Figure 22.7. The
data points for which good agreement is found correspond to snow samples having
low wetness levels, $< 3\%$, for which the contribution Δ' is small anyway. For
the samples in which m_v is more appreciable, there is a significant disagreement
between the measurements and the model given by eq. (22-8).

The errors in density estimates are caused by the model underestimating the
incremental increase $\Delta\epsilon'_{ws}$ for the higher wetness cases. Shown in Figure 22.8 is
a plot of $\Delta\epsilon'_{ws}$ as a function of m_v computed on the basis of eq. (22-8), and the

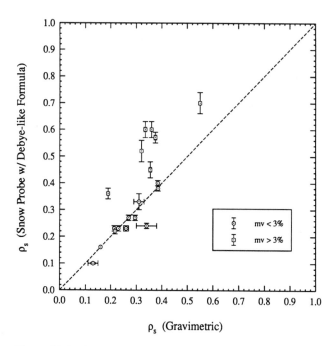

FIGURE 22.7. Comparison of snow density results obtained via
snow probe (in conjunction with Debye-like relation) and
gravimetric measurements. Data points represented with squares
were from snowpacks having volumetric wetness levels of
> 3%; with circles, < 3%.

measurement points were calculated from $\Delta\epsilon'_{ws} = \epsilon'_{ws} - \epsilon'_{ds}$, with ϵ'_{ws} being the
value measured by the probe and ϵ'_{ds} determined from eqs. (22-12) and (22-13).
The curve drawn through the data points is generated using a simple polynomial
fit, given by

$$\Delta\epsilon'_{ws} = 0.187m_v + 0.0045m_v^2. \tag{22-16}$$

A very good fit can also be obtained via eq. (22-8) by modifying the term $0.02m_v^{1.015}$
to $0.08m_v^{1.015}$; however, an arbitrary adjustment defeats the purpose of using a
model based on physical arguments. The Debye-like model of eq. (22-8) has essentially the same frequency dependence as the real part of the dielectric constant
of water, and its empirically derived coefficients—which effectively reduce the
value of this quantity from the theoretical value of the pure material—account
for the water being distributed in particle form within a host having a dielectric
constant somewhere between those of air and ice. With this understanding, there
does not appear to be any reason why a model which works well between 3 and (at
least) 15 GHz should need to be significantly modified to work at 1 GHz; physically speaking, the only difference between 3 GHz and 1 GHz is that the dielectric
constant of water increases from ≈80 to ≈87.

The literature contains certain pertinent experimental results that should be
considered. Tiuri et al. [11] report measurements also at 1 GHz. In these measure-

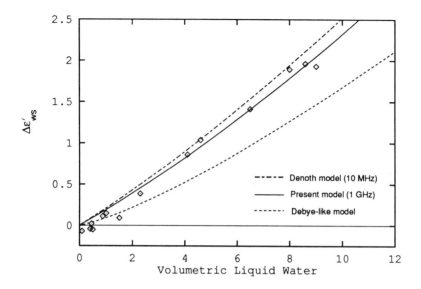

FIGURE 22.8. $\Delta\epsilon'_{ws}$ versus m_v. Shown are data and best-fit curve, plus a model obtained from a comparable study at 10 MHz, and Debye-like model.

ments, a dilatometer technique was employed for wetness measurements primarily, but a capacitor technique was used for some of the samples corresponding to the lowest m_v levels. The dielectric measurements were made with the Snow Fork developed at the Helsinki University of Technology. The relationship they report is

$$\Delta\epsilon'_{ws} = 0.089 m_v + 0.72 m_v^2. \tag{22-17}$$

This function, when plotted, closely resembles the Debye-like model eq. (22-8) evaluated at 1 GHz. Most recently Denoth [12] reported measurements made at 10 MHz, in which dielectric measurements were made using a simple plate capacitor and liquid water measurements were made using a freezing calorimeter. The relation he reports is

$$\Delta\epsilon'_{ws} = 0.206 m_v + 0.0046 m_v^2. \tag{22-18}$$

Denoth observes that this relation should continue to be valid up to approximately 2 GHz, since ϵ' of the constituents of wet snow—ice, air, and water—are all exactly or nearly frequency-independent in this range. In particular, for water, as seen in Figure 22.9, the real part of the dielectric constant of water, ϵ'_w, at 1 GHz differs from that of 10 MHz by only 1.1%. Also noted on the figure is the region through which the Debye-like model of [2] was reported to be valid (although above 15 GHz, the empirical coefficients shown in eq. (22-8) are slightly modified as a function of frequency). The best-fit function for the 10 MHz data, given by eq. (22-18), is also shown in Figure 22.10. The close agreement between the results at 10 MHz and 1 GHz tends to bear out

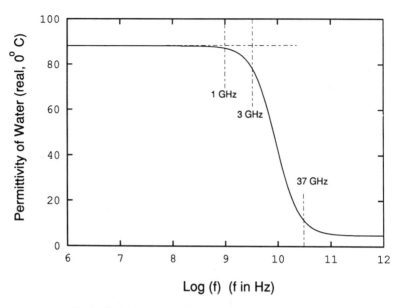

FIGURE 22.9. Real part of permittivity of water at 0°C.

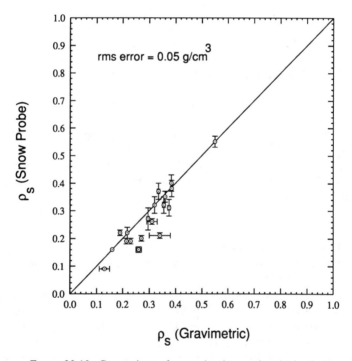

FIGURE 22.10. Comparison of snow density results obtained via snow probe (with associated empirical algorithm) and gravimetric measurements.

Denoth's prediction and suggests that at these frequencies, where scattering is an unimportant factor in calculating the dielectric constant, ϵ'_{ws} is directly relatable to ϵ'_w. In effect, a model like eq. (22-8) *should* be expected to work, but, in its present form, does not appear to. Regarding the discrepancy between our results and those of [11], a possible explanation is that they used a dilatometer to measure wetness (we found the dilatometer to give very unsatisfactory performance), whereas our standard (which was also used in [12]) was the freezing calorimeter technique, whose accuracy and precision have been demonstrated.

22.4. APPLICATION

Figure 22.11 is a nomogram, based on these equations which have been found to be valid in the specified ranges. It consists of contours of constant m_v and ρ_{ds}, respectively, in a two-dimensional representation bounded by the two parameters which are directly obtained by the snow probe: resonant frequency and bandwidth (3-dB) of the resonance spectrum. With the measurement of these two quantities, m_v and ρ_{ds} may be uniquely specified. Dry-snow density, ρ_{ds}, is related through eq. (22-12) to wet-snow density ρ_{ws}.

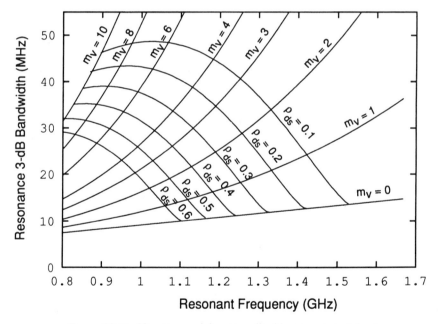

FIGURE 22.11. Nomogram giving snow liquid water content (m_v) and equivalent dry-snow density (ρ_{ds}) in terms of two parameters directly measured by the snow probe: resonant frequency (f) and resonance (3-dB) bandwidth (Δf).

As an example of the utility of the snow probe for elucidating snowpack character and behavior, we present in Figure 22.12 snow wetness data measured for an 0.88-m-deep snowpack over a diurnal cycle. During the period shown, from

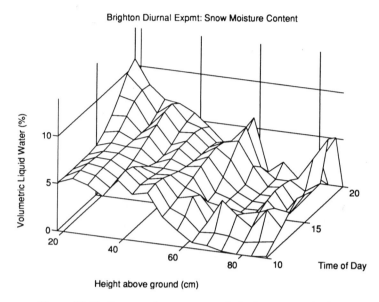

Brighton Diurnal Expmt: Snow Moisture Content

FIGURE 22.12. Snow moisture measured via the snow probe in a 0.88-m snowpack over a diurnal cycle, shown as a function of time and height above the ground.

10 A.M. to 8 P.M., the temperature rose from freezing to 6°C and down to −3°C again at 8 P.M. The lowest 16-cm of the pack was solid ice; therefore measurements start at 18 cm above ground and were made at roughly 5-cm intervals. Among the interesting features in the figure, even at 10 A.M., after subfreezing night temperatures, and while the surface is still completely dry, there is appreciable moisture deeper down in the snowpack. At the top surface, between the hours of 6 and 8 P.M., there was significant wetness which then quickly froze at about 8 P.M.

22.5. SUMMARY

This report has described the development and validation of an electromagnetic sensor and associated algorithm for the purpose of rapid ($\approx$20 s) and nondestructive determination of snow liquid water content and density. The sensor is similar in principle to an existing device known as a Snow Fork, but offers additional advantages in spatial resolution and accuracy owing to a novel coaxial cavity design. Also, the algorithm employed with that device [1] for relating complex dielectric constant to snow parameters does not agree with the results of the present study. We have consequently developed new relations for that purpose.

Direct methods of snow wetness determination were evaluated for their suitability as standards against which the snow probe could be tested. The dilatometer, though simple in principle, was found to give very unfavorable performance. The freezing calorimeter, which has, as a system, been brought to a high degree of sophistication in our lab, was found capable of delivering accuracy better than $\pm 1\%$ and excellent precision.

The snow probe determines the dielectric constant directly. Empirical and semiempirical models are used with this information to compute liquid water volume fraction and density. To test the suitability of existing models and/or allow the development of new models, the snow probe was tested against the freezing calorimeter and gravimetric density determinations. Originally, the relations set forth by Hallikainen et al. [2] were employed to translate measured dielectric constant to snow parameters. The equation relating ϵ''_{ws} to m_v and frequency was found to be entirely valid. However, the equation predicting $\Delta\epsilon'$ in terms of m_v and frequency was found to underestimate this quantity, leading to substantial errors in the estimates of ρ_{ds}. A purely empirical relation, given in eq. (22-16), was obtained instead and will be used in our parameter retrieval algorithm for the snow probe. Through the use of these functions, in association with the complex dielectric measurements of the snow probe, the following specifications are established: liquid water content measurement accuracy $\pm 0.66\%$ in the wetness range from 0% to 10% by volume; wet snow density measurement accuracy ± 0.05 g/cm^3 in the density range from 0.1 to 0.6 g/cm^3.

Two examples of pertinent experimental results were compared against ours: Denoth's measurements [12] at 10 MHz are very similar to ours; those reported by Tiuri et al. [11] at 1 GHz differ considerably from ours but agree well with the model given by eq. (22-8), our own starting point. Denoth [12], noting the diversity of empirical relations for the dielectric constant of wet snow, suggests that, because of the influence of the shape of the water component and the stage of metamorphism of the snow sample, "a valid simple relation between $\Delta\epsilon'$ or ϵ'' and liquid water content W may not exist." This may be the case, but the results of the present study are very consistent with results presented in [12] for $\Delta\epsilon'$ and in [2] for ϵ''. Accurate snow measurements of dielectric constant and liquid water content are notoriously difficult to make. This has doubtlessly been a factor in the diversity of results and is motivation for the development of instruments such as has been the focus of this investigation.

REFERENCES

[1] A. Sihvola and M. Tiuri, "Snow fork for field determination of the density and wetness profiles of a snow pack," *IEEE Trans. Geosci. Remote Sensing*, vol. Ge–24, pp. 717–721, 1986.

[2] M. Hallikainen, F. T. Ulaby, and M. Abdelrazik, "Dielectric properties of snow in the 3 to 37 GHz range," *IEEE Trans. Antennas Propagat.*, vol. AP–34, pp. 1329–1339, 1986.

[3] R. T. Austin, *Determination of the Liquid Water Content of Snow by Freezing Calorimetry*, Univ. of Michigan Radiation Lab Report 022872–2, 1990.

[4] E. B. Jones, A. Rango, and S. M. Howell, "Snowpack liquid water determinations using freezing calorimetry," *Nordic Hydrol.*, vol. 14, pp. 113–126, 1983.

[5] W. H. Stiles and F. T. Ulaby, *Microwave Remote Sensing of Snowpacks*, NASA Contractor Report 3263, 1980.

[6] S. C. Colbeck, "The layered character of snow covers," *Revs. Geophys.*, vol. 29, pp. 81–96, 1991.

[7] D. A. Ellerbruch and H. S. Boyne, "Snow stratigraphy and water equivalence measured with an active microwave system," *J. Glaciol.*, vol. 26, pp. 225–233, 1980.

[8] A. Denoth et al., "A comparative study of instruments for measuring the liquid water content of snow," *J. Appl. Phys.*, vol. 56, no. 7, 1984.

[9] M. A. H. Leino, P. Pihkala, and E. Spring, "A device for practical determination of the free water content of snow," *Acta Polytechnica Scandinavica*, Applied Physics Series No. 135, 1982.

[10] R. E. Collin, *Foundations for Microwave Engineering*, New York: McGraw-Hill, 1966.

[11] M. E. Tiuri, A. H. Sihvola, E. G. Nyfors, and M. T. Hallikainen, "The complex dielectric constant of snow at microwave frequencies," *IEEE J. Oceanic Engr.*, vol. OE-9, pp. 377–382, 1984.

[12] A. Denoth, "Snow dielectric measurements," *Adv. Space Res.*, vol. 9, no. 1, pp. 233–243, 1989.

23

Matti Fischer
Ebbe Nyfors
Pertti Vainikainen
Helsinki University of Technology, Espoo, Finland

On the Permittivity of Wood and the On-Line Measurement of Veneer Sheets

Abstract.   The paper reviews the dielectric properties of wood with special emphasis on experimental results of the effects on the permittivity of drying and remoistening wood. A sensor array for the on-line noncontact mapping of moisture and mass per area of veneer sheets is also described. The sensor is used to sort sheets according to dry mass and moisture after drying.

23.1.   INTRODUCTION

The permittivity of wood depends on the moisture, which makes the measurement of moisture by microwaves possible. The permittivity is, however, also affected by several other factors. The most important of these are the binding of water, temperature, density, ionic conductivity, species of the tree, moisture history of the wood, polarization, and frequency. In order to be able to measure the moisture accurately, the effect of these phenomena must be understood [1].

The binding of water lowers the Debye-relaxation frequency. In wood the mean relaxation frequency f_d is typically in the VHF or UHF range [2], whereas for free water it is approximately 23 GHz at room temperature. Because both the moisture and the temperature affect the mean strength of binding, both affect f_d. The effect of the dry density is straightforward. The ionic conductivity is important in the VHF range and below, causing the imaginary part of permittivity ϵ'' to increase at lower frequencies. Even some influence of the Maxwell-Wagner effect can be seen in the real part of permittivity ϵ' at low frequencies. The species of tree has some extra effect on the permittivity not accounted for by the difference in density, but at least for Finnish birch trees and conifers the differences are small. The moisture history of the wood has a rather important effect. The permittivity of a sample of wood once dried and later remoistened is in some cases different from the permittivity of the same sample at the same moisture before drying. The drying time and the time lapsed since the drying was finished are also important. As shown below, the permittivity changes for several days following fast drying. These phenomena complicate the calibration of sensors. Because of the anisotropy of wood, the permittivity depends on the orientation of the electric field with respect to the grains. The maximum permittivity occurs with the electric field aligned with the grains. The effect of the above phenomena depends on the measurement frequency. Depending on whether the measurement frequency is higher or lower than f_d, the change in temperature may cause ϵ_r'' to either increase or decrease.

23.2. CHOICE OF FREQUENCY

The optimal frequency for a moisture sensor depends on the kind of sensor, what is expected from the sensor, and on the phenomena affecting the permittivity. Usually it is advisable to use a frequency high enough to avoid the ionic conductivity. If inexpensive, noncontact measurement of veneer sheets is demanded and the measurement area should be tens of centimeters. A frequency just above the ionic conductivity limit may be preferred. For the sensor array described below, a frequency of 360 MHz was chosen.

23.3. MEASUREMENT OF MOISTURE
AND DENSITY

Because the permittivity is a complex quantity two independent measurements can be performed with one sensor. That means, e.g., measuring both the resonant frequency and the quality factor of a resonator or the phase shift and the attenuation of a transmission sensor. From the two measurements two quantities can be calculated, e.g., the moisture and the density or the mass per area. In many cases a simple way of calculating has been used. The value $K = \epsilon_r''/(\epsilon_r' - 1)$ is a function of moisture, but often almost independent of density [3]. From the moisture and, e.g., the resonant frequency the dry density can be calculated. In wood the dependence of K on moisture is, however, not monotonous. A maximum value occurs at about 15% of moisture, but the dependence of K on moisture is also affected by the history effects mentioned above.

23.4. SENSOR ARRAY FOR MOISTURE
AND DENSITY MAPPING OF VENEER SHEETS

At the Radio Laboratory of the Helsinki University of Technology stripline resonator sensors have been developed for various applications. An array of sensors with two shaped-center conductors, which are a quarter of a wavelength long, was developed for the fast mapping of moisture and dry mass per area of dried veneer sheets (Figure 23.1). The resonant frequency is 360 MHz and the electric field of the resonant mode (even mode) is parallel to the veneer sheets. Figure 23.2 presents the sensor and the electric field pattern of the sensor. At the plane of the

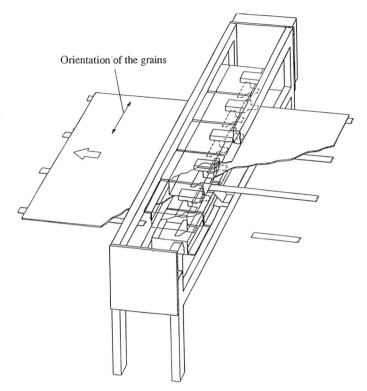

FIGURE 23.1. A stripline resonator sensor array for the fast mapping of the moisture and mass per area of veneer sheets.

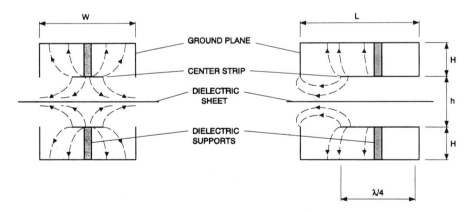

FIGURE 23.2. A stripline resonator sensor and the electric field pattern.

sheets, the electric field lines spread rather evenly to all directions. Furthermore in the factory the orientation of the grains of the veneer sheets is always perpendicular to the direction of motion (Figure 23.1). Thus the orientation of the grains does not have significant effect on the results. According to perturbation theory [4] for the even mode (for low-loss materials),

$$\frac{\Delta f_r}{f_r} \approx -\frac{\epsilon_r' - 1}{2} S, \tag{23-1}$$

$$\Delta(\frac{1}{Q_l}) \approx \epsilon_r'' S, \tag{23-2}$$

where f_r is the resonant frequency, Q_l is the loaded quality factor, and S is the filling factor. The spot size is 100×300 mm^2. The measurement time per sensor is 10 ms. Thus real-time mapping of the properties of the sheets is possible with full coverage at a production line speed of 2.5 m/s. The sensor is used for sorting the sheets according to moisture and density.

In calibrating the sensor array some troubles were detected. The calibrations made in the laboratory, and even with the sensor array in the factory using veneer samples of known moisture, gave erroneous results for the on-line measurements. Because clear indications of hysteresis-like effects in remoistened samples had been found previously [1, p. 88], a series of laboratory tests was conducted.

23.5. LABORATORY TESTS WITH VENEER SAMPLES

The first series of tests was done to further study the differences between green veneer and once dried and remoistened veneer. Samples were slowly dried in the open air and measured with short intervals with a sensor identical to the individual sensors in the array. For the measurement of the permittivity of veneer sheets the sensor was calibrated with plastic sheets. Different plastics (PVC, polyethene, polystyrene) of different thicknesses were used in order to determine the filling factor for veneer sheets. Each sample of veneer was dried to a different moisture level, and then gradually remoistened by spraying of water on the surface and letting the moisture get evenly distributed while storing in a plastic bag. The samples were pieces of spruce, 400×400 mm^2 large and 3.2 mm thick. The results, which are shown in Figures 23.3a and 23.3b, show no sign of hysteresis. There are some diffrences between the samples for high moistures, but each sample followed the same pattern in both directions. This contradicted earlier results, but meant that the reason for the differences between laboratory and on-line measurements must be found somewhere else.

The results of the laboratory measurements and the calibration measurements at the plant were similar but different from the results of the on-line measurements. After thoroughly studying the processes, no differences could be found other than the speed of drying. Therefore another series of laboratory tests was performed, where the veneer samples were dried in less than half an hour. After the temperature had stabilized about 5 min at room temperature, the samples were measured in short intervals. A dry sample was used as a reference in the measurements to be able to rule out effects of any drift in the measurement system. Samples were stored in plastic bags to keep the moisture constant, but the samples were also weighed after each measurement. The samples were marked

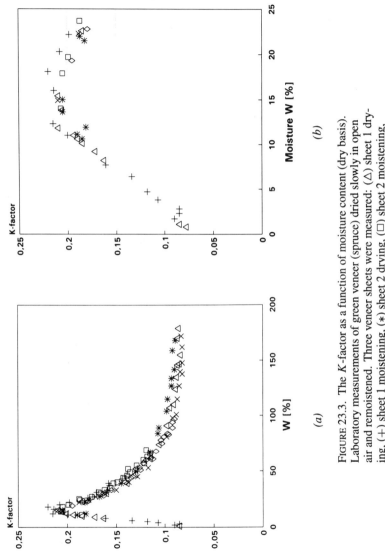

(a)

(b)

FIGURE 23.3. The K-factor as a function of moisture content (dry basis). Laboratory measurements of green veneer (spruce) dried slowly in open air and remoistened. Three veneer sheets were measured: ($\triangle$) sheet 1 drying, ($+$) sheet 1 moistening, ($*$) sheet 2 drying, ($\square$) sheet 2 moistening, ($\times$) sheet 3 drying, ($\diamond$) sheet 3 moistening. The density of dry wood was about 350 kg/m^3 for all sheets. (a) Moisture content 0–200%; (b) Moisture content 0–25%. No hysteresis can be noted.

so that they were placed similarly into the sensor every time. That way the effect
of the direction of the grain on the permittivity could be eliminated. The results
showed that the permittivity is changing after the completion of the drying. The
permittivity stabilizes only after a couple of days. Figure 23.4 shows K for two
sheets dried to 22% moisture. After five days the value of K has increased by
more than 20%. Half of this change occurred during the first hour. Figures 23.5
and 23.6 show separately ϵ_r' and ϵ_r'' for two samples. It can be seen that both
increase with time, but the increase is larger for ϵ_r''. Probably the bound water is
slowly redistributed during a period of time after the moisture has changed. The
results shown in Figures 23.4–23.6 are for 22% moisture, but the effect on K as a
function of moisture is to shift the maximum point (Figures 23.3a and 23.3b) from
15% moisture to about 10%, when measured immediately after fast drying. Below
the maximum point the changes after drying seem to be negligible. This reduces
the available measurement range quite substantially, when using the K-factor. By
using a more complicated model to describe the relationship between the moisture
and the measured quantities, it is possible to increase the measurement range. The
model used for the sensor array described above is of the type

$$W = C_1 K + C_2 K^2 + C_3(\epsilon_r' - 1) + C_4(\epsilon_r' - 1)^2 + C_5\epsilon_r'' + C_6(\epsilon_r'')^2 + C_7, \qquad (23\text{-}3)$$

where W is the dry basis moisture content and $C_1, \ldots, C_7$ are constants. The type
of the model was chosen rather arbitrarily only to give more degrees of freedom

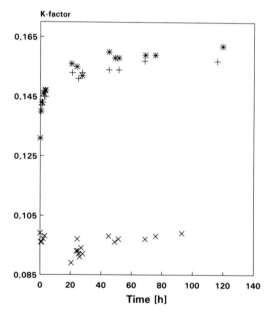

FIGURE 23.4. The K-factor of veneer as a function of time after
the completion of fast drying ($\leq$ 30 min): two dried samples
(22%), (+), (*), and one reference sample (3.5%), ($\times$).

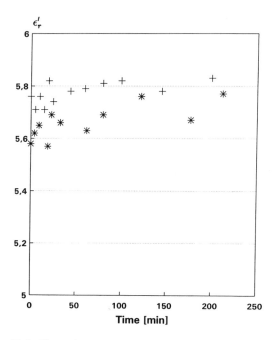

FIGURE 23.5. The real part of the permittivity of a veneer sample as a function of time after the completion of fast ($\leq$ 30 min) drying.

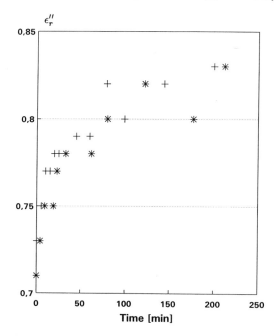

FIGURE 23.6. The imaginary part of the permittivity of a veneer sample as a function of time after the completion of fast ($\leq$ 30 min) drying.

for the fitting process. At low moisture content the dominating terms are $C_1 K$ and C_7, implying that the use of only K is possible for dry wood. This model is no longer fully density-independent even at low moisture contents, but with this model the measurement range could be increased to about 12%, which is enough for this application.

23.6. CONCLUSIONS

When using microwaves for the measurement of moisture in wood, bound water is of primary importance. Slow changes must be taken into account, when the drying process is fast. Therefore all calibrations should be performed at the plant under the same conditions as the on-line measurements will be made. In measuring dried wood, the ratio $K = \epsilon_r''/(\epsilon_r' - 1)$ or a more complicated function can be used, depending on the required measurement range.

REFERENCES

[1] E. Nyfors and P. Vainikainen, *Industrial Microwave Sensors*, Artech House, 1989.

[2] W. Trapp, "Das Dielectrische Verhalten von Holz und Zellulose im grossen Frequenz-und Temperaturbereich," Thesis, Technischen Hochschule, Braunschweig, 1954.

[3] M. Kent, "Complex permittivity of fish meal: A general discussion of temperature, density and moisture dependence," *J. Microwave Power*, vol. 12, no 4, pp. 341–345, Dec. 1977.

[4] R. Harrington, *Time-Harmonic Electromagnetic Fields*, McGraw-Hill, 1961, p. 324.

24

Jerzy Kaliński
Wiltech Industrial Metrology, Warsaw, Poland

On-Line Moisture Content Monitoring in Sheet Materials by Microwave Methods and Instrumentation

Abstract. Some aspects of water content measurement in a dielectric sheet material-water mixture by microwave methods and instrumentation are discussed. The results obtained by on-line water content measurement and control in initially dried cardboard are reported.

24.1. INTRODUCTION

The possibility of contactless, nondestructive, and immediate measurement of the water content in dielectric materials is of significant importance for various branches of industry. Indirect methods of moisture content measurement using electromagnetic radiation as a measuring means are commonly used in industrial applications. Direct methods of moisture content determination are also used, but because they are robust and time-consuming they are used as reference methods, which are necessary for the preliminary calibration of the indirect (microwave) measuring equipment.

In indirect measurement, however, the water content measurement results are influenced by additional factors related to the tested material characteristics, process technology, and measurement conditions. In water content measurement by microwave techniques these additional factors are: material temperature, density and granulometric spectrum, material layer thickness, surface geometry, and distance from the radiating port of the sensor, to name a few. The influence of these factors on the measured moisture content value depends on the conversion function selected for obtaining the electric quantity, which is proportional to the moisture content of the material under test. Moisture content may be converted into an electrical parameter, e.g., signal attenuation and/or phase shift, input impedance of a reflective sensor, or change in the resonant frequency and Q-factor of a resonant sensor loaded with the moist material. These influences are carefully minimized by means of well-chosen solutions of the material-forming unit, measuring microwave sensor, signal converter unit, and signal processing program.

The dependence of one or more electrical quantities on water content in the material under test is used as a basic conversion function [1]. In the multiparameter method of moisture content measurement, other kinds of radiation (infrared, gamma rays, or beta rays) can also be applied, particularly for determining other important parameters of the material under test (density, layer thickness, fat content, etc.).

24.2.   MATERIAL

The measurement possibilities and means which are necessary for on-line water content control of the sheet material by microwave radiation depend on kind of material to be tested (mainly its weight per unit area), required moisture content range, and production technology, e.g., installation used to transport the material such as paper, cardboard, textiles, films, plate, and so on, to the measuring point and cutting or rolling machine. Sheet materials are generally very inconvenient for on-line moisture content measurement, because of small amounts of water contained and because of user requirement for easy access to the measuring space of the sensor from both sides of the tested material band.

The typical transmission sensor designed for attenuation and/or phase shift measurement is not useful in this case because of the limited sensitivity of a simple microwave attenuation or phase-shift converter that can be used in industrial conditions. The closed resonators or slotted waveguide sensors [2] are also not usable because of the limited possibilities of introducing the on-line controlled material into the measurement space of the sensor and the troublesome problems with the displacement of this material throughout such a sensor. The slotted waveguides or open resonant cavity sensors are simple and demand only one-side access to the sheet material under test, but their sensitivity is insufficient to attain the required accuracy of the measurement [3].

The difficulties in the measurement of sheet materials increase when the moisture content decreases and approaches 0%. In cardboard technology the useful moisture content range is limited (after an initial drying) to values between a few and a dozen or so percent of water content.

24.3.   MOISTURE CONTENT MASKING EFFECTS

The results of cardboard moisture content measurement can be affected by changes and variations in

- Material weight per unit area
- Material temperature
- Material surface geometry
- Origin and composition of the material (waste paper, rags)
- Position of the material band with respect to the radiating surface of the sensor (distance, angle)

The influence of the weight per unit area can be minimized or eliminated by proper adjustment of the calibrated controls of the measuring instrument for a given weight value. A multiparameter method of measurement of microwave quantities [1] or an independent measurement of weight per unit area value by means of an isotopic method and instrumentation can also be used. In the last two cases the result of moisture content measurement should be obtained from a special programmed microprocessor.

The influence of the material temperature upon the cardboard moisture content measurement can be neglected in many cases, as the material temperature at the output of drying equipment is rather constant during the year. If this is not the case, the material temperature should be monitored and its influence compensated or taken into account in the microprocessor program. The influence of material surface geometry can often be neglected during the regular run of process technology. The distance between the radiating surface of the sensor and the strip of material, as well as the angle between them, is fixed during an initial calibration of the whole system, and they are then held constant during its use [3].

For a one-side measuring system connected with a reflection-type sensor, the important factor is the distance between the radiating plane of the sensor and the reflector of microwave radiation located behind the strip of material under test, because of extensive standing-wave effect. This factor is particularly important when the moisture content of the material should be monitored across the whole width of the band and the sensor is to be translocated across the strip.

24.4. INSTRUMENTATION

A microwave moisture content meter for initially dried cardboard with a horn antenna sensor has been developed [4] and installed over the material band as shown in Figure 24.1. The sensor contains two horn antennas spaced λ/m apart and connected in series by two circulators as shown in Figure 24.2 [4,5]. The distances between the first and second antennas of the sensor and the common reflector of microwave radiation are respectively

$$d \qquad \text{and} \qquad d - \frac{\lambda}{m} = d - \left(\frac{\lambda}{n} + \frac{\lambda}{8} \right),$$

and enable a reduction of the influence of antennas–reflector distance change (Δd) on the output signal of the sensor. The n-value should be chosen to ensure the

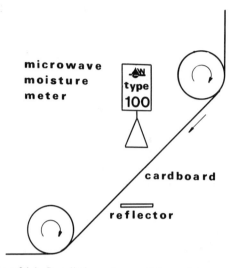

FIGURE 24.1. Installation scheme of the moisture content meter over the cardboard band.

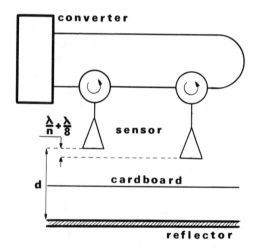

FIGURE 24.2. Block diagram of the microwave sensor with two horn antennas and the reflector behind the cardboard band.

"in-phase" summing of the output signals of both circulator antenna circuits. The additional $\lambda/8$ distance is introduced to obtain an "opposite phase" summing of output signal increments of the sensor, occurring by distance antennas–reflector change Δd. The compensation of output signal versus distance change is then based on the standing-wave effect, resulting in opposite increment sign of output signal of each antenna-reflector circuit versus distance, when the electric length difference between antennas and reflector equals $\lambda/4$.

The e_{out} dependence on Δd for single circulator antenna circuit (a), for two such circuits connected in cascade, with λ/n distance difference between antennas and reflector (b) and for two circulator antenna circuits with additional $\lambda/8$ displacement of the antenna (c) is shown in Figure 24-3.

The attenuation A of microwave radiation introduced by such sensor is

$$A(\lambda/n) = A_1 + A_2,$$

where
A_1 = attenuation introduced by first circulator antenna circuit
A_2 = attenuation introduced by second circulator antenna circuit.

The change ΔA of this attenuation by distance change Δd between antennas and reflector is

$$\Delta A(\lambda/n) = \Delta A_1 + \Delta A_2,$$

with additional $\lambda/8$ displacement become

$$\Delta A(\lambda/n + \lambda/8) = \Delta A_1 - \Delta A'_2,$$

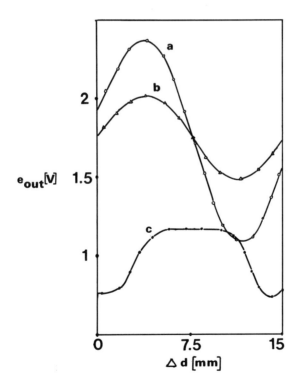

FIGURE 24.3. Dependencies of the microwave converter output signal on the distance between antennas and reflector.

where

ΔA_1 = attenuation change versus distance increment Δd of first circulator antenna circuit

ΔA_2 = attenuation change versus distance increment Δd of second circulator antenna circuit

$\Delta A_2'$ = attenuation change versus distance increment Δd of second circulator antenna circuit with additional $\lambda/8$ displacement

It is obvious that when

$$\Delta A_1 = \Delta A_2'$$

the sensor output signal dependence on distance increments Δd will be minimized.

The microwave converter circuit was constructed with stripline technology and contains microwave signal generator (MESFET) and detector (Schottky diode) as well as a system of PIN diode switches and microstrip ferrite isolators. On the output of this converter two signals proportional to the measured attenuation value and to the reference level are produced alternately [6]. The reference signal level is proportional to the instantaneous values of microwave generator output and microwave detector ac/dc conversion efficiency.

The dc electronic circuitry contains an automatic microwave converter sensitivity controller and drift-free chopped dc amplifier. An extremely high resolution of changes in the input signal, independent of the initial input signal level, has been ensured. The measured attenuation range from 0 to 30 dB, and the easily adjustable full-scale deflection attenuation increment measuring range of 2–20 dB, have been achieved (related to the attenuation of the empty measuring space). The conversion function $E = f(M)$, where E is the output voltage in volts and M is the material moisture content in percent, with the weight per unit area as a parameter, is shown in Figure 24.4.

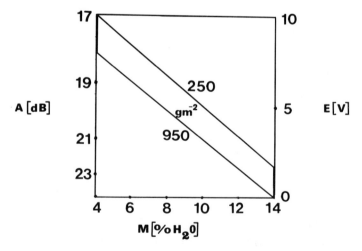

FIGURE 24.4. Dependence of the output signal of the moisture content meter on the water content in the cardboard band.

24.5. APPLICATION

The microwave moisture content measuring system described above, connected to an isotopic weight per unit area gauge and suitably programmed microcomputer has been installed in Paper Mill Jeziorna by Warsaw. An automatic control of moisture content in the initially dried cardboard strip, with the weight per unit area in the range of 250–950 gm^{-2} and an accuracy of about 0.5% H_2O, has been obtained.

REFERENCES

[1] A. Kraszewski and S. Kuliński, "An improved microwave method of moisture content measurement and control," *IEEE Trans. IECI*, vol. IECI-23, pp. 364–370, Nov. 1976.

[2] A. Kraszewski, "Microwave aquametry—a review," *J. Microwave Power*, vol. 15, pp. 209–220, 1980.

[3] A. Młodzka-Stybel, WILTECH Report No 23, pp. 8–14, 1992.

[4] J. Kaliński, "Microwave converter for moisture content measurement of sheet material," Polish Patent Appl., in process, 1993.

[5] J. Kaliński, "Microwave converter with reflection sensor," Polish Patent Appl., No. P-261694, 1986.

[6] J. Kaliński, "Method and arrangement for automatic reference signal level control by the absorption or reflection coefficient measurement of electromagnetic radiation," Polish Patent Appl., P-261870, 1986.

25

Albert Klein
Indutech GmbH, Simmersfeld, Germany

On-Line Microwave Moisture Monitoring with "Micro-Moist"

Abstract. The microwave moisture meter "Micro-Moist" uses a microwave transmission measurement in the S-band by measuring attenuation and phase shift. With an additional gamma ray transmission measurement to compensate for the influence of bulk density and layer thickness, a three-parameter measurement can be made. Utilizing all the above parameters the accuracy can be improved, and under specific conditions an additional disturbing parameter can be measured. Some examples of applications with different transducers are given to show that the instrument meets the requirements of industrial on-line moisture measurement.

25.1. INTRODUCTION

Process control and process optimization require reliable and accurate moisture measuring techniques. The techniques mainly used in the past, such as conductive, capacitive, or infrared techniques, rarely meet the requirements of today because they are strongly influenced by several disturbing parameters. Comparative investigations on German coal have shown that microwaves are less influenced by these disturbing parameters [1]. Figure 25.1 gives an overview of these results. Furthermore, a comparison of the attenuation and the phase shift of the transmitted signal has shown that phase measurements are less influenced by several disturbances like salt content, particle size distribution, and temperature [2] (see Figure 25.1). These results have been recently reconfirmed for Australian coal [3,4].

25.2. MICROWAVE MOISTURE DETERMINATION

25.2.1 Physical Background

Microwave moisture determination is based on the dielectric properties of water. The water molecule has a permanent dipole moment and therefore the dielectric

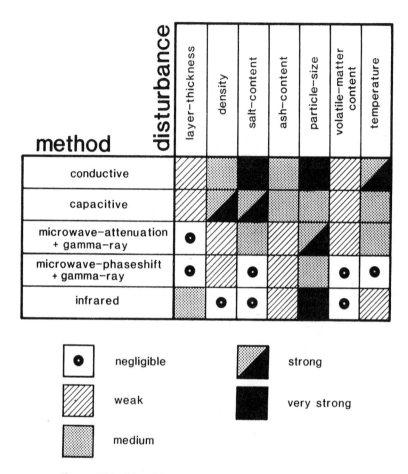

FIGURE 25.1. Disturbing paramaters on different methods for moisture determination of coal.

constant of water is much higher than the permittivity of most of the dry materials. Furthermore, caused by the relaxation process, in the microwave range water has characteristic dielectric losses. Therefore, both dielectric constant ϵ' and dielectric loss ϵ'' of the complex permittivity

$$\epsilon = \epsilon' - j\epsilon'' \qquad (25\text{-}1)$$

can be used as a measure of the moisture content.

For determining the dielectric properties of a nonmagnetic material at microwave frequencies a variety of techniques are available, which can be classified as reflection, transmission, and resonator methods. All three measuring techniques are used for moisture determination, too, because each has specific advantages and disadvantages. For contactless moisture determination of, e.g., bulky materials transported on a conveyor belt, the transmission technique usually is used.

The dielectric loss influences mainly the absorption, whereas the real part of the dielectric constant affects mainly the propagation velocity of the microwaves, transmitted through a material layer; i.e., attenuation and delay time of microwaves that have penetrated a material layer with a known or constant layer thickness and density are a measure of the moisture content [2].

In the past only attenuation was used for determining moisture, but instruments are now available which measure delay time. However, because of the short delays in the picosecond range a direct measure of the delay time is difficult. Therefore usually the phase shift of the microwaves caused by the sample layer is measured, which is proportional to the delay time; i.e., this type of moisture meter is principally a network analyzer.

25.2.2　Two-Parameter Techniques

To compensate for the influence of layer thickness and density, a second parameter must be measured. To eliminate both of these influences, attenuation and phase shift can be normalized on the mass per unit area of the sample. The mass per unit area can be determined with a balance for off-line measurements. For on-line conditions, an additional gamma ray transmission sensor can be used [2].

To avoid the need for an additional device for determining the mass per unit area, combinations of attenuation and phase measurements are possible. Because the attenuation and the phase shift, both normalized on the mass per unit area, which may be nonlinear functions of moisture, are not dependent on layer thickness and bulk density, the quotient of these two values must be independent of these factors, too. In this quotient, the mass per unit area can be canceled; i.e., the quotient of attenuation and phase shift is independent of layer thickness and bulk density. It is a measure for moisture only, if the attenuation is not proportional to the phase shift [5]. Assuming that the normalized attenuation and phase shift, respectively, are linear functions of moisture, one obtains a formula given previously [6]. These combinations are more sensitive for disturbing parameters than the phase shift normalized on the mass per unit area only, because the result is additionally influenced by the disturbances of the attenuation. Furthermore, the sensitivity to error is increased, because both signals depend on moisture. In the S-band a low accuracy was obtained with these formulas applied to data given in [1]. However, in the X-band the validity of the formulas could be reconfirmed [7–10], because in the X-band the attenuation is more sensitive for moisture and less influenced by disturbing parameters, e.g., the electrolytic conductivity, whereas the phase shift is less sensitive for moisture, i.e., relatively more sensitive for the mass per unit area.

25.2.3　Three-Parameter Techniques

Microwave moisture measurement can be improved by measuring further disturbing parameters, e.g., the temperature to compensate this influence. A three-parameter technique has been suggested for compensating the influence of particle size variations of bulk materials, whereby attenuation and phase shift, both normalized on the mass per unit area, are combined [2]. In general, with this technique all disturbances of varying material properties, which affect attenuation and phase shift to a different extent, can be compensated or reduced. Apart from the particle size, this method can be used for reducing the influence of, e.g., salt content and pH value. Assuming that practically only one material property is varying, with this three-parameter technique an additional measure for this property can be derived. Because in most cases these disturbing parameters are only varying in a limited range, it is not necessary to use the formula given in [2]. Usually a linear combination of phase shift P and attenuation A, both normalized on the mass per unit area m'', is sufficient; i.e., the moisture content M is calculated with the formula

$$M = a P/m'' + b A/m'' + c, \qquad (25\text{-}2)$$

where a, b, and c are calibration coefficients. The second material parameter can be calculated with the same formula by using another set of calibration coefficients.

The mass per unit area need not be determined in absolute units. A signal proportional to m'' is sufficient, because the missing proportional factor is automatically covered by the a and b coefficients, respectively, which are determined during the calibration procedure. Because the attenuation and phase shift, respectively, are combined with the mass per unit area in a fixed manner, the quotients A/m'' and P/m'' can be handled as one parameter; i.e., a reduction of one parameter is obtained, which simplifies the calibration. For example, if only the phase shift combined with the mass per unit area is used, the linear function allows the instrument to be started with a simple two-point calibration. For a recalibration only the moisture reading must be compared with the laboratory result to determine the new coefficients from the old ones. The basic readings, phase shift, and mass per unit area do not have to be available.

25.3. THE MOISTURE METER MICRO-MOIST

25.3.1 Description of the Instrument

With the Micro-Moist microwave moisture meter, it is possible to use the three-parameter method: It measures the attenuation and the phase shift of the microwaves transmitted through the sample layer. The instrument is optimized for penetrating thick layers or very moist materials. A relatively low frequency is used; the frequency is swept in the S-band. An input for a scintillation counter is provided to detect the gamma rays transmitted through the material layer for determining its mass per unit area. Temperature compensation is also provided. Standard analog and digital outputs are available. Figure 25.2 shows the principle of the measuring setup. In addition to a sliding average, a batch mode is available. Here the signal is averaged over a definite period, which can be specified by an external input. This feature can also be used for calibration of one-parameter measurements to get the average reading over a sampling period. For more parameter measurements special calibration software is provided, by which each parameter is averaged separately. This data reduction is very helpful during calibration. The laboratory results can be given via the keyboard to the instrument, and then the calibration coefficients are calculated automatically by the instrument. In addition to these features a PC can be connected via RS-232 interface for data logging, storing, visualization, and calibration. The software for this is available.

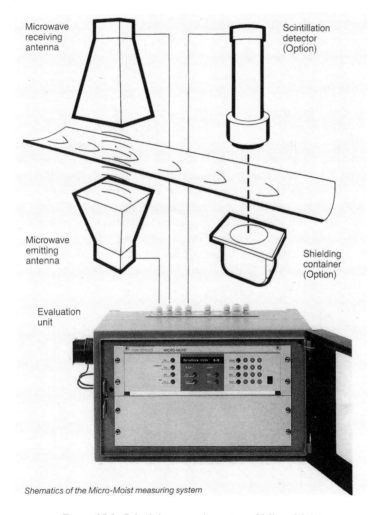

Shematics of the Micro-Moist measuring system

FIGURE 25.2. Principle measuring setup of Micro-Moist.

25.3.2   Field Experience
        with Micro-Moist

Micro-Moist has been on the market since 1988. Since then about 100 instruments have been installed in the field. Micro-Moist was first introduced for the coal industry because of our experience with this product. The instrument is now installed for many applications in various other industries. Several transducers are available. The instrument can be installed on conveyor belts using horn and spiral antennas. A standard chute is available, which is used for granular materials such as wheat, dried lignite, or fertilizer. A flow cell with a 50-mm diameter has been designed for liquid products, especially in the food industry. A similar flow cell is under test in combination with a small conveyor screw, which can be installed at a transfer station of two conveyor belts as a bypass. Measurements of inhomogeneous materials such as clay particles or the filtercake of a slurry can then be made more accurately.

25.3.3 Results

A typical application in the coal industry is the moisture determination of the coal feed to a coking oven to enable the control of the heating oven. This coal has a top size of 10 mm. The moisture content is in the range of 7–12%. The measurement is performed on a conveyor belt. The load on the 1200-mm belt is about 500 t/hr at a belt speed of 1.68 m/s. The moisture measurement utilizes phase shift combined with the additional gamma ray compensation. The microwaves are transmitted by horn antennas. Other options such as a temperature sensor or tachometer were not employed. The result of a 4-month test period is illustrated in Figure 25.3. The highest deviation between laboratory and Micro-Moist is 0.8% moisture. Of the 62 samples, 7 show a deviation of more than 0.3%, 2 of more than 0.6%. The one standard deviation is $s = 0.28\%$ moisture. In this value the sampling and analysis error are included [11].

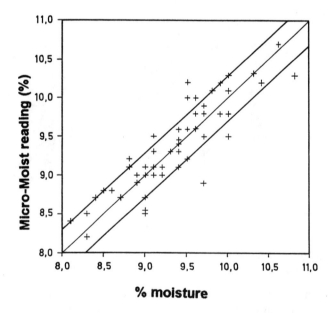

FIGURE 25.3. Micro-Moist reading versus lab moisture content of a coking coal.

An application without density compensation is the moisture measurement of butter. Horn or spiral antennas are installed on the extruded butter with a dimension of 180 × 180 mm. Unsalted and salted butter with a nominal salt content of 2% were measured. The moisture range is between 14% and 18%. The objective is to control the moisture content of the butter to 16% moisture. The result, shown in Figure 25.4, is obtained with a phase measurement. The influence of the salt content is practically negligible. A standard deviation of 0.2% was achieved. Using attenuation this would produce an error of 2.5%. As we use a combination of phase shift and attenuation, the moisture determination is improved negligibly; however, as we additionally use the attenuation, it is possible to determine the salt content.

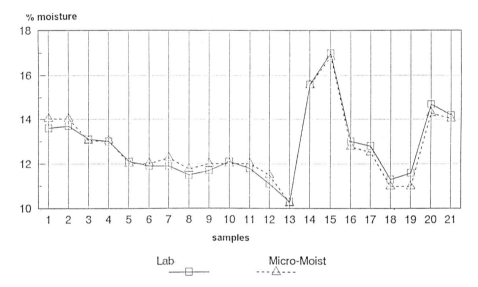

FIGURE 25.4. Tracking plot of the Micro-Moist reading and the lab moisture of butter.

An example for a chute application is the moisture determination of dried lignite with 10–17% moisture in a briquette factory. The main stream is transported on a chain conveyor. From the chain conveyor the lignite falls in the measuring chute with 250 mm × 350 mm (cross section) × 350 mm length. The microwaves penetrated through the chute at the 250-mm axis. Horn antennas are used. The material is transported back to the main stream by a conveyor screw. The material's temperature is about 80°C. Nuclear density compensation is used. The result is shown in Figure 25.5. The standard deviation is less than 0.3% moisture. This result is verified as long-time stability over a period of more than one year now.

As an example for the flow cell, an application on cream cheese in a dairy can be given. The flow cell is installed in a 50-mm pipe after a separator with the purpose of controlling the separator. The solid content is 46% to 50%; i.e., the moisture content is between 50% and 54%. For this measurement only the phase

shift is used. The result over a test period is shown in Figure 25.6. The standard
deviation is $s = 0.37\%$ moisture.

% moisture

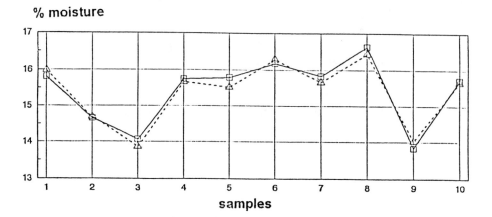

FIGURE 25.5. Tracking plot of the Micro-Moist reading and the
lab moisture of dried lignite.

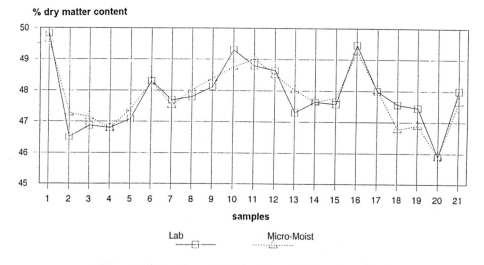

FIGURE 25.6. Tracking plot of the Micro-Moist reading and the
lab result of cream cheese.

25.4. CONCLUSION

The microwave moisture meter Micro-Moist enables a complex microwave transmission measurement by measuring attenuation and phase shift. Additionally the
gamma ray transmission measurement can be used; i.e., by using all features a

three-parameter measurement can be performed. This allows an improved accuracy and under special conditions a measure of a further, normally disturbing parameter. The results presented and the acceptance of the instrument in the market show that this instrument is a reliable solution for on-line moisture measurement.

REFERENCES

[1] A. Klein, "Comparison of rapid moisture meters," *Miner. Process.*, vol. 28, pp. 10–16, 1987.

[2] A. Klein, "Microwave determination of moisture in coal: comparison of attenuation and phase measurement," *J. Microwave Power*, vol. 16, pp. 289–304, 1981. Summary of: Untersuchung der dielektrischen Eigenschaften feuchter Steinkohle im Hinblick auf die Anwendbarkeit des Mikrowellenverfahrens zur Wassergehaltsbestimmung. Dissertation. Aachen, 1978.

[3] N. Cutmore, D. Abernethy, and T. Evans, "Microwave technique for the on-line determination of moisture in coal," *J. Microwave Power Electromag. Energy*, vol. 24, pp. 79–90, 1989.

[4] N. Cutmore, T. Evans, and A. McEwan, "On-conveyor determination of moisture in coal," *J. Microwave Power Electromag. Energy*, vol. 26, pp. 237–242, 1991.

[5] W. Meyer and W. Schilz, "A microwave method for density independent determination of the moisture content of solids," *J. Phys. D. Appl. Phys.*, vol. 13, pp. 1823–1830, 1980.

[6] A. Kraszewski and S. Kulinski, "An improved microwave method for moisture content measurement and control," *IEEE Trans. Ind. Electron Control Instr.*, vol. 23, pp. 364–370, 1976.

[7] M. Kent and W. Meyer, "A density-independent microwave moisture meter for heterogeneous foodstuffs," *J. Food Engng.*, vol. 1, pp. 31–42, 1982.

[8] E. Kress-Rogers and M. Kent, "Microwave measurement of powder moisture and density," *J. Food Engng.*, vol. 6, pp. 345–356, 1987.

[9] A. Kraszewski, "Microwave monitoring of grain—further considerations," *J. Microwave Power Electromag. Energy*, vol. 23, pp. 236–246, 1988.

[10] K. Kupfer, "Disturbances by the application of microwave moisture measurements," *Technisches Messen*, vol. 60, pp. 19–28, 1993. Summary of: Feuchtemessung an Zuschlagstoffen für die Betonherstellung unter Verwendung der Mikrowellenmeßtechnik, Dissertation. Weimar 1990.

[11] A. Klein and W. Pesy, "Experiences with the microwave moisture meter Micro-Moist," *Miner. Process.*, vol. 30, pp. 549–557, 1989.

S. K. Aggarwal
*Patent, Trade Marks and Design Offices,
Woden ACT, Australia*

R. H. Johnston
University of Calgary, Calgary, Alberta, Canada

26

Determination of Water Content in Oil Pipelines Using High Frequencies

Abstract. An experimental system for on-line measurement of water content in oil pipelines has been developed. The instrumentation measures the phase and attenuation of an electromagnetic wave traveling through the fluid mixture flowing in a specially designed applicator. The results show that the phase shift of the EM wave at 500 MHz is a strong function of the water content of the mixture and a weak function of the salt content and the temperature. To test the water/oil meter, an apparatus is constructed to supply flowing water/oil mixtures of varying ratios to the applicator. The measured data is compared with the results obtained by analyzing the applicator theoretically.

26.1. INTRODUCTION

Monitoring of water in crude oil and in various petroleum products is required for a number of reasons. If the water content in the fluid produced by oil wells can be determined, then enhanced oil recovery techniques can be used to maximum advantage, as well as being of benefit to an oil company for production control and accounting purposes. Also during the refining process, it is necessary to control the water level of the crude oil since water containing various salts, etc., may have adverse effects on the processing equipment.

The difference in dielectric properties of water and oil is well known and understood, and these properties have been used by earlier researchers for determining water content in oil/water mixtures [1,2], in mixtures with water contents up to 10%.

The method described in this paper also makes use of the difference in the dielectric properties of water and oil to determine the water content of oil. The attenuation and phase shift of an electromagnetic wave at 500 MHz traveling through the fluid mixture is measured. A specially designed applicator carries the electromagnetic wave in a coaxial transmission line–like structure through the flowing fluid. The attenuation and phase data can be used to determine the water content and water salinity at a given temperature.

26.2. MEASUREMENT SYSTEM

The measurement system required to perform the measurements as shown in Figure 26.1 is reasonably simple. It consists of a signal generator, which produces an EM wave at 500 MHz. The EM wave from the signal generator travels to a signal splitter, which divides the signal into two separate paths. One of these paths contains the applicator, and the signal, after being attenuated and phase-shifted by the fluid mixture in the applicator, travels to a vector voltmeter through a coaxial Tee. The coaxial Tee provides for a connection for the high-impedance vector voltmeter probe and matched termination for the signal. The other path contains a reference coaxial line (approximately the same electric length as the main line containing the applicator). The vector voltmeter measures the attenuation and phase shift of the EM wave traveling through the fluid mixture relative to the EM wave traveling through the reference line.

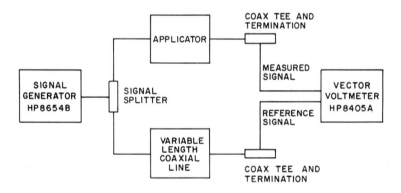

FIGURE 26.1. Instrumentation used for measurement.

26.2.1 Applicator

The applicator is constructed from a forged steel cross which provides the basic structure that is needed to withstand the pressure and the temperature of the water/oil mixtures. The two outlets of the cross that carry the mixture are unaltered, while the other two outlets are converted to a transmission line structure to carry the electrical signal. Details of the applicator are shown in Figure 26.2.

In actual two-phase flow systems, it is well known that the fluids may separate from each other. In a pipeline carrying heavy oil and water, the oil can coat the pipe (probably due to a greater viscosity and possibly a greater tendency to wet the steel pipe than water). The oil will thus travel more slowly than the water due to its

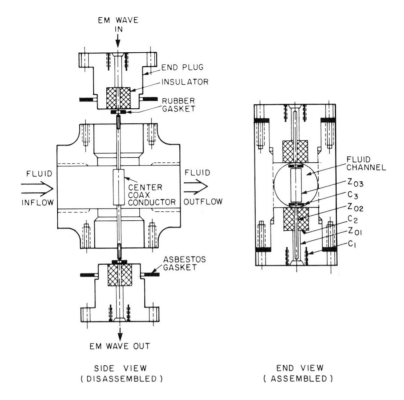

FIGURE 26.2. An exploded cross-sectional side view and an assembled cross-sectional end view of the applicator. Equivalent circuit elements are shown in the assembled applicator.

greater viscosity and due to the location in the cross-sectional area that it occupies. The average velocity of the water is greater than that of the oil, and the velocity differential will cause an error in translating the fractional volume measurement to a fractional flow rate reading. If the water and oil are thoroughly mixed in the form of small droplets, then the two fluids should have the same average flow rates even when the fluid velocity is not constant across the pipe cross section. Therefore, static fluid mixers are placed upstream of the cross to reduce the above source of error. The input and output pipes of the applicator have flanges attached to facilitate its connection to a measurement or pipeline system.

26.2.2 Electrical Analysis

The applicator can be represented by a series of coaxial line sections cascaded together as shown in Figure 26.3. S_1, S_2, and S_3 represent sections of transmission lines in the applicator, while C_1, C_2, and C_3 represent the capacitances due to discontinuities between these coaxial line sections. Z_0, γ, and l represent the characteristic impedance, propagation constant, and the length, respectively, for

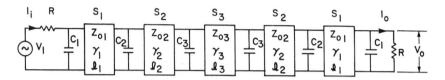

FIGURE 26.3. The equivalent circuit diagram of the applicator.

a coaxial line section. R represents the impedance of the input voltage source V_i, as well as the load impedance at the output end, and in this case is equal to 50 ohms.

If one knows the values of Z_0, γ, l, and the discontinuity capacitances, the circuit diagram in Figure 26.3 can be analyzed by using transmission matrices for each component.

The characteristic impedance for a section of coaxial line can be found from the formula

$$Z_0 = \frac{138}{\sqrt{\epsilon_r}} \log \left(\frac{D}{d} \right), \tag{26-1}$$

where D and d represent the diameter of the outer and the inner conductor, respectively, and ϵ_r represents the relative permittivity of the insulating material between the outer and the inner conductors of the coaxial line section.

The propagation constant of an EM wave in a coaxial line can be found from the expression

$$\gamma = j\omega\sqrt{\epsilon_0\mu_0} \, \sqrt{\epsilon_r}. \tag{26-2}$$

The values of capacitances C_1, C_2, and C_3 depend upon the dimensions of the coaxial line sections at the points of discontinuity. There are three types of discontinuities between these sections, which are shown in Figure 26.4 together with their equivalent circuits.

The discontinuities shown in Figures 26.4a and 26.4b are classified as single step. If the dimensions of the discontinuities are known, the polynomials discussed in [3] can be used to find the values of discontinuity capacitances. The double-step discontinuity shown in Figure 26.4c may be broken into two parts similar to those shown in Figures 26.4a and 26.4b. The total capacitance can be found by adding these capacitances for the individual cases.

In order to find the values of Z_3, γ_3, and C_3, the relative permittivity of the water and oil mixture in the measurement chamber of the applicator must be known. A number of theories have been presented for the permittivity of dielectric mixtures of different geometrical configurations [4,5], but not many theories are available for predicting the permittivity of a mixture of liquid dielectrics when one of the components is highly lossy. Ramu and Rao [6,7] have presented the following expression for calculating the complex relative permittivity ϵ_r^* of liquid dielectric mixtures and they report that it is accurate, even when the fractional volume of the dispersed fluid goes up to 0.4–0.45:

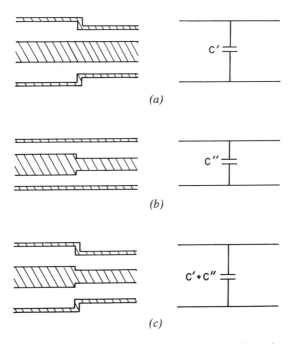

FIGURE 26.4. Coaxial line discontinuities and their equivalent circuits.

$$\epsilon_r^* = \epsilon_r' - j\epsilon_r'' = \frac{(k_1 k_3 + k_2 k_4)\epsilon_1' - (k_2 k_3 - k_1 k_4)\epsilon_1''}{k_3^2 + k_4^2}$$

$$- j \frac{(k_2 k_3 - k_1 k_4)\epsilon_1' + (k_1 k_3 + k_2 k_4)\epsilon_1''}{k_3^2 + k_4^2}, \tag{26-3}$$

where

$$k_1 = 2\epsilon_1'(1 - v_2) + \epsilon_2'(1 + 2v_2), \tag{26-4}$$

$$k_2 = 2\epsilon_1''(1 - v_2) + \epsilon_2''(1 + 2v_2), \tag{26-5}$$

$$k_3 = \epsilon_1'(2 + v_2) + \epsilon_2'(1 - v_2), \tag{26-6}$$

and

$$k_4 = \epsilon_1''(2 + v_2) + \epsilon_2''(1 - v_2), \tag{26-7}$$

and ϵ_1' and ϵ_1'' represent the dielectric constant and the loss factor, respectively, for the continuum fluid, ϵ_2' and ϵ_2'' are the dielectric constant and loss factor, respectively, for the dispersed fluid, and v_2 represents the fractional volume of the dispersed fluid in the continuum fluid.

Assuming that the oil is the continuum and the water is the dispersed fluid, respectively, eqs. (26-3) to (26-7) can be modified to give

$$\epsilon_{rm}^* = \frac{(k_1 k_3 + k_2 k_4)\epsilon_{ro}}{k_3^2 + k_4^2} - j\frac{(k_2 k_3 - k_1 k_4)\epsilon_{ro}}{k_3^2 + k_4^2}, \qquad (26\text{-}8)$$

where

$$k_1 = 2\epsilon_{ro}(1 - v) + \epsilon'_{rw}(1 + 2v), \qquad (26\text{-}9)$$

$$k_2 = \epsilon''_{rw}(1 + 2v), \qquad (26\text{-}10)$$

$$k_3 = \epsilon_{ro}(2 + v) + \epsilon'_{rw}(1 - v), \qquad (26\text{-}11)$$

$$k_4 = \epsilon''_{rw}(1 - v), \qquad (26\text{-}12)$$

where ϵ_{rw} and ϵ_{ro} are the relative permittivities of water and oil, respectively, and v is the fractional value of the water.

By substituting values into eq. (26-8) and then applying those to eq. (26-2), one may calculate the phase shift of a wave traveling through a water/oil mixture of a length and frequency as a function of the fractional volume of the water content. The result is shown in Figure 26.5, assuming that oil is the host material for a frequency of 500 MHz and a fluid length of 34.22 mm. Alternatively if water is assumed to be the host material using eqs. (26-3) and (26-2), a different phase-shift profile is obtained and is also plotted in Figure 26.5. It is to be noted that the assumption of which fluid is the host greatly affects the resulting phase shift or, conversely, would greatly affect the water content estimation obtained from a given phase shift. It is suggested [6,7] that the predominant material is the host material.

The calculation of the exact phase shift of a wave through an applicator is more complicated than indicated above and should account for transmission line sections, source, and load impedance that are connected on each side of the dielectric mixture. This analysis follows. Various circuit elements, as shown in Figure 26.3, can be represented by a matrix of ABCD parameters. The required matrices are

$$\begin{bmatrix} 1 & R \\ 0 & 1 \end{bmatrix}$$

for a resistor,

$$\begin{bmatrix} 1 & 0 \\ j\omega C & 1 \end{bmatrix}$$

for a capacitor, and

$$\begin{bmatrix} \cosh \gamma l & Z_0 \sinh \gamma l \\ (\sinh \gamma l)/Z_0 & \cosh \gamma l \end{bmatrix}$$

for a transmission line section.

It can be shown from Figure 26.3 that

$$\begin{bmatrix} V_i \\ I_i \end{bmatrix} = \begin{bmatrix} T_{11} & T_{12} \\ T_{21} & T_{22} \end{bmatrix} \begin{bmatrix} V_o \\ I_o \end{bmatrix}, \qquad (26\text{-}13)$$

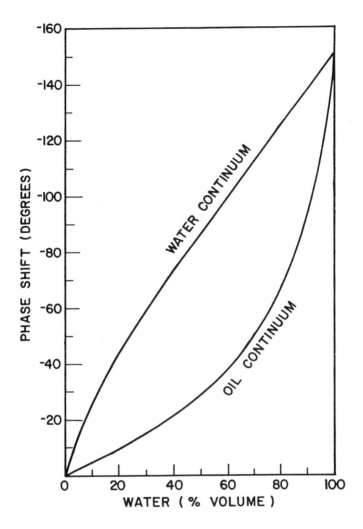

FIGURE 26.5. Theoretical phase shift of an EM wave in an oil/water mixture, $\epsilon_{rw} = 79.95 - j2.20$, $\epsilon_{ro} = 2.55$, $l = 34.22$ mm, $f = 500$ MHz.

where V_i and I_i represent input voltage and current, respectively, to the applicator, while V_o and I_o represent the load current and the voltage across the load resistor, respectively. T is a transmission matrix (also the ABCD matrix) representing the applicator and can be obtained by multiplying matrices for each element shown in Figure 26.3. The output current I_o can be written as

$$I_o = \frac{V_o}{R}. \tag{26-14}$$

Substituting eq. (26-14) into eq. (26-13) gives us

$$V_i = T_{11}V_o + T_{12}\frac{V_o}{R}$$

or

$$\frac{V_o}{V_i} = \frac{1}{T_{11} + T_{12}/R}, \tag{26-15}$$

which can be solved to find the theoretical phase shift and attenuation of the EM wave for a given mixture of water and oil in the applicator. The theoretical phase shift and attenuation of the EM wave at 500 MHz are found as a function of water contained in water/oil mixtures. Equation (26-15) has also been solved for phase shift and attenuation when 0.15 mol/l of salt is dissolved in water. Theoretical results based on the above analysis are compared with the measured data.

26.3.   MEASUREMENT PROCEDURE

For testing and calibration of the system, it is necessary to have a source of flowing water/oil mixtures with known content. This is supplied by the apparatus as shown in Figure 26.6. An outlet high on the tank feeds oil to a positive displacement pump. The lower outlet on the tank supplies water to another positive displacement pump. Each pump is driven by a dc motor. If the wiper of the rheostat is moved up, the upper motor slows down and the lower motor speeds up, and vice versa. The speed of each motor is approximately proportional to applied voltage; therefore the sum of the two motor speeds is approximately constant. The total fluid output of the pumps is approximately constant, and the mixture ratio can be varied from 0% to 100% by moving the wiper of the rheostat R_2 from one extreme position to the other. The series rheostat (R_1) allows some control of the total fluid flow and limits the motor start-up currents.

The combined water and oil mixture is applied to the input of the applicator where it is mixed and then it goes back to the tank. The oil and water should separate out of the mixture quickly, as the oil has a specific gravity significantly less than 1 and the oil chosen (a light white mineral oil) does not emulsify with water easily. Different oils were tried, and some difficulty was experienced in finding an oil that separates quickly without emulsifying, especially over a prolonged period of contact with the water. In addition, impurities (such as rust) also seem to stabilize oil/water emulsions. The temperature of the oil and water is thermostatically controlled.

A typical series of measurements starts with the fluid reservoir one-half full of oil and one-half full of water. First the electronic instrumentation is checked by replacing the applicator by a coaxial line of the same electrical length. The phase meter is adjusted to give a zero reading. The test is initiated by switching on the two dc motors with the speed control rheostat set for either an all-oil or all-water mixture. After vector voltmeter readings stabilize, the fluid sampling tap on the applicator is opened to provide a fluid sample to a graduated cylinder (sometimes three samples are taken). As soon as the readings and samples are

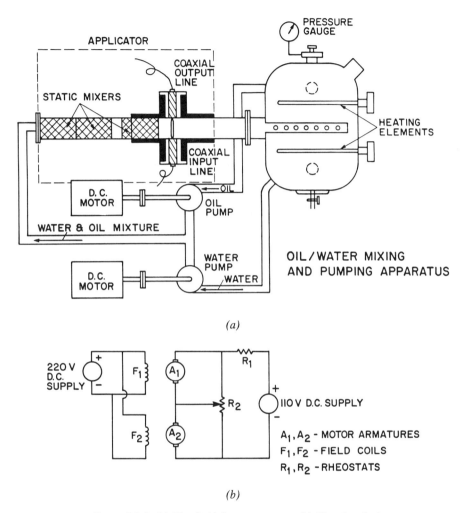

FIGURE 26.6. (a) The fluid flow apparatus. (b) The electrical connections for the dc motors.

taken, the dc motors are turned off so as to minimize the mixing of the oil and water in the reservoir. The measure sequence is again initiated with a new setting of the rheostat (R_2) to give a new mixture ratio, about 11 sets of measurements are taken over the 0% to 100% water content range.

After a series of measurements the fluid samples in the graduated cylinders are examined. Typically, the oil and water separate after a few minutes. It is estimated that the water/oil mixture ratio can be read with a repeatability of about ±1%, while the absolute accuracy is about ±3%. The fact that the oil to air surface is concave and the water/oil interface is convex makes precise determination of the water/oil ratios difficult to estimate with absolute accuracy.

26.4. RESULTS

Measurements were made with pure water/oil and salt water/oil mixtures. The concentration of sodium chloride in salt water was 0.15 mol/l. A series of measurements were made at 20°C, 50°C, and 90°C. The results are shown for the measurement made at 50°C.

The phase-shift data from four measurement runs using pure water and oil at 50°C are shown in Figure 26.7. The solid line represents a best fit including all runs. The data are well behaved at this temperature (and at 90°C). At 20°C (not shown) a significantly larger scatter of ±7% occurs. Almost every point lies within ±3% of the solid line sketched in Figure 26.7.

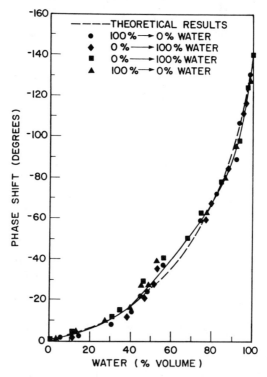

FIGURE 26.7. The phase shift in mixtures of pure water and oil at 50°C. The results are normalized with respect to 100% oil in the applicator. Runs shown by circles and diamonds are performed together; runs shown by squares and triangles are performed together.

The phase shift as a function of water content for 0.15 vol/l saline solution at 50°C is shown in Figure 26.8. The solid line represents the best fit of the points from the four measurement runs. The measurement results with salt water at 90°C are good without significant scatter, but the results at 20°C show a ±8% scatter.

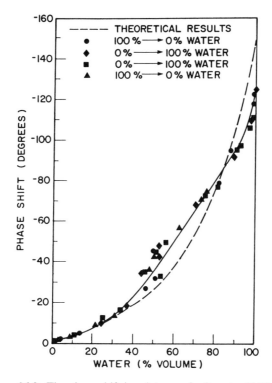

FIGURE 26.8. The phase shift in mixtures of salt water (0.15 mol/l of NaCl) and oil at 50°C. The results are normalized with respect to 100% oil in the applicator. Runs shown by circles and diamonds are performed together; runs shown by squares and triangles are performed together.

An illustration of the attenuation data for pure water/oil mixtures are plotted in Figure 26.9. The measured points are very close to the solid lines sketched in these figures. Attenuation is small and decreases with temperature. The attenuation data for salt water/oil mixtures is illustrated in Figure 26.10. The data are well behaved at 50°C and 90°C. The attenuation with salt water is large compared to the attenuation in pure water mixtures and increases with temperature and salinity.

The authors do not have a conclusive explanation for the scatter noted in phase shift and attenuation data at 20°C, but it is believed that it may be due to the formation of an emulsion between oil and water. At elevated temperatures, the emulsion may break more easily and the measured data do not show the previous scatter.

It is to be noted that the measured results bear a close resemblance to the calculated values for all water contents, even though the theory is expected to be applicable to water contents of up to 45% only. The theoretical analysis is based on the assumption that water is included in oil for all water contents. Theoretical analysis treating the oil as the included material with water as the host material gives poor agreement with the measured data regardless of the water content.

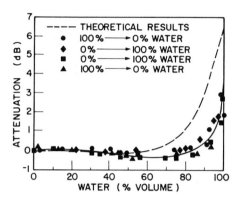

FIGURE 26.9. The attenuation in mixture of pure water and oil at 50°C. The results are normalized with respect to 100% oil in the applicator. Runs shown by circles and diamonds are performed together; runs shown by squares and triangles are performed together.

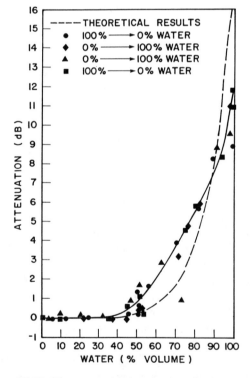

FIGURE 26.10. The attenuation in mixtures of salt water (0.15 mol/l of NaCl) and oil at 50°C. The results are normalized with respect to 100% oil in the applicator. Runs shown by circles and diamonds are performed together; runs shown by squares and triangles are performed together.

We feel that the forged steel used in the cross and the stainless steel center conductor have an affinity for oil compared with the water. Therefore even while the majority of the fluid is water, a thin coating of oil on the steel conductors has a strong influence on the currents traveling in the conductors and hence the wave traveling through the medium. In this way the oil acts as the host material even when it occupies a small fractional volume.

The phase-shift and attenuation data obtained from the solid lines in Figures 26.7–26.10 are plotted in Figure 26.11. This graph can be used to determine the water content and salinity of the mixture at 50°C for the given values of phase shift and attenuation.

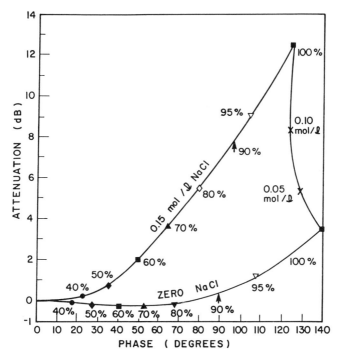

FIGURE 26.11. The phase shift and attenuation in mixtures of water and oil at 50°C. The results are normalized with respect to 100% oil in the applicator.

The phase-shift data for pure water and for salt water mixtures at 50°C are comparable as seen by comparing Figures 26.7 and 26.8, and it is found that if a single phase-shift curve is chosen for all data points in these figures the error in measuring the water content from the phase data would be less than ±5% in most cases.

26.5. CONCLUSIONS

A meter for measuring the water content of a pipeline carrying water and oil has been developed and tested. Measurement of water and salt content of the mixture can be made by measuring the phase shift and attenuation of the EM wave passing through the mixture of oil and water. If the salinity of the mixture is not required, the phase-shift data alone can be used to determine the water content of the mixture with slightly reduced accuracy. At all temperatures the curves are steep at the high water content and the measured points lie very close to the solid lines, thus making the measurement technique potentially accurate at high water levels. It is also apparent that the repeatability of the system is especially good at high temperatures, which may be due to the absence of oil/water emulsions. Typical oil field applications have a fluid production temperature of 50°C and above. The applicator is compact and easy to integrate with existing pipeline equipment.

An important feature of this applicator is the apparent behavior where oil acts as the continuous phase even at low oil content. This suggests that the applicator conductor materials are oleophilic. It is also apparent that under some conditions water can be the continuous phase and it will be necessary to use the top curve of Figure 26.5 to determine water content. It is desirable to pick applicator conductor materials that ensure that either oil is the continuous phase or water is the continuous phase. Thus the metallic parts of the applicator should be constructed of materials that are oleophilic and hydrophobic (or alternately hydrophilic and oleophobic).

REFERENCES

[1] G. S. P. Castle and J. Roberts, "A microwave instrument for the continuous monitoring of the water content of crude oil," *Proc. IEE*, vol. 62, no. 1, pp. 103–108, Jan. 1974.

[2] D. A. Doughty, "Determination of water in oil emulsions by a microwave resonance procedure," *Anal. Chem.*, vol. 49, no. 6, pp. 690–694, May 1977.

[3] P. I. Somolo, "Calculating coaxial transmission-line capacitances," *IEEE Trans. Microwave Theory Techn.*, p. 454, Sept. 1963.

[4] L. K. H. van Beek, "Dielectric behaviour of heterogeneous systems," *Prog. Dielectrics*, vol. 7, pp. 69–114, 1967.

[5] W. R. Tinga, W. A. G. Voss, and D. F. Blossey, "Generalized approach to dielectric mixture theory," *J. Appl. Phys.*, vol. 44, no. 9, pp. 3897–3902, Sept. 1973.

[6] Y. N. Rao and T. S. Ramu, "Determination of the permittivity and loss factor of mixture of liquid dielectrics," *IEEE Trans. Electrical Insulation*, vol. EI-7, no. 4, pp. 195–199, Dec. 1972.

[7] T. S. Ramu and Y. N. Rao, "On the evaluation and conductivity of mixtures of liquid dielectrics," *IEEE Trans. Electrical Insulation*, vol. EI-8, no. 2, pp. 55–60, June 1973.

27

M. Kent
R. H. Christie
A. Lees
Torry Research Station, Aberdeen, Scotland

A Portable Fat Meter Suitable for Live Salmonid Fish

Abstract. Various practical design considerations for a microwave hand-held instrument for the determination of fat content of whole fish were investigated. Calibrations of meter reading against fat content are presented for two versions of the instrument operating at 10 and 2 GHz. Its use is demonstrated for a number of species of pelagic fish and notably for salmonids.

27.1. INTRODUCTION

A rapid and nondestructive method for the measurement of the fat content of various fish species has long been desired by the fish processing industry where a knowledge of this variable is often crucial for product quality. Recently a suitable instrument was described [1] using microwave radiation at 10 GHz and a microstrip sensor [2]. The operation of the stripline sensor can be viewed in different ways. In one explanation, the fringing field of the microstrip is supposed to interact with the lossy material placed in contact with the sensor. In this model, the loss measured would be some function of the dielectric loss factor in a similar way to transmission through a lossy medium [3]. This, however, does not explain the reported relative independence of the insertion loss from the loss factor and its apparently strong dependence on the real part of the permittivity [2]. A more reasonable explanation might be that the system acts as an inefficient stripline antenna, the efficiency of which as a radiator increases with increasing real permittivity of the contact material [4]. The effect of this permittivity must be to change the effective electrical dimensions of the strip and thus its performance as an antenna. There is, however, no model known to the authors supporting this view. The use of the stripline as a sensor remains in this respect entirely empirical. Using miniaturized components it had been possible to make the whole device portable.

The technology used is simple but effective, employing a comparator circuit to monitor the insertion loss of the sensor when this is placed in contact with the material under test. Figure 27.1 shows the basic design [1], where those components within the dotted box are those in the hand-held part of the instrument, comprising (a) low-power microwave source (in the work described here at 2 or 10 GHz), (b) isolator, (c) directional coupler to provide a reference signal, (d) attenuator or isolator for improved performance, (e) stripline sensor, (f) detectors, (g) preamplifiers, (j) digital display unit. The exterior components coupled by cabling to the head are (h) further stage of amplification allowing adjustment for calibration purposes and (i) signal processing to provide output to (j) in terms of calibrated fat content.

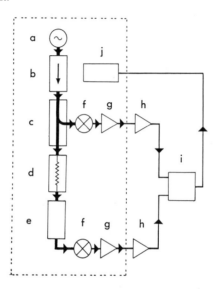

FIGURE 27.1. Schematic diagram of instrument (see text for details).

The device responds mainly to the water present in the fish tissue and not the fat but, as is well known, in fish these are highly correlated [5]. Seasonal variations in the fat content occur as a natural consequence of the breeding cycle involving changes in availability of food and other factors. Because fish must at all times maintain a slightly negative buoyancy, water and protein are displaced by fat, the increase of which tends to increase buoyancy. The phenomenon is definitely not one of dilution of one component by addition of another but in pelagic species involves actual loss of protein and water [5]. It is probable that the water lost is partly that "bound" to the protein and loss of the latter is the vehicle for the water loss [6].

The prototype instrument was calibrated for herring (*Clupea harengus*), but later work attempted to provide a calibration for sprats (*Sprattus sprattus*). Due to the smaller size of this species, a second instrument was designed with a smaller sensing head [6]. This later work also revealed that there was a marked seasonal variation in calibration and that this was probably due to (a) seasonal changes in protein content affecting the proportion of "bound" to "free" water in the tissue as indicated above and (b) the changes in fat distribution that occur through the season.

The solution to this was to produce a similar instrument operating at a lower frequency. This would have several direct and indirect advantages over the 10-GHz versions.

1. Because the instrument actually responds to the water content through its dielectric properties, measuring at a lower frequency (e.g., 2 GHz) would ensure that more, if not most, of the water present would be included in the response, since this frequency would be well below the dipolar relaxation frequency of water, ≈ 20 GHz at 25°C. It has been shown in fact that the correlation between dielectric properties and water content becomes optimum at around 2 GHz as the frequency is varied [7].

2. The longer wavelength of measurement would probably mean a greater penetration of the tissue by the microwave field, increasing the effective volume of tissue being "seen" by the sensor and enabling better averaging of fat and tissue inhomogeneities.

3. The sensor responding at 2 GHz to most of the tissue water might also lead to the elimination of the species dependence of the 10-GHz versions, which could arise from different structure and composition of the flesh from species to species.

4. The same factor could enable more successful use of the instrument on salmonids, which, having thicker skins than the smaller fish studied, did not show suitable dependence of meter reading on fat content at 10 GHz.

5. A lower-frequency microwave source would be more efficient and would result in a lower current drain on the batteries, achieved in the 10-GHz version only by intermittent operation (i.e., the source was switched on only for the actual measurement).

With all these considerations in mind, a third version of the instrument was constructed to operate at 2 GHz. The commercial version of this is shown in Figure 27.1.

27.2. MATERIALS AND METHODS

All pelagic fish used in this work were obtained from the ports of Ayr and Tarbert in the Clyde fishery and were landed within one day of catching. These were herring (*Clupea harengus*), mackerel (*Scomber scombrus*), and sprats (*Sprattus sprattus*). Salmonids were obtained, recently killed, from various fish farms in Scotland. These were Atlantic salmon (*Salmo salar*) and rainbow trout (*Salmo gairdneri*, also known as *Onchoryhynchus mykiss*). Some measurements were also made on live salmon prior to killing.

Instrument readings were obtained in the case of the pelagic fish and the trout by placing the sensing head firmly on the surface of the fish, just above the lateral line in the central portion of the body (Figure 27.2).

FIGURE 27.2. Instrument being used on whole fish.

For salmon a different approach was used. Being a much larger fish, it was decided to average several readings taken along both sides of the fish at equal intervals between the gills and the tail. Samples of flesh for composition analysis were also taken from these same positions, all lying above the lateral line. The mean fat content of these samples was then correlated with the mean meter reading. The live fish studied were measured immediately on removal from the fish cage, being restrained by one person with another taking the readings. They were then killed for later composition analysis.

The decision to measure only at points above the lateral line was taken to avoid the large fat deposits that accumulate in the belly walls. Similar deposits occur along the dorsal region, so this area was also avoided. Water content determinations were performed in triplicate by drying 10-gram samples of the flesh for 48 hr at 105°C.

Fat content measurements were also made in triplicate using a total of 40 g of tissue and utilizing a method involving dissolving the fat in tetrachlorethylene and measuring the change in density of the solvent (type 53100 oil meter, A/S Fos,

Denmark). For these analyses the flesh was macerated and blended for 2 min in a food processor (Magimix 2000) prior to the addition of the solvent.

27.3.   RESULTS AND DISCUSSION

As a confirmation of the interdependence of the fat and water contents the results for linear regression of water versus fat for the data sets of each species studied are present in Table 27.1. It is seen that a high degree of correlation exists, and it is readily shown that the gradients of the regression lines are significantly different from what would be expected if fat was accumulated independently. They are also very different from unity, which would be the case if fat replaced water weight for weight. The relationship between fat and water contents justifies the calibration of the instrument in terms of fat content even though it is responding mainly to the water content.

TABLE 27.1. Regression Equations for Fat versus Water for Various Fish Species

Species	Fat % $= a - bX$ (water %)	No. of Samples	Correlation Coefficient
Herring	$F = (92.86 \pm 1.07)$ $-(1.17 \pm 0.015) \times W$	138	0.989
Mackerel	$F = (92.77 \pm 2.11)$ $-(1.17 \pm 0.03) \times W$	104	0.960
Sprats	$F = (97.07 \pm 1.20)$ $-(1.22 \pm 0.019) \times W$	103	0.989
Salmon	$F = (73.14 \pm 1.20)$ $-(0.93 \pm 0.02) \times W$	179	0.960
Trout	$F = (64.88 \pm 7.49)$ $-(0.81 \pm 0.10) \times W$	39	0.920

Comparisons between species, however, should not be attempted using the results in Table 27.1. The methods for the determination of fat and water contents were different for the different species as circumstance dictated. For example, the salmon data were obtained on muscle only, whereas for sprats the data were for whole but eviscerated fish. However, the results for herring are close to the published data [5], and those for the salmon flesh are close to what would be expected for a system maintaining constant density. The table is presented merely to show the high degree of correlation between fat and water contents.

In Figure 27.3a results using the original 10-GHz instrument are shown, while in Figure 27.3b the results are plotted from which the calibration curves for the 2-GHz device, shown in Figure 27.4, were obtained. In the latter case, over most of the range, the results appear relatively independent of species. The departure from this generalization for mackerel is probably due to greater inhomogeneity of fat distribution in this species, especially as the fat content increases.

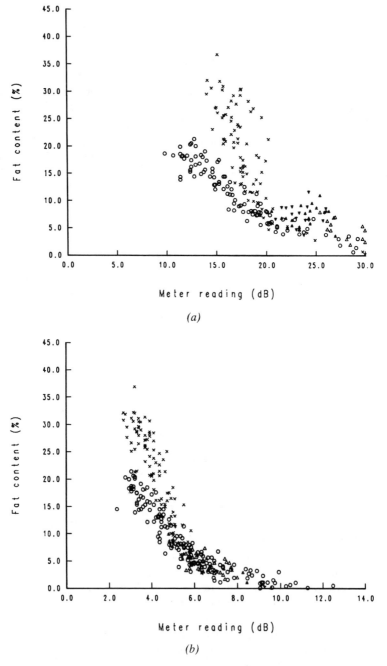

FIGURE 27.3. (a) Data collected using the earlier 10-GHz version of the instrument: herring = o; mackerel = x; salmon = *; trout = Δ. (b) As for (a) but for the 2-GHz version on the same fish demonstrating the improvement in the calibrations.

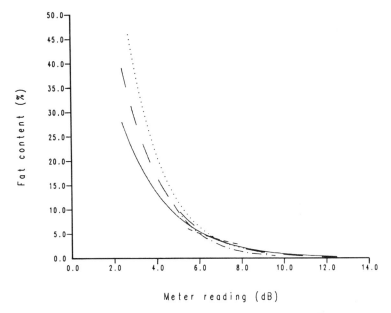

Figure 27.4. The calibration data of Table 27.2 plotted on a linear scale for the various species: herring _____; mackerel .... ; salmon -.-.-.-.-. ; trout ——— ; all species ___ ___ ___ .

These data can be approximated by the expression

$$\text{Log} F = a - bM,$$

where F is the fat content, M is the meter reading in decibels, and a and b are constants. This expression allows a linear regression of log F on M to be performed, from which the curves in Figure 27.4 are derived. The regression parameters are shown in Table 27.2 for herring, mackerel, salmon, and trout. Only a narrow spread of data was available for trout, and so the calibration is more inaccurate than for other species. No data were collected for sprats, the sensor being too large.

TABLE 27.2. Regression Analysis for Logarithm of Fat Content versus Meter Reading in dB

Species	Constant (a)	Gradient (b)	Correlation Coefficient
Herring	$1.91 \pm .05$	$-0.196 \pm .008$	.909
Mackerel	$2.35 \pm .05$	$-0.258 \pm .010$	.926
Salmon	$2.40 \pm .21$	$-0.283 \pm .033$	.893
Trout	$1.54 \pm .21$	$-0.140 \pm .032$	.584
All species	$2.13 \pm .03$	$-0.226 \pm .006$	.914

Although the regression results in Table 27.2 show significant differences between each species, Figures 27.2 and 27.3 suggest that for fat contents below about 10% (or water contents greater than 70%) no great penalty would be incurred by assuming the same calibration for all species. Pooling all the results gives the general calibration for all species (Table 27.2).

Further work has now been carried out using the instrument on minced meat products with some success [8].

27.4. CONCLUSIONS

The results show that earlier problems associated with the use of 10-GHz frequency of operation in an instrument for the measurement of fat content in fish can be reduced by lowering the frequency to 2 GHz. The instrument so designed has been found to be usable with species such as the salmonids that previously presented difficulties. This could be important to salmon growers who wish to monitor the growth of their stock and the effects of various feedstuffs upon them. The instrument can easily be used on immobilized or anesthetized fish and has the advantage of being nonpenetrative and nondestructive.

REFERENCES

[1] M. Kent, "Hand-held instrument for fat/water determination in whole fish," *Food Control*, vol. 1, pp. 47–53, 1990.

[2] M. Kent, "The use of strip-line configuration in microwave moisture measurement. II." *J. Microwave Power*, vol. 8, pp. 189–194, 1973.

[3] I. J. Bahl and S. S. Stuchly, "Analysis of a microstrip covered with a lossy dielectric," *IEEE Trans.*, vol. MTT-28, no. 2, pp. 104–109, 1980.

[4] D. J. Steele and M. Kent, "Microwave stripline techniques applied to moisture measurement in food materials," in *Proc. 13th Microwave Power Symposium*, Ottawa, 1978.

[5] T. D. Iles and R. J. Wood, "The fat/water relationship in North Sea herring (*Clupea harengus*) and its possible significance," *J. Marine Biol. Assoc. UK*, vol. 45, pp. 353–366, 1965.

[6] M. Kent, A. Lees, and R. H. Christie, "Seasonal variation in the calibration of a microwave fat/water meter for fish flesh," *Int. J. Food. Sci. Technol.*, vol. 27, pp. 137–143, 1992.

[7] M. Kent, "Measurement of dielectric properties of herring flesh using transmission time domain spectroscopy," *Int. J. Food. Sci. Technol.*, vol. 25, pp. 26–38, 1990.

[8] M. Kent, A. Lees, and A. Roger, "Estimation of the fat content of minced meat using a portable microwave fat meter," *Food Control*, vol. 4, pp. 222–225, 1993.

28

Van Nguyen Tran
Yang Shen
School of Engineering and Technology,
Deakin University, Geelong, Victoria, Australia

Microwave Moisture Meter Using a TEM Cell for Continuous Operation

Abstract. A microwave moisture meter has been developed using a TEM cell. The 4-GHz solid-state oscillator assembly at 4 GHZ is etched on a PCB including a two-port power divider. One port feeds a signal into the cell, while the other is used as a reference signal to a balanced mixer for measuring the phase change through each sample. Stripline dual-directional couplers are used at the input and output of the cell to monitor the forward and reflected signals, from which the attenuation through a sample is computed. The moisture content is related to this attenuation, with some variation due to packing or bulk density. Bulk density dependence is corrected by the phase information obtained from the balanced mixer. The whole system is under the control of an Intel microprocessor. It controls all functions of the meter, processes the data, and provides measuring information on an LCD display to a printer via a Centronics parallel interface or to a remote printer via an RS-232 serial port. A small printer is also available within the meter to provide a brief moisture content record. Moisture measurements are made separately in discrete samples or continuously by letting a material under test flow through the TEM cell.

28.1. INTRODUCTION

There has been considerable interest in developing a moisture meter using microwaves. Its major advantage is in a much faster response than the hot-air oven method and therefore such a meter can be used for on-line measurements. Stuchly and Hamid [1] review the state of the art in microwave sensors for measuring non-electrical quantities up to 1972. Kraszewski [2] discusses microwave aquametry up to May 1980. Klein [3] studies attenuation and phase shift through a coal sample over a wide frequency range and proposes to use attenuation with the additional weight per area information to eliminate the influence of bulk density. Meyer and Schilz [4] provide analysis to show that the ratio of attenuation over phase shift through a sample is constant within 5% standard deviation over the whole range of densities. Kent and Meyer [5] find that the claim for bulk density-independent moisture measurement based on attenuation and phase is not entirely justified, but an instrument based on two parameters is better than one based on a single parameter. Accuracy better than ±0.6% moisture content is achieved for heterogeneous materials. Okamura [6] reports a moisture meter developed in the X-band for measuring high moisture contents in grains. Recently, Cutmore et al. [7] developed microwave gauges for on-line analysis in the coal, mineral, and oil industries.

There are several moisture meters available on the market using the microwave absorption principle. In one system, a horn is used to transmit microwave energy in the X-band through the material under test. Another horn is used to detect the resulting signal passing through the sample. The sample volume is fixed and simultaneously weighed to determine the bulk density. This information seems to overcome at least some of the bulk density dependency. The meter is material specific, and a database of calibrations for different materials must be stored in the meter.

We designed and constructed a moisture meter using a solid-state oscillator at 4 GHZ and stripline technology with a TEM cell as the sample holder.

28.2. PRINCIPLE OF THE MOISTURE METER

The meter is intended to be used in a network to record the moisture contents of liquids or granular materials from locations which may be miles apart. For instance, in one application, the wheat authority may wish to obtain the moisture levels of grains in different silos scattered throughout a state or country. The moisture meter is preferably set up remotely, with little if any attendance from personnel. The measurements must be quick and on-line. In another application, the rice authority may want to divert grain into different bins, each having a designated moisture content range as it is being delivered to a depot. The food industry may wish to grade products, such as beef cubes or onion flakes, by moisture contents as they are being bagged. Each of these applications calls for a moisture meter accurate to better than 0.5%, which must provide on-line moisture information for quick action. Our study of attenuation and dielectric properties of several granular and liquid materials from 1 to 10 GHz convinced us that a microwave sensor would be a suitable choice. The frequency to 4 GHz is chosen because there we find a steep linear attenuation slope versus moisture contents. Furthermore, there are available solid-state oscillators and balanced mixers at very low cost and a well-developed stripline and surface-mount technology in dual-directional couplers for forward and reflected signal detection. After evaluating many different types of sample holders, we selected an open TEM cell. A TEM cell shown in Figure 28.1a is a section of an unbalanced transmission line propagating the TEM mode. Unlike a coaxial line, the TEM cell in our case consists of an inner conductor placed within two outer side or vertical rectangular plates. Thus, the top is left open so that a sample can be loaded from a hopper or a feed tank into the space between the inner conductor and the plates. For discrete measurements, a trapdoor is required at the

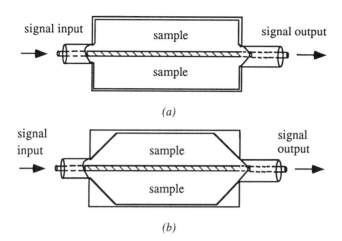

FIGURE 28.1. (a) TEM cell used as a sample holder; (b) TEM cell with tapered section.

bottom. For continuous measurements, the feed tank and trapdoor remain open so that a material can flow through the cell without interruption.

The impedance of the cell is matched to a standard coaxial line by shaping the inner conductor using a TDR. The outer plates are connected at each end, to an N-type connector or a coaxial line. It is essential that the impedance of the TEM cell containing a sample should be as close to the characteristic impedance of the coaxial line as possible. In this regard, some TEM cells are provided with tapered sections (Figure 28.1b). In some designs, a TEM cell is provided with one movable outer side plate which can be controlled by the microprocessor using a stepper motor. The movable outer side plate helps prevent higher-order modes by narrowing the space between itself and the inner conductor. It is also used to match the characteristic impedance of the cell to the coaxial line.

A wide cell width is required for large particles and a narrow cell width for small ones. Thus liquids and fine granular materials require a short and narrow cell. The cell length is also tailor-made for different moisture ranges. A long cell is used for low moisture contents and a short cell for high moisture contents.

The microwave signal is generated by a bipolar oscillator using surface-mount technology and printed circuit board design. It has excellent thermal, frequency, and amplitude stability. However, stability is not highly essential in this case because the attenuation and phase at the output of the cell are relative values of the same signal. The signals (forward and reflected) at the input and output are monitored separately by a stripline dual directional coupler. The phase information is obtained by multiplying a reference signal from the oscillator with the signal passing through the sample in the cell. Multiplication is achieved by the solid-state balanced mixer (Mini-Circuits ZAM-42). The power divider at the output of the oscillator provides these two signals: one signal passing through a length of semi-rigid coaxial line for phase reference, and the other through the sample in the cell.

28.3. PREPARATION OF LOOK-UP TABLES

Calibration tables consisting of average attenuation values corresponding to designated moisture contents must be prepared. These tables, one for each material, are stored in read-only memory (ROM) for looking up the moisture content whenever an attenuation value is determined. Thus the accuracy of the meter cannot be better than that of the look-up tables. These tables are prepared by measuring the attenuation of samples of materials by the meter over a range of moisture contents and at a selected phase value. A linear interpolation algorithm is used to estimate the moisture content if a measured attenuation value does not fall on a specific value in the appropriate table. Extrapolations for an attenuation value which falls outside the table range is made from the end value and the one nearest to it. If an attenuation value falls well outside the limits of the table, an error message is given. An alternative to look-up tables is to use expressions for fit curves, but these would require a more powerful processor and a high-level language.

28.4. MICROPROCESSOR CONTROL

The microprocessor system uses an Intel processor and provides controls for the operation of the moisture meter as shown in Figure 28.2. It has on board a ROM containing look-up tables of the attenuation values at different moisture levels for each product. Once a product is specified and a measurement is made, the table is consulted to find a moisture content corresponding to the measured signal level. If one is not found, an extrapolated routine is used to compute the moisture value. The microprocessor controls the movement of the stepper motor, which in turn moves the movable plate for matching or tuning. It also controls another stepper motor

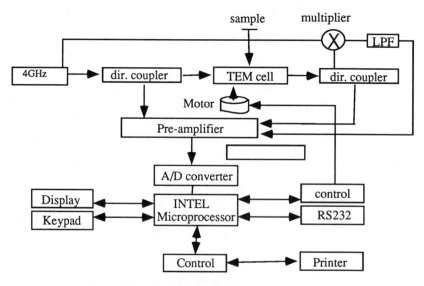

FIGURE 28.2. Block diagram showing all parts of the moisture meter.

for closing or opening the bottom plate. Data acquisition at the input and output of the TEM cell is achieved by first converting the analog signal level into a digital form for the microprocessor to read at the output of each ADC. It also obtains a moisture content from an appropriate look-up table. Finally, it outputs the result to one of the many different ports provided such as the local built-in printer, the RS-232 serial port, the Centronics port, or the LCD display. The microprocessor and its peripherals are mounted on a PCB with a standard STD bus, which includes a CPU, ROM, RAM, A/D converters, timer, and I/O ports. A keypad and an LCD display provide a local interface with the microprocessor system. This is an option to provide an on-site personnel with information such as moisture content, option prompts, date, and time. Such an option is also useful for entering or preparing new calibration tables. It is also used for fault diagnosis. When used in the stand-alone mode, all commands are generated from a keypad. The moisture result is displayed on an LCD or printed by a printer connected to one of the output ports. The built-in printer provides a copy of the measured result which includes sample type, operator-entered sample code, date, time, and moisture content.

The program controlling the meter is written in the assembler language for speed and stored in EPROM. The command set includes

a. Select material to be measured and carry out a measurement

b. Print the result

c. Transfer data to a remote host

d. Switch between remote and stand-alone mode

e. Set time and data

f. Calibrate the meter for a specific product

28.5. RESULTS

The meter was used to measure set moisture contents of different types of rice. The results obtained are most encouraging as set out below:

a. Inga brown short-grain rice
    Set moisture contents   15%   17%
    Measured values   14.9% ± 0.1   17.1% ± 0.4

b. Inga white short-grain rice
    Set moisture contents   15%   17.5%
    Measured values   15.2% ± 0.5   17.5% ± 0.2

c. Inga paddy rice
    Set moisture contents   11.5%   18%
    Measured values   11.6% ± 1   17.8% ± 1

It can be seen that the set results and the measured results are in close agreement. The accuracy depends very much on the preparation of the look-up tables and the accuracy in the determination of the moisture contents of the standard materials.

The meter was also tested against other materials such as beef cubes, chopped onions, and onion flakes. The worst recorded results are for onion flakes with a standard deviation of ±2.5% because of much variation in bulk density. Figure 28.3 shows the histogram for 1/4-inch beef cubes. A measured result of 3.9% ± 0.2 corresponds closely with the hot-air oven result of 3.7%. Figure 28.4 shows the histogram for chopped onions; a measured result of 6.5% ± 0.1 corresponds with the hot-air oven result of 6.3%.

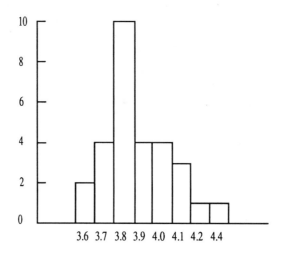

FIGURE 28.3. Histogram of readings vs. moisture content of beef cubes.

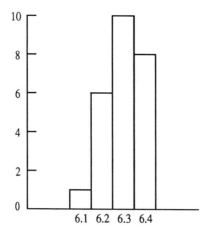

FIGURE 28.4. Histogram of readings vs. moisture content of chopped onion.

28.6. CONCLUSION

A microwave moisture meter has been developed from the measured dielectric data of a range of products. The frequency of 4 GHz was selected as a good compromise between sample sizes and dielectric properties. A TEM cell at a frequency in the X-band could make the sample size too small to be useful for large particles. A cell at a frequency much lower than 4 GHz would cause measurements to be influenced by ionic conductivity. The microprocessor controls the operation of the meter and provides a number of flexible options which are not possible otherwise. The meter can be used in the stand-alone mode or in the remote mode for discrete sampling or continuous flow measuring.

REFERENCES

[1] S. S. Stuchly and M. A. K. Hamid, "State of the art in microwave sensors for measuring non-electrical quantities," *Inst. J. Electron.*, vol. 33, p. 617, 1972.

[2] A. Kraszewski, "Microwave aquametry—a review," *J. Microwave Power*, vol. 15, no. 4, p. 209, 1980.

[3] A. Klein, "Microwave determination of moisture in coal: comparison of attenuation and phase measurement," *J. Microwave Power*, vol. 16, nos. 3, 4, p. 289, 1981.

[4] W. Meyer and W. M. Schilz, "Feasibility study of density independent moisture measurement with microwaves," *IEEE Trans. MTT.*, vol. 29, p. 732, 1981.

[5] M. Kent and W. Meyer, "A density independent microwave moisture meter for heterogeneous foodstuffs," *J. Food Engineering*, vol. 1, p. 31, 1982.

[6] S. Okamura, "High moisture content measurement of grain by microwaves," *J. Microwave Power*, vol. 16, nos. 3, 4, p. 253, 1981.

[7] N. Cutmore, D. Abernethy, and T. Evans, "Microwave technique for the on-line determination of moisture in coal," *J. Microwave Power Electromag. Energy*, vol. 24, no. 2, p. 79, 1989.

T. Lasri
B. Dujardin
Y. Leroy
Université des Sciences et Technologie de Lille,
Villeneuve d'Ascq, France

Y. Vincent
G. Mallick
France Maintenance Automatismes, Vandoeuvre, France

29

Present Possibilities for Moisture Measurements by Microwaves

Abstract.   Knowledge of the moisture content and permittivity of materials is of great importance during manufacturing processes. The aim of the work is to investigate the possibility of using microwave techniques for nondestructive control during these industrial processes. We demonstrate the feasibility of these techniques to determine the moisture content of materials such as fireproof materials, explosives, stonework sand, wood, caster sugar, urban muds, etc. For this purpose, measurements of microwave reflection/transmission coefficient are made on these materials by low-cost systems made of microwave components and integrated circuits.

29.1.   INTRODUCTION

Nowadays microwave technology has advanced to the point where robust and low-cost instruments are feasible, due to the availability of new components and integrated circuits. Therefore one can think of building on-line or in situ sensors which are needed at all stages (production, storage) of industries like agricultural, food, pharmaceutical, etc. With reference to this, an overview of the actual needs and perspectives of microwave aquametry is given by Kraszewski [1].

In this paper, we first present a few examples of microwave measurements (amplitude/phase of the reflection/transmission coefficient) done at 2.45 GHz and 10 GHz leading to the measurement of moisture content. We also point out that such a method can be applied to other examples of nondestructive measurements. We present also the way in which a very simple and low-cost microwave circuit can be built around such a smart sensor.

29.2.   EXAMPLES OF MOISTURE CONTENT MEASUREMENTS BY MICROWAVES

Different calibrations have been carried out on a lot of materials in terms of the reflection and transmission coefficients of samples versus moisture content.

Table 29.1 points out the cases that have been considered. Some details about several of these experiments are given below.

TABLE 29.1. Examples of Materials Tested

MATERIAL	MOISTURE CONTENT (H%)	FREQUENCY (GHz)
Fireproof materials	$21\% < H < 60\%$	2.45
Pyrotechnics products	$43\% < H < 46\%$	10
Stonework sand	$0\% < H < 20\%$	2.45
Caster sugar	$H < 2\%$	2.45
Wood	$H < 20\%$	10
Barley	$10\% < H < 66\%$	10
Urban muds	$70\% < H < 82\%$	10
Textiles	$H < 80\%$	10

29.2.1 Fireproof Materials

The corresponding boards move between a first array of applicators and a second array of receiving antennas, connected through a microwave switch to a transmission coefficient measurer (Figure 29.1). Figure 29.2 shows examples of measurement results for different thickness of material obtained with the system in Figure 29.1

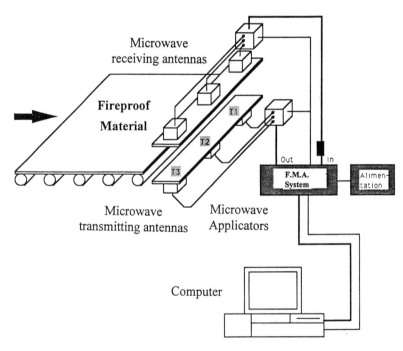

FIGURE 29.1. Synoptic of the system for measuring moisture content in fireproof materials.

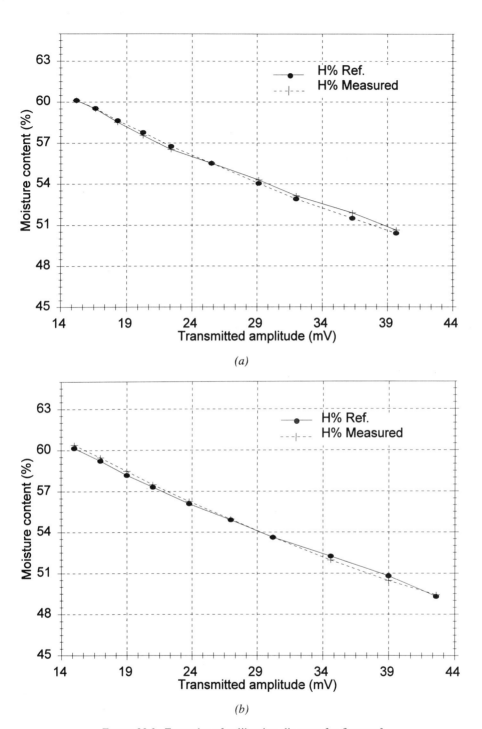

Figure 29.2. Examples of calibration diagram for fireproof materials at 2.45 GHz: (a) material thickness 23.3 mm; (b) material thickness 24.9 mm.

(moisture content versus the transmitted amplitudes). In both cases the calibration curve (determined by gravimetry) is presented with solid lines, and examples of measurements are given with dotted lines. In fact, first the calibration is carried out by gravimetry (weighing before and after drying); the measurement data are verified by the same method. In these cases the accuracy observed is better than 0.35%. Note also that for the two examples in these figures, the transmission coefficient depends slightly on sample thickness. This result can be due to near-field effects of the antennas (rectangular waveguide apertures).

29.2.2 Pyrotechnic Products

Another example of measurement corresponds to pyrotechnic products such as propergol during lamination. The basic scheme of the process is shown in Figure 29.3. The material inserted between two rollers of a rolling mill sticks on one of the rollers, which is rough. The corresponding reflection coefficient is measured by a horn connected to this system, which provides the amplitude and phase of the reflection coefficient. These data are deduced from the real and imaginary parts of the correlation products of these signals given by the balanced mixer.

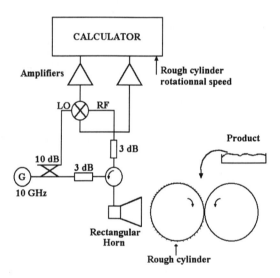

FIGURE 29.3. Basic scheme of the lamination process.

The output signal obtained versus time during the lamination of the product (propergol) is presented in Figure 29.4 [2]. This signal, obtained after filtering, exhibits different characteristic parts:

- *ab* corresponds to the empty rolling mill;
- *bc* points out the signal evolution during laminating of propergol;
- *cd* is obtained with the rolling mill empty again.

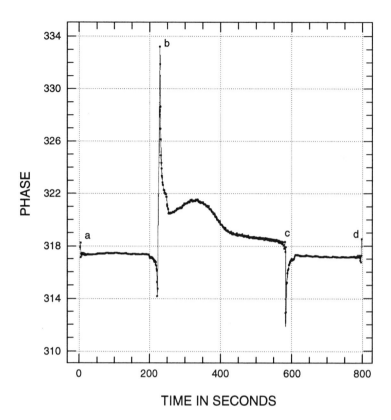

TIME IN SECONDS

FIGURE 29.4. Evolution of the signal during lamination
(after filtering).

A lot of experiments have shown that during the last stage of lamination (the end
of part bc), the phase θ of the reflection coefficient and the moisture content can
be modeled empirically as a function of time:

$$\theta = A + \frac{B}{C + t} \qquad \text{and} \qquad \frac{1}{H} = \frac{1}{H_0} + Kt,$$

where $H_0 =$ moisture content at the origin
             $H =$ moisture content at time t
$A, B, C, K =$ constant parameters which depend on the product;
             these parameters have no physical meaning but
             describe the process correctly.

So we find

$$H = \frac{H_0(\theta - A)}{(1 - K H_0 C)\theta + K H_0 B - A}.$$

This method is described in detail by Mallick et al. [2].

Briefly speaking, in practical applications, for a well-known material, with a reproducible thickness, the fact that a given phase θ is reached means that we have a desired moisture content H.

29.3.   A MICROWAVE SENSOR

Extensive laboratory studies have shown the possibility of building measurement systems devoted to such applications much simpler than automatic network analyzers [3,4,5]. So in order to carry out these measurements, a simple setup devoted to industrial measurements has been developed.

Several solutions go through well-known devices such as six ports, reflectometers, etc. We have chosen the principle of an interferometer [3]. The basic scheme is shown in Figure 29.5. The monochromatic signal e is divided into two components, which go either through a reference branch, including a reference impedance (R.I.), or through a measurement branch including the device under test (DUT). After that, the two signals merge into a quadratic detector (S). The case which is considered corresponds, for example, to the measurement of the transmission coefficient of two rectangular apertures placed on both sides of the material under test.

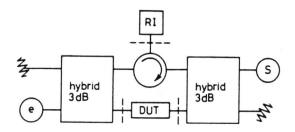

FIGURE 29.5. Basic scheme of the interferometer.

The layout of the hybrid circuit working at 2.45 GHz is shown in Figure 29.6. The source is connected to port 1, the R.I. to port 2, and the detector to port 4. The DUT is inserted between ports 3 and 5, and finally in 6 is added a control detector.

The principle of the interferometry implies that the signals transmitted in both branches exhibit nearly the same level. In most cases, the power transmitted through the DUT is much lower than the power arising from the reference arm, which includes a reflection coefficient near 1. In consequence, we have to adjust the level in the two branches to be nearly equal. This function is achieved by means of controlled voltage attenuators placed in the reference arm of the device.

Such a device has proven to be quite accurate (better than 0.5 dB in amplitude and $2°$ in phase) and stable. For example, an on-line measurement system of

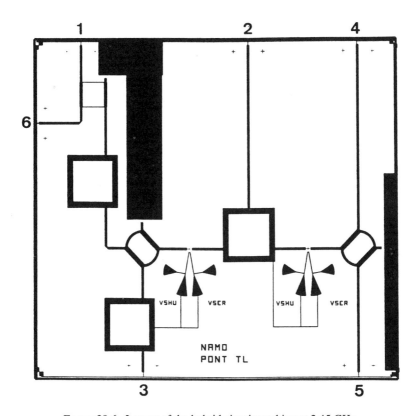

FIGURE 29.6. Layout of the hybrid circuit working at 2.45 GHz.

moisture in caster sugar has been implemented in sugar refinery. The reproducibility of the moisture measurement is 0.1%.

29.4. OTHER EXAMPLES OF NONDESTRUCTIVE CONTROL

If microwaves can work in moisture measurement because of a variation of permittivity of the material, they work also for nondestructive control. The next example concerns the detection of knots in wood. We present in Figure 29.7 several diagrams observed for a knot, for the beginning of a branch, and for several knots along the same board. This measurement works only when moisture content does not hide the heterogeneity effect ($H < 20\%$).

Another example concerns the detection of holes in a polyester board loaded with fiberglass. Other examples of nondestructive control are related to holes in refractory bricks, density measurements of cellulose boards (Figure 29.8), and thickness of polypropylene (Figure 29.9). Such data are obtained after calibration with known material.

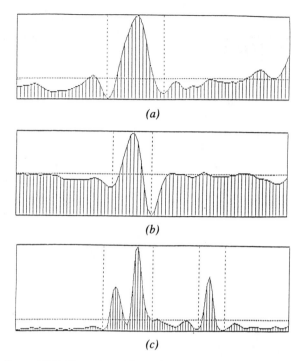

FIGURE 29.7. Detection of knots in wood: (a) characteristic
pattern for a knot; (b) pattern for the beginning of a branch;
(c) pattern of three knots along the same board.

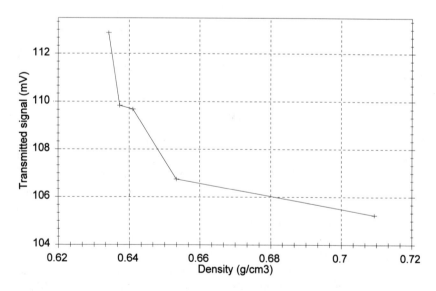

FIGURE 29.8. Density measurements of cellulose boards.

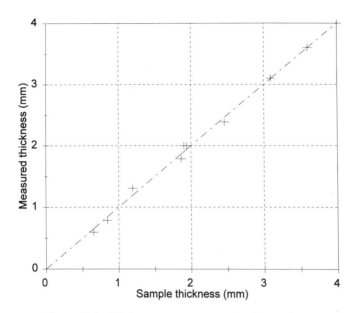

FIGURE 29.9. Thickness measurements of polypropylene.

29.5. CONCLUSION

We have demonstrated the interest of microwaves for applications to military or telecommunications in the field of nondestructive control and moisture measurements. The feasibility of systems, now made quite easy from the development of hybrid and integrated circuit technologies, allows us to think that we are able to fulfill the demand for low-cost microwave sensors. Moreover, the applications we mentioned demonstrate that a lot of industrial requirements could be fulfilled by such developments.

ACKNOWLEDGMENTS

This work is supported by the Ministère de la Recherche et de l'Espace, contract 92 B 0274.

REFERENCES

[1] A. Kraszewski, "Microwave aquametry—needs and perspectives," *IEEE Trans. Microwave Theory Tech.*, vol. 39, pp. 828–835, May 1991.

[2] G. Mallick, M. Rosso, J. M. Tauzia, E. Giraud, O. Brenachot, and G. Roussy, "Capteur microonde intelligent pour la mesure de l'humidité des poudres

pyrotechniques au cours de leur laminage," *L'onde Electrique*, vol. 73, no. 3, pp. 53–57, Mai–Juin 1993.

[3] T. Lasri, B. Dujardin, and Y. Leroy, "Microwave sensor for moisture measurements in solid materials," *IEE Proc.-H*, vol. 138, no. 5, pp. 481–483, 1991.

[4] M. Craman, J. David, and R. Crampagnes, "Mesures de constantes diélectriques complexes à l'aide d'un dispositif autonome utilisant un calculateur compatible P.C.," *Rev. Phys. Appl.*, pp. 469–474, Mai 1990.

[5] R. J. King, K. V. King, and K. Woo, "Microwave moisture measurement of grains," *IEEE Trans. Instrum. Meas.*, vol. 41, no. 1, pp. 111–115, Feb. 1992.

30

Ashley Robinson
Marek Bialkowski
*University of Queensland, Brisbane,
Queensland, Australia*

Single- and Multiple-Frequency Phase Change Methods for Microwave Moisture Measurement

Abstract. Three methods of measuring the phase change of a transmitted signal across a sample for use in determining moisture content are examined. These methods overcome the problem of the 360° limit on the phase measurements by continuous monitoring of the phase change of a single-frequency signal or by using multiple frequencies. Methods of calibration for a six-port network analyzer or a quadrature mixer to obtain measurements in real time are also described.

30.1. INTRODUCTION

Previous work [1] on microwave moisture measurement has shown that the real part of the relative permittivity of a material varies in proportion to the square of the moisture content, i.e., $(\% \ moist)^2 \propto \epsilon_r$, and the loss tangent is proportional to the moisture content. As the phase change of a transmitted signal is proportional to the square root of the real part of the dielectric constant (for a small imaginary part), it is therefore also proportional to the moisture content. The proportionality of the loss tangent to the moisture content leads to the result that attenuation of a transmitted signal is proportional to the moisture content. Thus, the problem of moisture content measurement can be reduced to the measurement of the phase change of a transmitted signal. For increased accuracy of moisture content measurement, extra parameters such as the transmitted signal attenuation and mass per area to take into account the effects of other variables, e.g., sample density, can be included.

This paper presents methods of phase measurement that meet the following requirements:

- The measurement system is cheap and rugged for use in industry.
- Measurements are simple from the operator's point of view.
- Measurements are accurate.
- Measurements are completed in real time.
- The phase measurement is not limited to the 0° to 360° range.

Single-frequency phase counting, multiple-frequency phase counting, and Fourier transform methods of phase measurement for moisture content measurements are investigated theoretically and experimentally. In each case measurements are assumed to be made using either a quadrature mixer or a six-port network analyzer.

30.2. SINGLE-FREQUENCY PHASE COUNTING METHOD

The phase change can be measured at a single frequency using a vector network analyzer, a six-port network analyzer, or a quadrature mixer. In all these cases, however, the phase measurement is limited to only 0° to 360°. To increase the range of possible phase measurements, the phase can be measured continuously and then the direction of movement of the phasor can be tracked over time. The only limitation in this approach is that the change in phase between any two successive measurements in time must be limited to a 180° range. This forms the basis of a single-frequency phase counting method.

The single-frequency phase counting method can now be used to measure the phase of the signal accurately in real time. A vector network analyzer can do this task but is too expensive and fragile for use in industrial measurements. An alternative is a six-port or a quadrature mixer which does not have this high price tag (Figure 30.1). A six-port network analyzer has similar accuracy to the vector network analyzer, and the quadrature mixer also provides reasonable accuracy, but both of these devices require a suitable calibration for real-time operation. In order to obtain real-time operation, a method of calibration and measurement as described in [2] is used.

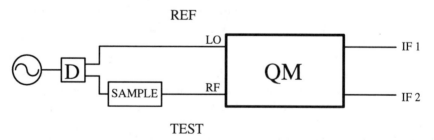

FIGURE 30.1. Quadrature mixer configuration.

For the six-port network in Figure 30.2, the signals incident on the detectors are linear combinations of signals a and b. For detectors operating in a square-law

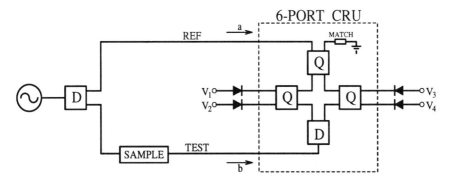

FIGURE 30.2. Six-port network analyzer configuration.

region the voltage at the output of the ith detector is given by

$$V_i = |A_i a + B_i b|^2 + V_{ri}, \qquad (30\text{-}1)$$

where A_i and B_i are complex constants and V_r is a residual voltage appearing at the detector's output when the RF power is off. Expression (30-1) can be rewritten in the form

$$V_i = (|A_i|^2|T|^2 + |B_i|^2)|b|^2 + 2\text{Re}\left\{A_i B_i^*|T|e^{j\phi}\right\}|b|^2 + V_{ri}, \qquad (30\text{-}2)$$

where T is a phasor $T = |T|e^{j\phi} = a/b$, the asterisk implies a conjugate value, and $j = \sqrt{-1}$. For real-time operation of the six-port the transmission coefficient phasor $T_m = b/a$ is formed by voltage differences:

$$T_m = \tilde{a}_x\,(V_1 - V_2) + \tilde{a}_y(V_3 - V_4). \qquad (30\text{-}3)$$

When the quadrature mixer is used, the phasor is

$$T_m = \tilde{a}_x V_{\text{IF1}} + \tilde{a}_y V_{\text{IF2}}. \qquad (30\text{-}4)$$

Note that the x and y components of T_m can be regarded as the real and imaginary components of the phasor T_m.

By using eq. (30-2) and by assuming that the signal b is kept constant, the x and y components of the vector T_m for the six-port can be rewritten as

$$\begin{aligned}
T_{mx} &= E_1|T|^2 + F_1|T|\cos(\phi - \phi_1) + G_1, \\
T_{my} &= E_2|T|^2 + F_2|T|\cos(\phi - \phi_2) + G_2,
\end{aligned} \qquad (30\text{-}5)$$

where E_i, F_i, G_i, and ϕ_i are real-value constants. It is apparent that in the calibration process these constants have to be determined.

By inspection, eq. (30-5) describes, in general, a family of ellipses, where circles $|T| = \text{constant}$ correspond to ellipses T_m. These ellipses are alike; i.e., the ratio of the length of the major axis to the minor axis is constant, but the positions of the ellipses' centers vary.

The ideal design of a real-time display six-port CRU requires hardware such that eq. (30-5) describes a family of concentric circles centered at the origin of the x-y coordinate system. In theory, this is equivalent to having constants E_i and G_i equal to zero, constants F_1 and F_2 equal, and phase difference $\phi_1 - \phi_2$ equal to 90°. These conditions can alternatively be achieved via calibration software. The details of the transformation of the family of ellipses in eq. (30-5) into a family of concentric circles can be found in [2]. A brief description of the method follows.

To transform the off-center, angled ellipse shown in Figure 30.3b to the ideal circle of Figure 30.3a, first the center of the ellipse is found using the maximum and minimum deviation of the x and y components of T_m, while the phase ϕ varies continuously from 0° to 360°. This first stage result is demonstrated in Figure 30.3c.

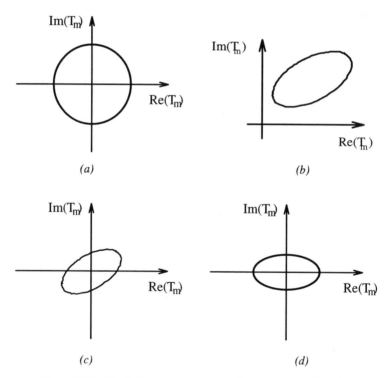

FIGURE 30.3. Single-frequency complex plane representation of the measured transmission coefficient T_m: (a) ideal case $T = e^{j\phi}$; (b) nonideal case $T = e^{j\phi}$; (c) after subtracting offset from (b); (d) after subtracting angular offset from (c).

The second stage of the correcting procedure is to find the tilt angle ϕ_0 by recording ϕ at which the distance to the origin is a maximum. (Note that ϕ_0 corresponding to the minimum distance could also be chosen, but the maximum is preferred because of a sharper appearance.) The result of the second stage is shown in Figure 30.3d.

Now, only axes $\mathrm{Re}\{T_m\}$ and $\mathrm{Im}\{T_m\}$ have to be scaled to obtain the result of Figure 30.3a.

During measurements, T_m is determined from measured voltages V_1, V_2, V_3, V_4. The magnitude and phase of T is then determined using a two-stage iterative procedure (described in [2]).

When a quadrature mixer is used, $\mathrm{Re}\{T_m\}$ and $\mathrm{Im}\{T_m\}$ are obtained from the two IF port voltages. It is assumed that the system is ideal enough to give circles as the phase rotates from $0°$ to $360°$. In practice this is achieved by the proper design of the mixer. The only correction needed in this case is the offset of the circle center. The required offset is measured and recorded by using an infinite attenuation calibration standard such as radar absorbant material or by simply pointing the transmit and receive antennas away from each other.

30.3. MULTIPLE-FREQUENCY PHASE COUNTING

For a frequency step from f_1 to f_2 the transmitted signal changes by an amount (TEM wave case)

$$\Delta\phi_{\text{tot}} = \Delta\beta L + \Delta\phi_i = \frac{2\pi(f_2 - f_1)L\sqrt{\epsilon_r}}{c} + \Delta\phi_i, \tag{30-6}$$

where L is the sample length, ϵ_r is the sample dielectric constant, c is the speed of light, and $\Delta\phi_i$ is the in-built phase characteristic of the transmission path. By knowing the change in phase over the frequency sweep, the relative permittivity can be calculated from the equation

$$\epsilon_r = \left(\frac{(\Delta\phi_{\text{tot}} - \Delta\phi_i)c}{2\pi(f_2 - f_1)L}\right)^2. \tag{30-7}$$

As in the single-frequency phase counting method, the change in phase between the two measurement frequencies must be less than π to avoid ambiguous results. This restriction, which is related to Nyquist's criterion, can be fulfilled by choosing a sufficiently small frequency step. In practice, the frequency is changed in a digital manner between two frequencies $f_{\min}$ and $f_{\max}$. The change in phase between each discrete frequency in the sweep is added to give the total phase change between the minimum and maximum frequencies.

The $\Delta\phi_i$ term in eqs. (30-6) and (30-7) at each frequency is independent of the sample. This term can be canceled by the calibration using known sample ϵ_r.

The most common practical problem with the multiple-frequency phase counting method is dislocation of the constant attenuation circle center when frequency is varied (Figure 30.4). This is due to nonidealities of the measuring system. The reasons are imperfect attenuation measuring system and reflections from the sample. This dislocation can be measured using an infinite attenuation calibration standard (Figure 30.5). After these measurements are taken, the system can be calibrated to give circles centered at the origin of the complex transmission coefficient plane (Figure 30.6). Still in this method an assumption is made that after removing the circles' center dislocation the hardware is ideal enough to get ϕ accurately. In practice, the error still remaining is another nonideality of the system which is exhibited by elliptical type of constant attenuation curves. In practice this contributes to the phase error by 10° to 15°.

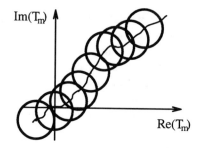

FIGURE 30.4. Typical measured results for T_m for the case of a lossless TEM transmission when the frequency is varied.

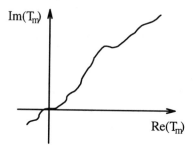

FIGURE 30.5. Infinite attenuation calibration measurement.

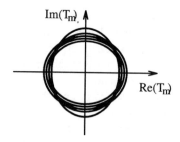

FIGURE 30.6. Results of Figure 30.4 after correcting for circle center dislocation.

30.4. FOURIER TRANSFORM METHOD

The Fourier transform can form the basis for measuring the phase and therefore the moisture content in the following manner.

The transfer function of the sample in the frequency domain is

$$T(f) = \pi_W(f) \times e^{j\beta L}, \tag{30-8}$$

where W is the sweep size $= f_{max} - f_{min}$. Over a frequency sweep this transfer function is a periodic signal with period inversely proportional to the square root of the relative permittivity; i.e., the phase or moisture content will be proportional to the frequency of the periodic signal. To obtain the periodic signal frequency, the Fourier transform with respect to the electrical length can be taken (Figure 30.7). The transfer function in the electrical length domain is

$$t(x) = W \operatorname{sinc}\left(Wx + \frac{L\sqrt{\epsilon_r}}{c}\right). \tag{30-9}$$

The phase change of the sample can be calculated from the peak position of the transfer function $t(x)$ using the equation

$$\phi = L\sqrt{\epsilon_r} = \frac{(\text{peak position})c}{W}. \tag{30-10}$$

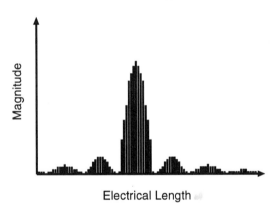

FIGURE 30.7. Plot of the transmission coefficient $t(x)$ as a function of electrical length of the sample.

30.5. RESULTS

The multiple-frequency phase counting and Fourier transform methods have been tested using computer simulations of a dielectric slab. These results are shown in Figure 30.8. The multiple-frequency phase counting method gives a mean error in the calculated relative permittivity of 0.13% when a single path through the dielectric block is used, and an error of 1.7% when multiple paths are simulated. For the Fourier transform method when a single path is used, the mean error in the calculated dielectric constant is 1.7%. When multiple paths through the slab were simulated, the Fourier transform method did not perform accurately and led to erroneous results. It was then concluded that a modified version of this method with additional digital signal processing frequency estimation techniques is required to make this method useful. These additional techniques have not been pursued further here. The simulations also showed that both methods gave more accurate results as the thickness of the sample increased, i.e., as the total phase change across the sample increased. This can be attributed, in part, to the increased attenuation of the multiple paths through the dielectric to allow the direct path to dominate the transmission coefficient.

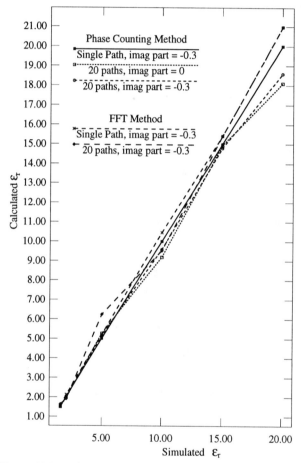

FIGURE 30.8. Computer simulation results for multiple-frequency phase counting and Fourier transform methods.

In experimental tests a six-port network analyzer and a quadrature mixer were used.

Tests of the six-port network analyzer have shown that phase measurements are typically accurate to within ±1° and attenuation measurements to within 1 dB from 0 to 40 dB. The analyzer is capable of taking 30 single-frequency measurements per second when a 386 PC is used. The quadrature mixer was less accurate (±10° phase error) in the attenuation range from 0 to 30 dB. The measurement speed for the quadrature mixer was similar to the six-port since the results are written to the computer monitor.

Brief moisture measurements on beach sand with moisture in the 0% to 15% moisture range and sample thickness aproximately 30 mm have been performed using the single- and multiple-frequency phase counting methods (see Figure 30.9). The single-frequency method gave accurate results for the entire moisture range (0.77% moisture maximum error—occurred for the 12% moisture sample). The multiple-frequency phase counting method gave accurate results for moisture content above 5%. However, when the moisture content of the sample was below 5%, moisture problems with the multiple paths through the sample occurred, resulting in erroneous results which are not included here. It is expected that as sample thickness increases, the multiple-frequency phase counting method would give better results since the internally reflected signals attenuation will be higher, which occurred in the simulation results.

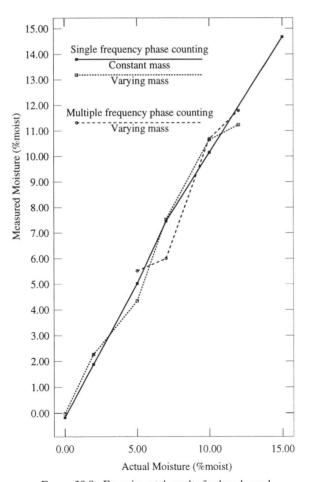

FIGURE 30.9. Experimental results for beach sand.

As a result of the experiments it has been concluded that the single-frequency phase counting method has the advantage that the measurements are simple, fast, and accurate when compared to the multiple-frequency method. Problems with the single-frequency method may, however, occur if the phase count is lost. This may be due to noise, unexpected abrupt change in sample characteristics, or power failure. In all these cases the phase would become ambiguous. The multiple-frequency method is more reliable, but is more time consuming as a number of frequency points must be measured to avoid ambiguous phase measurements.

30.6. CONCLUSIONS

Three methods for the measurement of the phase change of a transmitted signal over a sample to accurately determine its moisture content have been presented. It has been shown that these methods of phase measurement are accurate when multiple reflections inside the dielectric samples can be neglected. Practical moisture measurements show that the single- and multiple-frequency phase counting methods can give accurate results.

REFERENCES

[1] N. Cutmore, D. Abernathy, and T. Evans, "Microwave technique for the on-line determination of moisture in coal," *J. Microwave Power Electromag. Energy*, vol. 24, pp. 79–90, 1989.

[2] M. Bialkowski and A. Dimitrios, "A step-frequency six-port network analyser with a real-time display," *Arch. Elektron. Ubertragungstechnik*, vol. AEU, no. 3, pp. 193–197, Mai 1993.

[3] A. Robinson and M. Bialkowski, "An investigation into microwave moisture measurements," in *Asia Pacific Microwave Conf. Proc.*, Adelaide, pp. 571–574, 1992.

[4] A. Klein, "Microwave determination of moisture in coal: comparison of attenuation and phase measurements," *J. Microwave Power*, vol. 16, pp. 289–303, 1981.

[5] A. Kraszewski and S. Kulinski, "An improved method of moisture content measurements and control," *IEEE Trans. Ind. Electron. Control Instr.*, vol. IECI-23, pp. 364–370, 1976

Section V

A Reader's
Bibliography

Andrzej W. Kraszewski

USDA, ARS, Russell Research Center, Athens, GA

Fifteen Years of Literature in Microwave Aquametry

Editor's note: *This bibliography contains almost 600 patents, papers, and reports from generally accessible literature on moisture content determination, measurement, and monitoring by microwave methods that have been published during the past 15 years. It is a continuation of the work published in the* Journal of Microwave Power *(vol. 15, no. 4, pp. 298–310, 1980) 15 years ago. Generally, the bibliography covers papers classified under UDC codes 621.317.39, 621.385, 53.093, and 543.812.08 and patents in the international classification G01N22 group, as well as the U.S. classification 324/63 and 324/64 groups (former 324/58.5C group). The entries are listed alphabetically in chronological order and, according to the rule accepted in the previous bibliography, patents are listed according to the date of filing. Most of the references are drawn from English sources, but there are many important contributions from Europe and Japan. Soviet and Chinese literature has been largely inaccessible during this period and may be therefore underrepresented in the listing.*

At the end, a group of papers on the application of microwave power for drying samples of various materials for determination of moisture content is listed, as this use of microwave energy finds broad application in scientific and industrial laboratories around the world.

The author acknowledges the generous response to his appeal for cooperation in completing this bibliography. Special thanks go to Dr. Mike Kent of the Torry Research Station in Aberdeen, Scotland, to Dr. Albert Klein of Indutech GmbH in Simmersfeld, Germany, to Dr. Josef Mladek of the Food Science Institute in Prague, Czech Republic, and to Dr. Ebbe Nyfors of the Helsinki University of Technology, Finland, who, among many others, supplied the author with many valuable references. The invaluable help of our librarian, the late Barbara Gazda, in finding and providing the later references is gratefully acknowledged.

1977

[1] Ahlgrimm, H. J. Dielectric determination of the moisture content of food (in German). *Z. Lebensm.-Tech. Verfahrens*, 28: 305–311, 1977.

[2] Belayev, M. I., Ryzhova, N. V. and Yazikov, V. N. Dielectric properties of some edible fats (in Russian). *Maslod. zhirov. Prom.*, (4): 25–29, 1977.

[3] Dekermandji, G. and Joines, W. T. A new method for measuring the electrical properties of sea water and wet earth at microwave frequencies, *IEEE Trans. Instrum. Meas.*, IM-26 (2): 124–127, 1977.

[4] Doughty, D. A. Determination of water in oil emulsions by a microwave resonance procedure. *Anal. Chem.*, 49 (6): 690–694, 1977.

[5] Fitzky, H. G. Neuere Entwicklungen von Mikrowellen-Feuchtemessgeraten fur Labor und Betrieb. *G-I-T Fachzeitschrift Laboratorium*, 21: 1062–1070, Oct. 1977.

[6] Furusawa, M. and Nagasawa, T. Studies of the electric moisture meter for tobbaco. Performance of a microwave attenuation moisture meter. *Sci. Pap. Cent. Research Inst. Japan*, 119: 13–22, 1977.

[7] Hoberg, H. and Klein, A. Wassergehaltsbestimmung von feinkornigen Steinkohlen nach dem Mikrowellen-verfahren. *Gluckauf*, 113 (17): 859–863, Sep. 1977.

[8] Iskander, M. F. and Hamid, M. A. K. A new strip transmission line for moisture content measurements. *J. Microwave Power*, 12 (1): 16–18, 1977.

[9] Kalinski, J. Self-adjustable microwave homodyne circuit for on-line simultaneous attenuation and phase shift measurement. *Proc. 7th Europ. Microwave Conf.*, 267–272, Copenhagen, Denmark, 1977.

[10] Kalinski, J. Microwave moisture-attenuation transducer based on comparison in time domain (in Polish). *Arch. Elektr.*, 26 (4): 679–690, 1977.

[11] Kraszewski, A. Prediction of the dielectric properties of two-phase mixtures. *J. Microwave Power*, 12 (3): 215–222, 1977.

[12] Kraszewski, A., Kulinski, S. and Stosio, Z. A preliminary study on microwave monitoring of moisture content in wheat. *J. Microwave Power*, 12 (3): 241–252, 1977.

[13] Kraszewski, A., Lenarcik, A. and Stosio, Z. Microwave instrument for moisture content determination in granular and powdered materials in laboratories (in Polish). *Pomiary Automatyka Kontrola (Measur., Automation and Control)*, 23 (1): 8–11, 1977.

[14] Miyai, Y., Kojo, T., Takahashi, J., Yoshioka, S., Nakamichi, K., Okada, K. and Takeda, S. Moisture content meter. U.S. Patent 4,104,584; filed Feb. 2, 1977.

[15] Najim, M., Zyoute, M., Najim, K. and Ouazzani, T. Complex permittivity of maroccan phosphate as function of moisture content. *Proc. 7th Europ. Microwave Conf.*, 261–266, Copenhagen, Denmark, 1977.

[16] Neelakantaswamy, P. S. and Ramaswamy, N. S. On microwave moisture meters for agricultural products. *Zeitschrift elektr. Inform. u. Energietechnik (Leipzig)*, 7 (4): 344–352, 1977.

[17] Nelson, S. O. Use of electrical properties of grain—moisture measurement. *J. Microwave Power*, 12 (1): 67–72, 1977.

[18] Okada, K., Nakamichi, K., Miyai, Y., Takahasi, I., Kosyo, T. and Yoshioka, S. Microwave moisture meter (in Japanese). TG SSD 77-55, IECE, Japan, 1977.

[19] Pakulis, I. E. Microwave moisture sensor chute. U.S. Patent 4,131,845; filed Oct. 3, 1977.

[20] Ramachandraiah, M. S. Microwave moisture meter (in French). Swiss Patent 604,163; filed Dec. 31, 1977.

[21] Schofield, J. W. Measurement of impurity concentration in liquids. U.S. Patent 4,104,585; filed Jan. 26, 1977.

[22] Schwan, H. P. and Foster, K. R. Microwave dielectric properties of tissue. *Biophys. J.*, 17: 193–197, 1977.

[23] Taro, M., Takahiro, Y., Makoto, S. and Shinya, N. Microwave hygrometer. U.S. Patent 4,103,224; filed May 31, 1977.

[24] Vestergaard, P., Moller, G. and Madsen, R. F. Method and apparatus for measuring the concentration of fluids. U.S. Patent 4,196,385; filed Dec. 27, 1977.

[25] Webman, I., Jortner, J. and Cohen, M. H. Theory of optical and microwave properties of microscopically inhomogeneous materials. *Phys. Rev.*, B15: 5712–5723, 1977.

[26] Wobschall, D. A theory of complex dielectric permittivity of soil containing water: semidisperse model. *IEEE Trans. Geosci. Electr.*, GE-15: 49–58, 1977.

1978

[27] Fitzky, H. G., Schmitt, F., Bollongino, N. and Rehrmann, H. Apparatus for determining the water content of isotropic materials by means of microwave absorption. U.S. Patent 4,203,067; filed Mar. 15, 1978.

[28] Fitzky, H. G., Schmitt, F., Bollongino, N. and Rehrmann, H. Apparatus for determining the water content of isotropic materials by means of microwave absorption. U.S. Patent 4,206,399; filed Mar. 27, 1978.

[29] Fitzky, H. G., Schmitt, F., Bollongino, N. and Rehrmann, H. Apparatus for determining the water content of isotropic materials. U.S. Patent 4,211,970; filed May 30, 1978.

[30] Ganzlin, G. and Soder, J. M. Sofortmessung der Feuchtigkeit in Hopfen (Er-probung der Feuchtemessung in Hopfen und Hopfenpellets mit einem Mikro-wellen-Feuchtemessgerat). *Brauwissenschaft*, 31: 307–310, 1978.

[31] Grant, E. H., Sheppard, R. J. and South, G. *Dielectric Behaviour of Biological Molecules in Solutions: Monographs on Physical Biochemistry.* Oxford: University Press, 1978.

[32] Heikkila, S. Method and apparatus for detecting grain direction in wood, particularly in lumber. U.S. Patent 4,500,835; filed June 1, 1978.

[33] Hoberg, H. and Klein, A. Quickly performed measuring method for ascertaining the concentration of the polar components in a material otherwise mainly non-polar. U.S. Patent 4,233,559 (FRG Patent 2,808,739); filed March 1, 1978.

[34] Kalinski, J. Microwave attenuation measurement with chopped subcarrier method for on-line moisture monitoring. *Proc. 1978 IMPI Microwave Power Symp.*, 142–144, Ottawa, Canada, June 1978.

[35] Kalinski, J. An industrial microwave attenuation monitor (MAM) and its application for continuous moisture content measurements. *J. Microwave Power*, 13 (3): 275–281, 1978.

[36] King, R. J. Microwave electromagnetic nondestructive testing of wood. *Proc. 4th Symp. on Nondestructive Testing of Wood*, 121–134, Vancouver, WA, Aug. 1978.

[37] Kraszewski, A. Continuous measurement and control of moisture content in food products using microwave instrumentation. *Proc. 1978 Conf. on Meas. and Control in Food Industry*, Praha, Czechoslovakia, Feb. 1978.

[38] Kraszewski, A. A model of the dielectric properties of wheat at 9.4 GHz. *Proc. 1978 IMPI Microwave Power Symp.*, Ottawa, Canada, June 1978.

[39] Kraszewski, A. Model of the dielectric properties of wheat at 9.4 GHz. *J. Microwave Power*, 13 (4): 293–296, 1978.

[40] Lin, C. K. An investigation of a microwave device for the determination of veneer moisture content and its application in veneer dryer for continuous recording and automatic controlling (in Chinese). *Yeh K'ohsueh*, 1: 56–62, 1978.

[41] Miyai, Y. A new microwave moisture meter for grains. *J. Microwave Power*, 13 (2): 163–166, 1978.

[42] Nelson, S. O. Frequency and moisture dependence of the dielectric properties of high-moisture corn. *J. Microwave Power*, 13 (2): 213–218, 1978.

[43] Okado, K., Nakamichi, K., Miyai, Y., Takahashi, J., Kojo, T. and Yoshioka, S. Microwave moisture meter. *Natl. Techn. Rep. Matsushita Electr. Ind.*, 24 (3): 520–525, 1978.

[44] Poley, J. P., Nooteboom, J. J. and de Waal, P. J. Use of VHF dielectric measurements for borehole formation analysis. *Log Anal.*, 19: 8–30, 1978.

[45] Schilz, W. Novel microwave techniques for industrial measurements. *Proc. 8th Europ. Microwave Conf.*, 166–175, Paris, France, Sept. 1978.

[46] Shilian, C. Measurement of water content in crude oil with microwave phase method (in Chinese). *J. Petroleum Inst. E. China*, no. 3, 1978.

[47] Steele, D. J. and Kent, M. Microwave stripline techniques applied to moisture measurement in food materials. *Proc. 1978 IMPI Microwave Power Symp.*, 33–36, Ottawa, Canada, June 1978.

[48] Strandberg, C. F. and Strandberg, R. C. Microwave moisture indicator and control. U.S. Patent 4,156,843; filed March 13, 1978.

[49] Ulaby, F. T., Batlivala, P. P. and Dobson, M. C. Microwave backscatter dependence on surface roughness, soil moisture, and soil texture, Part I: Bare soil. *IEEE Trans. Geosci. Remote Sensing*, GE-16: 286–295, 1978.

[50] Wang, J. R. and Schmugge, T. J. An empirical model for the complex dielectric permittivity of soils as a function of water content. *NASA Techn. Memo. 79659*, Goddard Space Flight Center, Greenbelt, MD, 1978.

[51] Wobschall, D. A frequency shift dielectric soil moisture sensor. *IEEE Trans. Geosci. Electr.*, GE-16 (2): 112–116, 1978.

[52] Wyslouzil, W. Microwave moisture-profile gauge. U.S. Patent 4,193,027; filed Oct. 2, 1978.

[53] Zander, L., Zander, Z. and Wasilewski, R. Dielectric properties of butter with high water content. Proc. 20th Int. Dairy Congress, E: 873, 1978.

1979

[54] Akyel C. and Bosisio R. G. Microprocessor controlled permittivity measurements on high-loss materials using an active Q-multiplier technique (in French). *Proc. 1979 IMPI Microwave Power Symp.*, Monte Carlo, Monaco, June 1979.

[55] Bastida E., Fanelli N. and Marelli E. Microwave instruments for moisture measurement in soils, sands and cements. *Ibid.*, 147–149, June 1979.

[56] Berliner, M. A. Measurements of moisture content (in Russian). *Metrologia i izmierit. tekhn. (Metrol. Measure. Eng.)*, 4: 187–252, *Itogi nauki i tekhn. (Rev. Sci. Technol.)*, Moscow, USSR, 1979.

[57] Chu, F. Y. and Balls, B. W. Moisture measurements: infra-red vs. microwave (in papermaking). *Pulp Paper Mag. Canada*, 80 (5): 71–73, 1979.

[58] Chudobiak, W. J., Syrett, B. A. and Hafez, H. M. Recent advances in broadband VHF and UHF transmission line methods for moisture content and dielectric constant measurement. *IEEE Trans.*, IM-28 (4): 284–289, Dec. 1979.

[59] Faxvog, F. R. and Krage, M. K. Microwave acoustic spectrometer. U.S. Patent 4,277,741; filed June 25, 1979.

[60] Freedman, R. and Vogiatzis, J. P. Theory of microwave dielectric constant logging using the electromagnetic wave propagation method. *Geophysics*, 44: 969–986, 1979.

[61] Hafez, H. M., Chudobiak, W. J. and Wight, J. S. The attenuation rate in fresh water at VHF frequencies. *IEEE Trans. Instrum. Meas.*, IM-28 (1): 71–74, 1979.

[62] Henry, F. Mesures simultaneés des variations de masse et de permittivité. *Proc. 1979 IMPI Microwave Power Symp.*, Monte Carlo, Monaco, June 1979.

[63] Hill, J. H. Delay line microwave moisture measuring apparatus. U.S. Patent 4,319,185; filed Dec. 5, 1979.

[64] Hoppe, W., Meyer, W. and Schilz, W. Vorrichtung zur Fechte messung mit Mikrowellen. FRG Patent P29 42 971.8, filed Oct. 24, 1979.

[65] Kent, M. and Prince, T. E. Compact microstrip sensor for high moisture content measurement. *J. Microwave Power*, 14 (4): 363–365, 1979.

[66] Klein, A. The dielectric properties of moist coal and consequences of these on results for moisture content determination using microwaves. *Proc. 1979 IMPI Microwave Power Symp.*, 156–158, Monte Carlo, Monaco, June 1979.

[67] Kraszewski, A. A model of the dielectric properties of grain at microwave frequencies. *IEE Conf. Publ.* (177): 174–177, Sept. 1979.

[68] Kwok, B. P., Nelson, S. O. and Bahar, E. Time-domain measurements for determination of dielectric properties of agricultural materials. *IEEE Trans.*, IM-28 (2): 109–112, 1979.

[69] Meyer, W. and Schilz, W. Verfahren zur Messung der relativen Feuchte eines Messguts mit Hilfe von Mikrowellen. FRG Patent P29 28 487.5, filed July 26, 1979.

[70] Meyer, W. and Schilz, W. A microwave method for measuring the relative moisture content in an object. British Patent 2,057,137 (U.S. Patent 4,361,801), filed July 14, 1979.

[71] Meyer, W. and Schilz, W. Microwave absorption by water in organic materials. *IEE Conf. Publ.* (177): 215–219, 1979.

[72] Nelson, S. O. Electrical properties of grain and other food materials. *J. Food Process. Preserv.*, 2: 137–154, 1979.

[73] Ozamis, J. M. and Hewitt, S. J. Microwave moisture measurement system. *Proc. 9th Europ. Microwave Conf.*, paper MS-2, Brighton, U.K., Sept. 1979.

[74] Paap, H. J. Microwave-gamma ray water in crude monitor. U.S. Patent 4,289,020; filed Dec. 26, 1979.

[75] Paap, H. J. Microwave water in crude monitor. U.S. Patent 4,301,400; filed Dec. 29, 1979.

[76] Partain, L. D. and Lakshminarayana, M. R. Resonant circuit sensor of multiple properties of objects. U.S. Patent 4,257,001; filed Apr. 13, 1979.

[77] Rubino, N. A microwave apparatus for the rapid determination of kinetics parameters of reactions in aqueous solution. *Proc. 1979 IMPI Microwave Power Symp.*, Monte Carlo, Monaco, June 1979.

[78] Sasaki, S. Apparatus for measuring a percentage of moisture and weighing of a sheet-like object. U.S. Patent 4,297,875; filed Oct. 26, 1979.

[79] Taneya, S., Handa, M., Hayashi, H. and Sone, T. Automatic control of moisture in continuous buttermaking process. *Milchwissenschaft*, 34: 197–200, 1979.

[80] Tiuri, M. and Heikkila, S. Microwave instrumnet for accurate moisture measurement of timber in sawmill. *Proc. 9th Europ. Microwave Conf.*, paper P-9, Brighton, U.K., 1979.

[81] Ulaby, F. T., Bradley, G. A. and Dobson, M. C. Microwave backscatter dependence on surface roughness, soil moisture, and soil texture. Part II: Vegetation-covered soil. *IEEE Trans. Geosci. Remote Sensing*, GE-17: 33–40, 1979.

[82] Walsh, J. E., McQueeney, D., Layman, R. D. and McKim, H. L. Development of a simplified method for field monitoring of soil moisture content. *Proc. 2nd Coll. of Planetary Water and Polar Processes*, 40–43, 1979.

[83] Wyslouzil, W. and Kashyap, S. C. Microwave moisture profile gauges for sheet materials. *Proc. 1979 IMPI Microwave Power Symp.*, 153–155, Monte Carlo, Monaco, 1979.

[84] Xu, Deming. Microwave moisture measurements (in Chinese). *Electron Technol.* (Shanghai, China), no. 4, 1979.

[85] Xu, D. and Zhaonian, L. A study of microwave properties of oil reservoirs (in Chinese). *J. Shanghai Univ. Sci. Technol.*, no. 1, 1979.

[86] Zurcher, J. F. and Gardiol, F. E. Measurement of moisture distribution in construction materials. *Proc. 1979 IMPI Microwave Power Symp.*, 27–29, Monte Carlo, Monaco, 1979.

1980

[87] Anderson, J. G. Paper/board moisture measurement by microwave loss. *Proc. 4th IFAC Conf.*, 75–84, Ghent, Belgium, June 1980.

[88] Bahl, I. J. and Stuchly, S. S. Analysis of a microstrip covered with a lossy dielectric. *IEEE Trans. Microwave Theory Techn.*, MTT-28 (2):104–109, 1980.

[89] Ballario, C., Bonincontro, A., Cametti, C. and DiBiasio, A. Dielectric properties of water-cellulose system at microwave frequencies. *J. Colloid Interface Sci.*, 78: 242–245, 1980.

[90] Bazhenov, V. A., Stepanchenko, E. S. and Tereschenko, A. K. Analytical expression of relationship between dielectric permittivity and moisture content of butter (in Russian). *Moloch. Prom.*, (2): 45–47, 1980.

[91] Berliner, M. A. and Spiridonov, V. I. Unified microwave moisture gauge. *Meas. Tech. (USSR)*, 23 (3): 277–281, 1980.

[92] Brodwin, M. and Benway, J. Experimental evaluation of microwave transmission moisture sensor. *J. Microwave Power*, 15 (4): 261–265, 1980.

[93] Brooke, R. L. High speed bulk grain moisture measurement apparatus. U.S. Patent 4,326,163; filed Jan. 30, 1980.

[94] Brown, D. R., Bozorgsmanesh, H. and Gozani, T. Microwave equipment for moisture content monitoring in coal. *Coal Mining Process.*, 17: 88–90, 1980.

[95] Chaloupka, H., Ostwald, O. and Schiek, B. Structure independent microwave moisture measurement. *J. Microwave Power*, 15 (4): 221–232, 1980.

[96] Chang, A. T., Atwater, S. G., Salomonsun, V. V., Estes, J. E., Simoneff, D. S. and Bryan, M. L. L-band radar sensing of soil moisture. *IEEE Trans.*, GE-18: 303–310, 1980.

[97] Flygare, W. H. and Balle, T. J. Method and apparatus for the spectroscopic observation of particles. U.S. Patent 4,369,404; filed Sept. 15, 1980.

[98] Freedman, R. and Montague, D. R. Electromagnetic propagation tool (EPT): comparison of log derived and in situ oil saturations in shaly fresh water sands. Soc. Petr. Eng., paper 9266, 55th Annual Fall Conf., Dallas, TX, 1980.

[99] Ghobrial, S. I. Effects of hygroscopic water on dielectric constant of dust at X-band. *Electron. Lett.*, 16 (10): 393–394, 1980.

[100] Hewitt, S. J., Ozamiz, J. M. and Yallup, A. E. Measurement of moisture content. U.S. Patent 4,311,957; filed Mar. 31, 1980.

[101] Hoppe, W., Meyer, W. and Schilz, W. Density-independent moisture measuring in fibrous materials using a double-cutoff Gunn oscillator. *IEEE Trans. Microwave Theory Techn.*, MTT-28 (12): 1449–1452, 1980.

[102] Ismatullaev, P. R. and Grinvald, A. B. Mesurement of the moisture content of cottonseeds and cottonseed products using microwave method. *Meas. Techn. (USSR)*, 23 (4): 361–363, 1980.

[103] Jacobson, R., Meyer, W. and Schrage, B. Density independent moisture meter at X-band. *Proc. 10th Europ. Microwave Conf.*, 216–220, Warsaw, Poland, 1980.

[104] Kalinski, J. Microwave instrumentation for industrial on-line measurement of improved accuracy. *Proc. 10th Europ. Microwave Conf.*, 521–525, Warsaw, Poland, 1980.

[105] Karmas, E. Techniques for measurement of moisture content of food. *Food Technol.*, 34 (4): 52–56, 1980.

[106] Klein, A. Microwave moisture determination of coal—a comparison of attenuation and phase measurement. *Proc. 10th Europ. Microwave Conf.*, 526–530, Warsaw, Poland, 1980.

[107] Kraszewski, A. Problems related to microwave monitoring of moisture content in grains. *Proc. MITEKO 80 Microwave Conf.*, 103–108, Pardubice, Czechoslovakia, 1980.

[108] Kraszewski, A. Microwave aquametry (invited paper). *Proc. 10th Europ. Microwave Conf.*, 48–58, Warsaw, Poland, 1980.

[109] Kraszewski, A. Microwave aquametry—a review. *J. Microwave Power*, 15 (4): 209–220, 1980.

[110] Kraszewski, A. Microwave aquametry—a bibliography. *J. Microwave Power*, 15 (4): 298–310, 1980.

[111] Kraszewski, A., Kulinski, S., Madziar, J. and Zielkowski, K. Microwave on-line moisture content monitoring in low-hydrated organic materials. *J. Microwave Power*, 15 (4): 267–276, 1980.

[112] Kulinski, S., Zielkowski, K., Madziar, J. and Kraszewski, A. Microwave on-line moisture content monitoring in organic materials. *Proc. 10th Europ. Microwave Conf.*, 221–226, Warsaw, Poland, 1980.

[113] Meyer, W. and Schilz, W. A microwave method for density independent determination of moisture content of solids. *J. Phys. D: Appl. Phys.*, 13: 1823–1830, 1980.

[114] Mladek, J. and Beran, Z. Sample geometry, temperature and density factors in the microwave measurement of moisture. *J. Microwave Power*, 15 (4): 243–250, 1980.

[115] Mudgett, R. E., Goldblith, S. A., Wang, D. I. C. and Westphal, W. B. Dielectric behavior of semi-solid food at low, intermediate and high moisture content. *J. Microwave Power*, 15 (1): 27–36, 1980.

[116] Nagy, L. L. and Myers, M. E. Method of detecting soot in engine oil using microwaves. U.S. Patent 4,345,202; filed Dec. 19, 1980.

[117] Nelson, S. O. Moisture-dependent kernel- and bulk-density relationships for wheat and corn. *Trans. ASAE*, 23 (1): 139–143, 1980.

[118] Nelson, S. O. Microwave dielectric properties of fresh fruits and vegetables. *Trans. ASAE*, 23 (6): 1314–1317, 1980.

[119] Nelson, S. O., Fanslow, G. E. and Bluhm, D. D. Frequency dependence of the dielectric properties of coal. *J. Microwave Power*, 15 (4): 277–282, 1980.

[120] Ostwald, O., Schiek, B. and Chaloupka, H. A new approach for a quantitative microwave moisture measurement. *Proc. 10th Europ. Microwave Conf.*, 211–215, Warsaw, Poland, 1980.

[121] Pascal, H., Rankin, D., Freedman, R. and Vogiatzis, J. P. The theory of microwave dielectric constant logging using the electromagnetic wave propagation method. *Geophysics*, 45: 1530–1533, 1980.

[122] Praxmarer, W. On moisture content determination in powdered high polymers by microwave method. *Plaste Kautschuk*, 27 (5): 252–253, 1980.

[123] Rau, R. N. and Wharton, R. P. Measurement of core electrical parameters at UHF and microwave frequencies. Soc. Petr. Eng., paper 9380, 1980.

[124] Schmugge, T. J., Jackson, T. J. and McKim, H. L. Survey of methods of soil moisture determination. *Water Resources Res.*, 16 (6): 961–979, 1980.

[125] Shiraiwa, T., Kobayashi, S., Koyama, A., Tokuda, M. and Koizumi, S. Microwave moisture gauge for limestone. *J. Microwave Power*, 15 (4): 255–260, 1980.

[126] Steinbrecher, D. H. Apparatus and method for moisture measurement. U.S. Patent 4,358,731; filed May 23, 1980.

[127] Tiuri, M., Jokela, K. and Heikkila, S. Microwave instrument for accurate moisture and density measurement of timber. *J. Microwave Power*, 15 (4): 251–254, 1980.

[128] Topp, G. C., Davis, J. L. and Allen, A. P. Electromagnetic determination of soil water content in coaxial transmission lines. *Water Resources Res.*, 16 (3): 574–582, 1980.

[129] Wang, J. R. and Schmugge, T. J. An empirical model for the complex dielectric permittivity of solids as a function of water content. *IEEE Trans. Geosci. Remote Sensing*, GE-18 (4): 288–295, 1980.

[130] Wharton, R. P., Hazen, G. A., Rau, R. N. and Best, D. L. Electro-magnetic propagation logging—Advances in technique and interpretation. *SPE 55th Annual Fall Tech. Conf.*, 1980.

1981

[131] Bernard, P., Vauclin, M., Taconet, O. and Vidal-Madjar, D. Possible use of active microwave remote sensing data for prediction of regional evaporation by numerical simulation of soil water movement in the unsaturated zone. *Water Resources Res.*, 17: 1603–1610, 1981.

[132] Bradley, G. A. and Ulaby, F. T. Aircraft radar response to soil moisture. *Remote Sensing Environment*, 11: 419–438, 1981.

[133] Davis, L. A., Brost, D. F. and Haskin, H. K. Chemical flood testing method. U.S. Patent 4,482,634; filed Dec. 31, 1981.

[134] Davis, L. A., Brost, D. F. and Haskin, H. K. Microwave means for monitoring fluid in a core of material. U.S. Patent 4,490,676; filed Dec. 31, 1981.

[135] Delacey, E. H. B. and White, L. R. Dielectric response and conductivity of dilute suspensions of colloidal particles. *J. Chem. Soc. Faraday Trans.*, 277: 2007–2039, 1981.

[136] Foster, K. R. and Schepps, J. L. Dielectric properties of tumor and normal tissue at radio through microwave frequencies. *J. Microwave Power*, 16: 108–119, 1981.

[137] Goldberg, I. B., Ho, W. W., Chung, K. E., McCoy, L. R. and Wagner, R. I. Microwave coal water slurry monitor. *Proc. 1981 Symp. on Instrum. and Control of Fossil Energy Processes*, San Francisco, CA, 1981.

[138] Ho, W. W., Harker, A. B., Goldberg, I. B. and Chung, K. E. Microwave meter for fluid mixture. U.S. Patent 4,423,623; filed Aug. 24, 1981.

[139] Iskander, M. F., Tyler, A. L. and Elkins, D. F. A time-domain technique for measurement of the dielectric properties of oil shale during processing. *Proc. IEEE*, 69: 760–762, 1981.

[140] Jackson, T. J., Chang, A. and Schmugge, T. J. Aircraft active microwave measurements for estimating soil moisture. *Photogrammetr. Eng.*, 47: 801–805, 1981.

[141] Kaatze, U. and Uhlendorf, V. The dielectric properties of water at microwave frequencies. *Z. Phys. Chem. Neue Folge*, 126 (5): 151–165, 1981.

[142] Kalinski, J. A chopped subcarrier method of simultaneous attenuation and phase shift measurement under industrial conditions. *IEEE Trans. Ind. Electron. Control Instrum.*, IECI-28 (3): 201–209, 1981.

[143] Karmas, E. Measurement of moisture content in food. *Cereal Food World*, 26: 332–334, 1981.

[144] Kent, M. Use of stripline sensors with density independent moisture meter (abstract). *16th Microwave Power Symp.*, IMPI, Toronto, Canada, 1981.

[145] Kent, M. and Meyer, W. Density independent moisture metering in fishmeal industry. *Proc. 11th Europ. Microwave Conf.*, 448–453, Amsterdam, Holland, 1981.

[146] Kent, M. and Stroud, G. D. Microwave attenuation of frozen *Nephrops norvegicus*. *J. Food Technol.*, 16: 647–654, 1981.

[147] Kimura, K., Endo, A., Shibata, T. and Morozumi, H. Microwave alcohol fuel sensor. U.S. Patent 4,453,125; filed July 15, 1981.

[148] King, R. W. P. and Smith, G. S. *Antennas in Matter*. MIT Press, Cambridge, MA, 1981.

[149] Klein, A. Microwave determination of moisture in coal: comparison of attenuation and phase measurement. *J. Microwave Power*, 16 (3/4): 289–304, 1981.

[150] Knoechel, R. Arrangement for measuring moisture content. U.S. Patent 4,546,311; filed Dec. 18, 1981.

[151] Knoechel, R. and Meyer, W. Continuous moisture determination in fluids and slurries (abstract). *16th Microwave Power Symp.*, IMPI, Toronto, Canada, 1981.

[152] Kraszewski, A. New developments in microwave techniques of moisture content monitoring and measurement (abstract). *Int. Conf. on Recent Developments in Determination of Water Content*, SIRA Institute, Chislehurst, Kent, England, 1981.

[153] Kraszewski, A., Stuchly, M. A. and Stuchly, S. S. In-situ dielectric measurements of industrial materials at microwave frequencies (abstract). *16th Microwave Power Symp.*, IMPI, Toronto, Canada, 1981.

[154] Mazzagatti, R. P. Oil-water monitor. U.S. Patent 4,429,273; filed Mar. 20, 1981.

[155] McClean, V. E. R., Sheppard, R. J. and Grant, E. H. A generalized model for the interaction of microwave radiation with bound water in biological materials. *J. Microwave Power*, 16 (1): 1–7, 1981.

[156] Meador, R. A. and Paap, H. J. Crude oil production stream analyzer. U.S. Patent 4,458,524; filed Dec. 28, 1981.

[157] Meyer, W. and Schilz, W. Feasibility study of density-independent moisture measurement with microwaves. *IEEE Trans. Microwave Theory Techn.*, MTT-29 (7): 732–739, 1981.

[158] Nelson, S. O. Review of factors influencing the dielectric properties of cereal grains. *Cereal Chem.*, 58 (6): 487–492, 1981.

[159] Nelson, S. O. Frequency and moisture dependence of the dielectric properties of chopped pecans. *Trans. ASAE*, 24 (6): 1573–1576, 1981.

[160] Nelson, S. O., Beck-Montgomery, S. R., Fanslow, G. E. and Bluhm, D. D. Frequency dependence of the dielectric properties of coal. Part II. *J. Microwave Power*, 16 (3/4): 319–326, 1981.

[161] Okamura, S. High-moisture content measurement of grain by microwaves. *J. Microwave Power*, 16 (3/4): 253–256, 1981.

[162] Reboul, J.-P. Microwave studies of Portland cement hydration. *J. Microwave Power*, 16 (1): 25–29, 1981.

[163] Riggin, M. T. Microwave moisture sensor. U.S. Patent 4,484,133; filed Dec. 23, 1981.

[164] Schilz, W. and Schiek, B. Microwave systems for industrial measurements. *Adv. Electron. Electron Phys.*, 55: 309–381, 1981.

[165] Steinbrecher, D. H. Microwave test apparatus and method. U.S. Patent 4,381,485; filed Feb. 23, 1981.

[166] Strandberg, Jr., C. F. and Strandberg, R. C. Microwave moisture measuring, indicating and control apparatus. U.S. Patent 4,399,403; filed Sept. 22, 1981.

1982

[167] Bastida, E. M., Fanelli, N., Marelli, E. and Ricca, A. M. Microwave instruments for industrial applications developed at CISE. *Proc. 4th Natl. Meeting on Electromagnet. Appl.*, 341–343, IROE, Florence, Italy, 1982.

[168] Bernard, R., Martin, P., Thony, J. L., Vauclin, M. and Vidal-Madjar, D. C-band radar for determining surface soil moisture. *Remote Sensing Environment*, 12: 189–200, 1982.

[169] Chew, W. C. A response of the deep propagation tool. Soc. Petr. Eng., paper 10989, presented at the Louisiana SPE Meeting, 1982.

[170] Chew, W. C. and Sen, P. N. Potential of a sphere in an ionic solution in thin double layer approximation. *J. Chem. Phys.*, 77: 2042, 1982.

[171] Davis, L. A. and Brost, D. F. Method and apparatus for measuring relative permeability and water saturation of a core of earthen material. U.S. Patent 4,486,714; filed Sept. 8, 1982.

[172] Davis, L. A. and Haskin, H. K. Two dimensional microwave chemical flood testing means and method. U.S. Patent 4,519,982; filed Mar. 15, 1982.

[173] Handa, T., Fukuoka, M. and Yoshizawa, S. The effect of moisture on dielectric relaxation in wood. *J. Appl. Polym. Sci.*, 27: 439–453, 1982.

[174] Helms, D. A., Hatton, G. J. and Williams, T. M. Water cut monitoring means and methods. U.S. Patent 4,499,418; filed Aug. 5, 1982.

[175] Huchital, G. Deep propagation tool. A new electromagnetic logging tool. Soc. Petr. Engng. Paper No. 10988, presented at the Louisiana SPE Meeting, 1982.

[176] Kent, M. and Meyer, W. A density independent microwave moisture meter for heterogeneous foodstuffs. *J. Food Eng.*, 1: 31–42, 1982.

[177] Khalid, K. B. Determination of dry rubber content in hevea latex by microwave technique. *Pertanika (Malaysia)*, 5 (2): 192–195, 1982.

[178] Klein, A., Schicker, W. and Schiek, B. Microwave moisture measurements with reduced sensitivity to particle size and shape. *Proc. 12th Europ. Microwave Conf.*, 593–598, Helsinki, Finland, 1982.

[179] Laine, J. and Paivanen, J. Water content and bulk density of peat. *Proc. Int. Symp.*, 422–430, IPS Commissions II and IV, Minsk, USSR, 1982.

[180] Mailander, M. P., Schueller, J. K. and Krutz, G. An evaluation of four continuous moisture sensors on a combine. ASAE Paper #82-1576, Am. Soc. of Agricultural Engineers, St. Joseph, Michigan, 1982.

[181] Mendelson, K. S. and Cohen, M. H. The effect of grain anisotropy on the electrical properties of isotropic sedimentary rocks. *Geophysics*, 47: 257, 1982.

[182] Meyer, W. and Schilz, W. High-frequency dielectric data on selected moist materials. *J. Microwave Power*, 17 (1): 67–81, 1982.

[183] Meyer, W. and Schilz, W. Microwave measurement of moisture content in process materials. *Philips Tech. Rev.*, 40 (4): 62–69, 1982.

[184] Nagy, L. L. and O'Rourke, M. J. Microwave probe for measurement of dielectric constants. U.S. Patent 4,503,384; filed April 28, 1982.

[185] Nelson, S. O. Factors affecting the dielectric properties of grain. *Trans. ASAE*, 25 (4): 1045–1049, 1056, 1982.

[186] Njoku, E. G. and O'Neill, P. E. Multifrequency microwave radiometer measurements of soil moisture. *IEEE Trans. Geosci. Remote Sensing*, GE-20 (4): 468–475, 1982.

[187] Pakulis, I. E. Apparatus and process for microwave moisture analysis. U.S. Patent 4,485,284; filed Jan. 11, 1982.

[188] Ptitsyn, S. D., Sekanov, Yu.P. and Batalin, M.Yu. Modeling the dielectric properties of grain (in Russian). *Mekhanizatsiya i Elektrifikatsia Sel'skogo Khozyaistva*, 12: 47–49, 1982.

[189] Rau, R. N. and Whorton, R. P. Measurement of core electrical parameters at ultrahigh and microwave frequencies. *J. Petro. Technol.*, 2689–2700, Nov. 1982.

[190] Shutko, A. M. and Reutov, E. M. Mixture formulas applied in estimation of dielectric and radiative characteristics of solids and grounds at microwave frequencies. *IEEE Trans. Geosci. Remote Sensing*, GE-20 (1): 29–32, 1982.

[191] Skoog, B. G., Askne, J. I. H. and Elgered, G. Experimental determination of water vapor profiles from ground-based radiometer measurement at 21.0 and 31.4 GHz. *J. Appl. Meteorol.*, 21 (3): 394–400, 1982.

[192] Tiuri, M. and Toikka, M. Radio-wave probe for in-situ water content measurement of peat. *Suo*, 33 (3): 65–70, 1982.

[193] Topp, G. C., Davis, J. L. and Annan, A. P. Electromagnetic determination of soil water content using TDR: 1. Application to wetting fronts and steep gradients. *Soil Science Soc. Am. J.*, 46: 672–678, 1982.

[194] Topp, G. C., Davis, J. L. and Annan, A. P. Electromagnetic determination of soil water content using TDR: 2. Evaluation of installation and configuration of parallel transmission lines. *Soil Science Soc. Am. J.*, 46: 678–684, 1982.

[195] Ulaby, F. T., Aslam, A. and Dobson, M. C. Effects of vegetation cover on the radar sensitivity to soil moisture. *IEEE Trans. Geosci. Remote Sensing*, GE-20: 476–481, 1982.

[196] Walker, C. W. E. Microwave moisture measurement of moving particulate layer after thickness leveling. U.S. Patent 4,475,080; filed May 10, 1982.

[197] Xu, D. A study of microwave bridge techniques in nonelectrical quantities detecting (in Chinese). *Electron. Meas. Technol.*, no. 5, 1982.

[198] Zagorskii, V. V., Nesterov, V. M., Zamotrinskaya, E. A. and Mikhailova, T. G. Dependence of dielectric permittivity of moist disperse materials on the temperature. *Sov. Phys. J.*, 25 (1): 62–65, 1982.

1983

[199] Bahar, E. and Saylor, J. D. A feasibility study to monitor soil moisture content using microwave signals. *IEEE Trans. on Microwave Theory Techn.*, MTT-31 (7): 533–541, 1983.

[200] Blanchard, A. J. and Chang, A. T. C. Estimation of soil moisture from SeaSat SAR data. *Water Res. Bull.*, 19: 803–810, 1983.

[201] Davis, L. A. Method and apparatus for measuring relative permeability and water saturation of a core. U.S. Patent 4,543,821; filed Dec. 14, 1983.

[202] Dietrich, K., Unkelbach, K. H., Nimtz, G. and Aichmann, H. Verfahren und Einrichtung zur Messung der Feuchte von Schuttgutern. BRD Patent 33 17 200 A1; filed May 11, 1983.

[203] Gardiol, F. E., Sphicopoulos, T. and Teodoridis, V. The reflection of open-ended circular waveguides—application to nondestructive measurements of materials. *Reviews of Infrared and Millimeter Waves*, 325–365. Plenum Press, New York, 1983.

[204] Hogg, D. C., Guiraud, F. O., Snider, J. B., Decker, M. T. and Westwater, E. R. A steerable dual-channel microwave radiometer for measurement of water vapor and liquid in the troposphere. *J. App. Meteor.*, 22 (5): 789–806, 1983.

[205] Iskander, M. F. and DuBow, J. B. Time- and frequency-domain techniques for measuring the dielectric properties of rocks: a review. *J. Microwave Power*, 18 (1): 55–74, 1983.

[206] Jain, V. K., Sanyal, S. N. and Rizvi, S. F. H. Dielectric constant of some Indian timbers at microwave frequencies. *IE (Indian) J.—EL*, 63: 230–231, April 1983.

[207] Kent, M. and Meyer, W. Dielectric relaxation of adsorbed water in microcrystalline cellulose. *J. Phys. D: Appl. Phys.*, 16: 915–925, 1983.

[208] Khalid, K. B. and Abdul Wahab, M. B. Microwave attenuation for fresh hevea latex. *J. Rubber Res. Inst. Malaysia*, 313: 145–150, 1983.

[209] King, R. J. and Stiles, P. Microwave nondestructive evaluation of composites. Review of Progress in Quantitative Nondestructive Evaluation, 3, *Proc. 10th Annual Review*, 1073–1081, Santa Cruz, CA, 1983.

[210] Nelson, S. O. Dielectric properties of some fresh fruits and vegetables at frequencies of 2.45 to 22 GHz. *Trans. ASAE*, 26 (2): 613–616, 1983.

[211] Nelson, S. O. Density dependence of the dielectric properties of particulate materials. *Trans. ASAE*, 26 (6): 1823–1825, 1829, 1983.

[212] Sakurai, T., Mizuno, H. and Shibata, Y. Apparatus for measuring the ratio of alcohol contained in mixed fuel. U.S. Patent 4,651,085; filed April 6, 1983.

[213] Schmugge, T. J. Remote sensing of soil moisture: recent advances. *IEEE Trans. Geosci. Remote Sensing*, GE-21 (3): 336–343, 1983.

[214] Sen, P. N. and Chew, W. C. The frequency dependent dielectric and conductivity response of sedimentary rocks. *J. Microwave Power*, 18 (1): 95–105, 1983.

[215] Steffens, D. SCANPRO moisture meters for paper pulp industry (in German). *Wochenblatt Papierfabrikation*, (8): 259–262, 1983.

1984

[216] Anderson, J. G. Advances in paper moisture measurement by microwave loss. *IFAC Proc. Ser.*, 8: 63–72, 1984.

[217] Ansoult, M., de Backer, L. W. and Declercq, M. Statistical relationship between dielectric constant and water content in porous media. *Soil Sci. Soc. Am. J.*, 48: 47–50, 1984.

[218] Chapoton, A., Lebrun, A. and Wattrelot, F. Devices for "in-situ" measurements of soil moisture using water dielectric properties at radio- and microwave frequencies (in French). *Proc. 2nd Int. Coll. on Signatures spectrales d'objects en teledetection*, paper P.III.3, Bordeaux, France, 1983. Published by INRA Publ., *Les Colloques de l'INRA*, no. 23, 1984.

[219] Dalton, F. N., Herkelrath, W. N., Rawlins, D. S. and Rhoades, J. D. Time domain reflectometry method: Simultaneous measurement of soil water content and electrical conductivity with a single probe. *Science*, 224: 989–990, 1984.

[220] Ishikawa, H. and Kiyobe, S. Microwave moisture sensor. U.S. Patent 4,620,146; filed April 20, 1984.

[221] Kalinski, J. and Rakowski, J. On-line measurement of material quality by microwaves. *Proc. Int. Symp. on Metrology for Quality Control in Production*, 94–99, Tokyo, Japan, 1984.

[222] Kent, M. and Kohler, J. Broadband measurements of stripline moisture sensors. *J. Microwave Power*, 19 (3): 173–179, 1984.

[223] Kent, M. and Meyer, W. Complex permittivity spectra of protein powders as a function of temperature and hydration. *J. Phys. D.: Appl. Phys.*, 17: 1687–1698, 1984.

[224] Klein, A. Microwave determination of moisture compared with capacitive, infrared and conductive measurement methods. Comparison of on-line measurements at coal preparation plants. *Proc. 14th Europ. Microwave Conf.*, 661–666, Liege, Belgium, 1984.

[225] Kuhne, M., Gast, T., Nolte, K. and Dittmann, R. Vorrichtung zur Messung der Feuchtigkeit von Rauchematerialien. BRD Patent 34 07 819 C1; filed March 2, 1984.

[226] Matzler, C. and Schanda, E. Snow mapping with active microwave sensors. *Int. J. Remote Sensing*, 5 (2): 409–422, 1984.

[227] Mladek, J. and Beran, Z. The basic aspects of moisture content measurement in food materials by microwave methods (in Czech). *Potravin. Vedy*, 2 (2): 149–160, 1984.

[228] Nelson, S. O. Density dependence of the dielectric properties of wheat and whole-wheat flour. *J. Microwave Power*, 19 (1): 55–64, 1984.

[229] Nelson, S. O. Moisture, frequency and density dependence of the dielectric constant of shelled, yellow-dent field corn. *Trans. ASAE*, 27 (5): 1573–1578, 1984.

[230] Nyfors, E. and Vainikainen, P. Sensor for measuring the mass per unit area of a dielectric layer. *Proc. 14th Europ. Microwave Conf.*, 667–672, Liege, Belgium, 1984.

[231] Osaki, S., Fujii, Y., Tomita, O. and Saiwai, K. Method and apparatus for measuring orientation of constituents of webs and sheets. U.S. Patent 4,581,575; filed May 30, 1984.

[232] Pomeranz, Y. and Bolte, L. C. Time-dependent moisture gradients in conditioned wheat determined by electrical methods. *Cereal Chem.*, 61: 559–563, 1984.

[233] Sadeghi, A. M., Hancock, G. D., Waite, W. P., Scott, H. D. and Rand, J. A. Microwave measurements of moisture distributions in upper soil profile. *Water Resources Res.*, 20 (7): 927–934, 1984.

[234] Sakurai, T., Mizuno, H. and Shibata, Y. Apparatus for measuring the ratio of alcohol contained in mixed fuel. U.S. Patent 4,651,085; filed April 5, 1984.

[235] Schmugge, T. J. Microwave remote sensing of soil moisture. *SPIE 481 Recent Adv. Civil Space Remote Sensing*, 249–257, 1984.

[236] Schulze, E. and Kuehn, W. Continuous and nondestructive determination of the water content in plants by absorption of microwaves (in German). *Angew. Bot.*, 58 (5/6): 465–474, 1984.

[237] Scully, J. P. and Ward, R. Water moisture measuring instrument and method. U.S. Patent 4,600,879; filed June 15, 1984.

[238] Smith, D. M. and Mitchell, J., Jr. "Aquametry." Part II, *Electrical and Electronic Methods.* Wiley, New York, 1984.

[239] Tiuri, M., Toikka, M., Tolonen, K. and Rummukainen, A. Capability of new radiowave moisture probe in peat resource inventory. *Proc. 7th Int. Peat Congress, Dublin, Ireland*, (1): 157, 1984.

[240] Topp, G. C., Davis, J. L., Bailey, W. G. and Zebchuk, W. D. The measurement of soil water content using a portable TDR hand probe. *Canad. J. Soil Sci.*, 64: 313–321, 1984.

[241] Tran, V. N., Stuchly, S. S. and Kraszewski, A. Dielectric properties of selected vegetables and fruits at 0.1–10.0 GHz. *J. Microwave Power*, 19 (4): 252–258, 1984.

[242] Ulaby, F. T. and Jedlicka, R. P. Microwave dielectric properties of plant materials. *IEEE Trans. Geosci. Remote Sensing*, GE-22, 530–535, 1984.

[243] Wallender, W. W., Sackman, G. L., Kone, K. and Keminaka, S. M. Soil moisture measurement by microwave forward-scattering. ASAE Paper No. 84-2523, 1984.

[244] Ward, R. L., Felver, T. and Pyper, J. W. Bound and free moisture in explosives and plastics. *J. Hazardous Mater.*, 9: 69–76, 1984.

1985

[245] Aggarwal, S. K. and Johnston, R. H. The effect of temperature on the accuracy of microwave moisture measurements on sandstone cores. *IEEE Trans. Instrum. Meas.*, IM-34 (1): 21–25, 1985.

[246] Ansoult, M., Debacca, L. W. and Declerq, M. Statistical relationship between apparent dielectric constant and water content in porous media. *Soil Sci. Am. J.*, 49 (1): 47–50, 1985.

[247] Bell, J. F. M. Moisture meter. U.S. Patent 4,788,853; filed Oct. 25, 1985.

[248] Bramanti, M. and Del Bravo, A. Method and apparatus for determining the dielectric constant of materials. U.S. Patent 4,754,214; filed Nov. 8, 1985.

[249] Dasberg, S. and Dalton, F.N. Time domain reflectometry field measurements of soil water content and electrical conductivity. *Soil Sci. Soc. Am. J.*, 49: 293–297, 1985.

[250] Dobson, M. C., Ulaby, F. T., Hallikainen, M. T. and El-Rayes, M. A. Microwave dielectric behavior of wet soil, II: Dielectric mixing models. *IEEE Trans. Geosci. Remote Sensing*, GE-23 (1): 35–46, 1985.

[251] Feng, S. and Sen, P. N. Geometrical model of conductive and dielectric properties of partially saturated rocks. *J. Appl. Phys.*, 55 (8): 3236–3243, 1985.

[252] Gardiol, F. "Open-ended waveguides: principles and applications." *Advances in Electronics and Electron Physics*, 139–187. Academic Press, New York, 1985.

[253] Hallikainen, M. T., Ulaby, F. T., Dobson, C. M., El-Rayes, M. A. and Wu, L. Microwave dielectric behaviour of wet soil. Part 1: Empirical models and experimental observations. *IEEE Trans. Geosci. Remote Sensing*, GE-23 (1): 25–34, 1985.

[254] Hallikainen, M. T., Ulaby, F. T. Dobson, C. M. and El-Rayes, M. A. Microwave dielectric behaviour of wet soil. Part 2: Dielectric mixing models. *IEEE Trans. Geosci. Remote Sensing*, GE-23 (1): 35–46, 1985.

[255] Hane, B. Method for measuring the moisture ratio of organic material and apparatus herefor. U.S. Patent 4,675,595; filed June 18, 1985.

[256] Hatton G. J., Helms, D. A. and Williams, T. M. Petroleum stream monitoring means and method. U.S. Patent 4,660,414; filed Sept. 12, 1985.

[257] Hubert, J. and Turek, B. Moisture content measurement in brown coal using microwaves (in Polish). *Gornictwo Odkrywkowe*, (4): 6–8, 1985.

[258] Jachowicz, R. and Sochon, J. Methods of disturbing parameters elimination at moisture content measurements in solids. *Proc. ISA Int. Symp. on Moisture and Humidity*, 659–669, Washington, D.C., April 1985.

[259] Jakkula, P. Method and apparatus for measuring the moisture content or dry-matter content of materials using a microwave dielectric waveguide. U.S. Patent 4,755,743; filed Oct. 14, 1985.

[260] James, W. L., Yen, Y.-H and King, R. J. A microwave method for measuring moisture content, density and grain wood angle of wood. Res. Note FPL-0250, Madison, WI: U.S. Dept. of Agriculture, Forest Service, Forest Products Lab., 1985.

[261] Jian, Z. et al. On dielectric constant of yarns in microwave frequency range (in Chinese). *J. Textile Sci. Technol. (E. China)*, no. 2, 1985.

[262] Kent, M., Christie, R. M. and Lees, A. Microwave and infra-red drying versus conventional oven drying methods for moisture determination in fish flesh. *J. Food Technol.*, 20: 117–127, 1985.

[263] King, R. J., James, W. L. and Yen, Y.-H. A microwave method for measuring moisture content, density and grain angle in wood. *Proc. 1st Int. Conf. on Scanning Technology in Sawmilling*, San Francisco, CA, 1985.

[264] Latorre, V. R. and King, R. J. In situ microwave measurements of dielectric properties and moisture in materials. *Proc. ISA Int. Symp. on Moisture and Humidity*, 919–925, Washington, D.C., 1985.

[265] Maley, L. E. Continuous moisture analysis instrumentation. *Proc. ISA Int. Symp. on Moisture and Humidity*, 649–657, Washington, D.C., 1985.

[266] Martinson, T., Sphicopoulos, T. and Gardiol, F. E. Nondestructive measurement of materials using a waveguide-fed series slot array. *IEEE Trans. Instrum. Meas.*, IM-34 (3): 422–426, 1985.

[267] Ness, J. Broad band permittivity measurements using the semi-automatic network analyzer. *IEEE Trans. Microwave Theory Techn.*, MTT-33 (12): 1222–1226, 1985.

[268] Nyvlt, V. and Mladek, J. On water binding in food products (in German). *Lebens-mittelindustrie*, 32 (H3): 112–114, 1985.

[269] Pyper, J. W. The determination of moisture in solids—a selected review. *Anal. Chim. Acta*, 170: 159–175, 1985.

[270] Pyper, J. W., Buettner, H. M., Cerjan, C. J., Hallam, J. S. and King, R. J. The measurement of bound and free moisture in organic materials by microwave methods. *Proc. ISA Int. Symp. on Moisture and Humidity*, 909–917, Washington, D.C., 1985.

[271] Schmidt, J. and Wunderlich, B. Moisture content in components of concrete by microwave measuring techniques (in German). *Messen, Steuern, Regeln*, 28 (5): 214–216, 1985.

[272] Shen, L. C. A laboratory technique for measuring dielectric properties of core samples at ultrahigh frequencies. *Soc. Petrol. Engrs. J.*, 25 (4): 502–514, 1985.

[273] Shen, L. C., Savre, W. C., Price, J. M. and Athavale, K. Dielectric properties of reservoir rocks at ultra-high frequencies. *Geophysics*, 50 (4): 692–704, 1985.

[274] Smith, G. S. and Nordgard, J. D. Measurement of the electrical constitutive parameters of materials using antennas. *IEEE Trans. Antennas Propagation*, AP-33 (7): 783–792, 1985.

[275] Sphicopoulos, T., Teodoridis, V. and Gardiol, F. E. Simple nondestructive method for the measurement of material permittivity. *J. Microwave Power*, 20 (3): 165–172, 1985.

[276] Teodoridis, V., Sphicopoulos, T. and Gardiol, F. E. The reflection from an open-ended rectangular waveguide applied on a layered dielectric medium. *IEEE Trans. Microwave Theory Techn.*, MTT-33: 359–366, 1985.

[277] Toikka, M., Tiuri, M. and Tolonen, K. Capability of a new radiowave moisture probe in tropical peat resource inventory. *Proc. IPS Symp. on Tropical Peats*, 215–223, Kingston, Jamaica, 1985.

[278] Topp, G. C. and Davis, J. L. Measurement of soil water content using time-domain reflectometry (TDR): a field evaluation. *Soil Sci. Soc. Am. J.*, 49: 19–24, 1985.

[279] Ulaby, F. T. and Wilson, E. A. Microwave attenuation properties of vegetation canopies. *IEEE Trans. Geosci. Remote Sensing*, GE-23, 746–753, 1985.

[280] Vainikainen, P. and Nyfors, E. Sensor for measuring the mass per unit area of dielectric layer: results of using an array of sensors in a particle board factory. *Proc. 15th Europ. Microwave Conf.*, 901–905, Paris, France, 1985.

[281] Wallender, W. W., Stackman, G. L., Kone, K. and Kaminaka, M. S. Soil moisture measurement by microwave forward-scattering. *Trans. ASAE*, 28 (4): 1206–1211, 1985.

[282] Westwater, E. R., Falls, M. J. and Decker, M. T. Remote sensing of atmospheric water vapor by ground-based microwave radiometry. *Proc. ISA Int. Symp. on Moisture and Humidity*, 259–268, Washington, D.C., 1985.

[283] Yukl, T. Dielectric constant change monitoring. U.S. Patent 4,947,848; filed Jan. 22, 1985.

[284] Zaghloul, H. and Buckmaster, H. A. The complex permittivity of water at 9.356 GHz from 10 to 40° C. *J. Phys. D: Appl. Phys. (GB)*, 18: 2109–2118, 1985.

1986

[285] Afsar, H. N., Birch, J. R., Clarke, R. N. and Chantry, G. W. The measurement of the properties of materials. *Proc. IEEE*, 74 (1): 183–199, 1986.

[286] Aggarwal, S. K. and Johnston, R. H. Oil and water content measurement of sandstone cores using microwave measurement techniques. *IEEE Trans. Instrum. Meas.*, IM-35 (4): 630–637, 1986.

[287] Bernard, R., Vidal-Madjar, D., Baudin, F. and Laurent, G. Data processing and calibration for airborne scatterometer. *IEEE Trans. Geosci. Remote Sensing*, GE-24: 709–716, 1986.

[288] Dalton, F. N. and van Genuchten, M. Th. The time-domain reflectometry method for measuring soil water content and salinity. *Geoderma*, 38: 237–250, 1986.

[289] Dobson, M. C. and Ulaby, F. T. Active microwave soil moisture research. *IEEE Trans. Geosci. Electr. Remote Sensing*, GE-24 (1): 23–26, 1986.

[290] Du, S. A new method for measuring dielectric constant using the resonant frequency of a patch antenna. *IEEE Trans. Microwave Theory Tech.*, MTT-34 (9): 923–931, 1986.

[291] Grant, J. P., Clarke, R. N., Symm, G. T. and Spyrou, N. M. A critical study of the open-ended coaxial line sensor for medical and industrial dielectric measurements. *Digest IEE Colloq. Industr. Appl. of Microwave*, 1986/73, 1986.

[292] Kalinski, J. On-line water content monitoring of building materials using microwaves. *Proc. MICROCOL*, 171–172, Budapest, Hungary, 1986.

[293] Kent, M. and Kress-Rogers, E. Two parameter microwave technique for measurement of powder moisture and density. Leatherhead Food RA Research Rept. No. 553, 1986.

[294] Ledieu, J. P., De Ridder P., De Clerck, P. and Dautrebande, S. A method of measuring soil moisture by time-domain reflectometry. *J. Hydrol.*, 88: 319–328, 1986.

[295] Paletta, F. and Ricca, A. Concrete moisture evaluation by microwaves. *Alta Frequenza*, 55 (4): 255–264, 1986.

[296] Parkhomchuk, P. and Wallender, W. W. Electromagnetic sensing of subsurface soil moisture. ASAE paper 86-2005, 1986.

[297] Parrent, G. B., Zeiders, G. W., Reilly, J. P. and Khazen, A. Apparatus and method for sensing multiple parameters of sheet material. U.S. Patent 4,789,820; filed July 11, 1986.

[298] Pissis, P. and Daoukaki-Diamanti, D. Dielectric study of adsorbed water in galactose. *Chem. Phys.*, 101: 95–104, 1986.

[299] Revus, D. E. and Boyer, R. E. Microwave oil saturation scanner. U.S. Patent 4,764,718; filed April 23, 1986.

[300] Scott, B. N. and Yang, Y. S. Microwave apparatus for measuring fluid mixtures. U.S. Patent 4,862,060; filed Nov. 18, 1986.

[301] Scott, W. R. and Smith, G. S. Dielectric spectroscopy using monopole antennas of general electrical length. *IEEE Trans. Antennas Propagation*, AP-34 (7): 919–929, 1986.

[302] Shafer, F. L, Smith, D. and Roberts, J. A. Dielectric response of germinating wheat seeds using a resonant cavity. *J. Microwave Power*, 16 (2): 167–177, 1986.

[303] Sihvola, A. and Tiuri, M. Snow fork for field determination of the density and wetness profiles of a snow pack. *IEEE Trans. Geosci. Remote Sensing*, GE-24 (5): 717–721, 1986.

[304] Sims, R. W. The measurement of water in homogeneous and heterogeneous solutions and solids by microwave absorption. *Adv. Instrum.*, 41: 353–376, 1986.

[305] Sowerby, B. D. and Cutmore, N. G. Moisture and density determination. U.S. Patent 4,817,021; filed Jan. 22, 1986.

[306] Swanson, C. V. Apparatus and method for using microwave radiation to measure water content of a fluid. U.S. Patent 4,812,739; filed Sept. 15, 1986.

[307] Ulaby, F. T. and Batlivala, P. P. Optimum radar parameters for mapping soil moisture. *IEEE Trans. Geosci. Electr. Remote Sensing*, GE-24 (2): 81–93, 1986.

[308] Vainikainen, P., Agarwal, R. P., Nyfors, E. and Toropainen, A. Electromagnetic humidity sensor for industrial applications. *Electron. Lett.*, 22 (19): 985–987, 1986.

[309] Vainikainen, P., Nyfors, E. and Fischer, M. Sensor for measuring the moisture content and the mass per unit area of veneer. *Proc. 16th Europ. Microwave Conf.*, 382–387, Dublin, Ireland, 1986.

[310] Von der Eltz, H. U. Moisture content measurements in continuous dyeing processes as a basis for optimal methods of working (in German). *Textilveredlung*, 21 (7/8): 261–266, 1986.

[311] Walker, C. W. E. Microwave moisture measurement using two microwave signals of different frequency and phase shift determination. U.S. Patent 4,727,311; filed Mar. 6, 1986.

[312] Wilson, M. A. Comparison of alternate approaches to measurement of moisture in grains and oil seeds. *Cereal Foods World*, 31 (6): 406–408, 1986.

[313] Xu, D. The development of the microwave detection of non-electrical properties in China. *IEEE Trans. Instrum. Meas.*, IM-35 (4): 651–654, 1986.

1987

[314] Baillie, L. A., Hsu, F. H. and Yang, Y. S. Multiphase fluid flow measurement systems and methods. U.S. Patent 4,776,210; filed June 3, 1987.

[315] Bell, J. P., Dean, T. J. and Hodnett, M. G. Soil moisture measurement by an improved capacitance technique. Part II: Field techniques, evaluation and calibration. *J. Hydrol.*, 93: 79–90, 1987.

[316] Bose, T. K., Chahine, R., Merabet, M., Akyel, C. and Bosisio, R. G. Computer-based permittivity and analysis of microwave power absorption in conductive dielectrics. *IEEE Trans. Electr. Insul.*, EI-22: 41–46, 1987.

[317] Botsco, R. J. Microwave methods and application in nondestructive testing. Sec. 19 in *Nondestructive Testing Handbook*, P. McIntire (ed.), American Society for NDT, 1987.

[318] Chouiki, S. M., Ferdinand, J. M., Smith, A. C. and Kent, M. Use of a microwave-attenuation sensor for moisture measurement inside an extrusion cooker. *J. Food Eng.*, 6: 113–121, 1987.

[319] De, B. R., Donoho, P. L., Revus, D. E. and Boyer, R. E. Microwave system for monitoring water content in a petroleum pipeline. U.S. Patent 4,902,961; filed April 8, 1987.

[320] Dean, T. J., Bell, J. P. and Baty, A. J. B. Soil moisture measurement by an improved capacitance technique. Part I: Sensor design and performance. *J. Hydrol.*, 93: 67–78, 1987.

[321] Flemming, M. A. and Plested, G. N. Microwave probe. U.S. Patent 4,829,233; filed Mar. 9, 1987.

[322] Florig, H. K. and Purta, D. A. Concentration detection system. U.S. Patent 4,767,982; filed June 1, 1987.

[323] Gentili, G. B. and Gori, F. Open ended coaxial probe for soil moisture measurements. *Proc. 1987 SBMO Int. Microwave Symp.*, 1: 151–156, Rio de Janeiro, Brazil, 1987.

[324] Halbertsma, J., Przybyla, C. and Jacobs, A. Application and accuracy of a dielectric soil water content meter. *Proc. Int. Conf. Meas. of Soil and Plant Water Status*, 1: 11–15, Utah State University, 1987.

[325] Hanyan, M., Qun, W. and Shatofan, D. Analysis of errors in the microwave moisture measurement (in Chinese). *J. Harbin Inst. Technol.*, 118–120, 1987.

[326] Kalinski, J. A simple moisture content to DC voltage microwave converter with an exceptionally high resolution. *Proc. 22nd JUREMA*, Zagreb, Yugoslavia, 1987.

[327] Kent, M. *Electrical and Dielectric Properties of Food Materials: A Bibliography.* Science and Technology Publishers, Hornchurch, Essex, UK, 1987.

[328] Kent, M. and Kress-Rogers, E. Microwave moisture and density measurements in particulate solids. *Trans. Inst. Measure. Control*, 8: 161–168, 1987.

[329] Klein, A. Comparison of rapid moisture meters (in German). *Aufbereitungs-Technik*, 28 (1): 10–16, 1987.

[330] Kraszewski, A. Microwave aquametry in grains. *Proc. 1987 SBMO Int. Microwave Symp.*, 1: 327–332, Rio de Janeiro, Brazil, 1987.

[331] Kraszewski, A. Microwave moisture monitoring in grains. *22nd Int. Microwave Power Symp.*, Summaries, 30–33, IMPI, Cincinnati, OH, 1987.

[332] Kress-Rogers, E. and Kent, M. Microwave measurement of powder moisture and density. *J. Food Eng.*, 6: 345–376, 1987.

[333] Maeno, Y. and Higashi, S. Apparatus and method for measuring physical quantities. U.S. Patent 4,890,054; filed Dec. 8, 1987.

[334] Manson, B. M. Microwave method of moisture content measurement in practice. *Proc. 2nd Int. Conf. on Scanning Technology in Sawmilling*, Oakland, CA, 1987.

[335] Nakayama, S. Simultaneous measurements of basis weight and moisture content of sheet materials by microwave cavity (in Japanese). *Jap. J. Appl. Phys.*, 26 (7): 1198–1199, 1987.

[336] Nelson, S. O. Models for the dielectric constants of cereal grains and soybeans. *J. Microwave Power*, 22 (1): 35–39, 1987.

[337] Osaki, S., Nagata, S. and Fujii, Y. Method of measuring orientation of sheet or web like material. U.S. Patent 4,781,063; filed Jan. 30, 1987.

[338] Rasmussen, V. P. and Campbell, R. H. A simple microwave method for the measurement of soil moisture. *Proc. Int. Conf. Meas. of Soil and Plant Water Status*, 1: 275-278, Utah State University, 1987.

[339] Riemschneider, B. Moisture measurement of materials in the textile industry (in German). *Textiltechnik*, 37 (4): 182–184, 1987.

[340] Simpson, J. R. and Meyer, J. J. Water content measurement comparing a TDR array to neutron scattering. *Proc. Int. Conf. Meas. of Soil and Plant Water Status*, 1: 111–114, Utah State University, 1987.

[341] Steel, P. Precision waveguide cells for the measurement of complex permittivity of lossy liquids and biological tissue at 35 GHz. *J. Phys. E: Sci. Instrum.*, 20: 872–876, 1987.

[342] Swanson, C. V. Apparatus and method for using microwave radiation to measure water content of a fluid. U.S. Patent 4,820,970; filed Jan. 3, 1987.

[343] Tiuri, M. Microwave sensor applications in industry. *Alta Frequenza*, 56 (12): 393–397, 1987.

[344] Topp, G. C. The application of time domain reflectometry to soil water content measurement. *Proc. Int. Conf. Meas. of Soil and Plant Water Status,* 1: 85–93, Utah State University, 1987.

[345] Toropainen, A., Vainikainen, P. and Nyfors, E. Microwave humidity sensor for difficult environmental conditions. *Proc. 17th Europ. Microwave Conf.,* 887–891, Rome, Italy, 1987.

[346] Vainikainen, P., Nyfors, E. and Fischer, M. Radiowave sensor for measuring the properties of dielectric sheets and its application to veneer moisture content and mass per unit area measurement. IEEE Instrum. and Meas. Tech. Conf., Boston, MA, 1987.

[347] Vainikainen, P., Nyfors, E. and Fischer, M. Radiowave sensor for measuring the properties of dielectric sheets: Application to veneer moisture content and mass per unit area measurement. *IEEE Trans. Instrum. Meas.,* IM-36 (4): 1036–1039, 1987.

1988

[348] Al-Rizzo, H. and Al-Hafid, H. T. Measurement of dielectric constant of sand and dust particles at 11 GHz. *IEEE Trans. Meas. Instrum.,* IM-37 (1): 110–113, 1988.

[349] Baillie, L. A. System for measuring multiphase fluid flow. U.S. Patent 4,813,270; filed Mar. 4, 1988.

[350] Berger, L., Krieg, G. and Schmitt, G. Method and apparatus for analysis by means of microwaves. U.S. Patent 4,972,699; filed July 8, 1988.

[351] Bruchler, L., Witono, H. and Stengel, P. Near surface soil moisture estimation from microwave measurements. *Remote Sens. Environm,* 26: 101–121, 1988.

[352] Carr-Brion, K. G. The on-line determination of moisture in bulk solids. *Proc. IMechE on On-Line Moisture Meas. of Bulk Solids for Process Control,* Westminster, 1988.

[353] Chouikhi, S. M. Use of RF wave reflection method for moisture content determination in powdered and granulated products. Proc. IMechE, Westminster, 21–24, 1988.

[354] De, B. R. Sample accommodator and method for measurement of dielectric properties. U.S. Patent 4,866,371; filed Sept. 28, 1988.

[355] Fischer, M., Vainikainen, P., Nyfors, E. and Kara, M. Fast moisture profile mapping of a wet paper web with a dual-mode resonator array. *Proc. 18th Europ. Microwave Conf.,* 607–612, Stockholm, Sweden, 1988.

[356] Flemming, M. A. and Plested, G. N. Material characterization. U.S. Patent 4,866,370; filed Jan. 22, 1988.

[357] Helms, D. A. and Marrelli, J. D. Petroleum stream microwave watercut monitor. U.S. Patent 4,947,129; filed Dec. 5, 1988.

[358] Hume, A. L. and Auchterlonie, L. J. Assessment of moisture content of fibres and fabrics by measurement of extriction cross-section in an open resonator. *Electron. Lett.,* 24 (17): 1097–1098, 1988.

[359] Jain, R. C. and Voss, W. A. G. Dielectric temperature data for water at three ISM Frequencies in polynomial form. *Electromag. Energy Rev.*, 1: 39–43, 1988.

[360] Kent, M. Microwave measurement for moisture content and bulk density. *Proc. IMechE*, 43–47, 1988.

[361] Khalid, K. The application of microstrip sensor for determination of moisture content in hevea rubber latex. *J. Microwave Power*, 23 (1): 45–51, 1988.

[362] Khalid, K. B., Maclean, T. S. M., Razaz, M. and Webb, P. W. Analysis and optimal design of microstrip sensors. *IEE Proc.*, 135, pt. H (3): 187–195, 1988.

[363] Khalid, K. B., Wahab, Z. B. A. and Kasmani, A. R. Microwave drying of hevea rubber latex and total solid content (TSC) determination. *Pertanika (Malaysia)*, 11 (2): 289–297, 1988.

[364] Klein, A. On-line microwave meter for determining the moisture content of building materials on conveyor belts. *Proc. IMechE*, 37–42, Westminster, 1988.

[365] Kolpak, M. M. Three phase fluid flow measuring system. U.S. Patent 4,852,395; filed Dec. 8, 1988.

[366] Korneta, A. and Milewski, A. The application of two- and three-layer dielectric resonators to the investigation of liquids in the microwave region. *IEEE Trans. Instrum. Meas.*, IM-37 (1): 106–113, 1988.

[367] Kraszewski, A. Microwave monitoring of moisture content in grain— further considerations. *J. Microwave Power*, 23 (4): 236–246, 1988.

[368] Kraszewski, A., You, T. S. and Nelson, S. O. Determination of moisture content in single soybeans by microwave measurements. *23rd Int. Microwave Power Symp.*, Summaries :1-2, IMPI, Ottawa, Ont., Canada, 1988.

[369] Kraszewski, A. W., You, T. S. and Nelson, S. O. Microwave resonator technique for moisture content determination in single soybean seeds. *Proc.18th Europ. Microwave Conf.*, 903–908. Stockholm, Sweden, 1988.

[370] Kupfer, K. and Morgeneier, K. D. General problems related to moisture measurement in liquid materials using microwave equipment (in German). *Betontechnik*, (2): 57–58, 1988.

[371] Lewis, R. W. Measurement apparatus and method utilizing multiple resonant modes of microwave energy. U.S. Patent 4,904,928; filed Dec. 9, 1988.

[372] Marabet, M. and Bose, T. K. Dielectric measurements of water in the radio and microwave frequencies by time domain reflectometry. *J. Phys. Chem.*, 92: 6144–6145, 1988.

[373] Maris, P. On-line moisture measurement of whole grain and flour. *Proc. IMechE*, 33–36, 1988.

[374] Nelson, L. D., Erb, L. A., Ware, R. H. and Rottner, D. Microwave radiometer and methods for sensing atmospheric moisture and temperature. U.S. Patent 4,873,481; filed Feb. 16, 1988.

[375] Oh, K. H., Ong, C. K. and Tan, B. T. G. Simple microwave method of moisture content determination in soil samples. *J. Phys. E: Sci. Instrum.*, 21: 937–940, 1988.

[376] Osaki S. Instrument for measuring high frequency characteristics of sheet-like materials. U.S. Patent 4,943,778; filed Dec. 20, 1988.

[377] Perry, E. M. and Carlson, T. N. Comparison of active microwave soil water content with infrared surface temperature and surface moisture availability. *Water Resources Res.*, 24 (10): 1818–1824, 1988.

[378] Powell, S. D., McLendon, B. D., Nelson, S. O., Kraszewski, A. and Allison, J. M. Use of a density-independent function and microwave measurement system for grain moisture measurements. *Trans. ASAE*, 31 (6): 1875–1881, 1988.

[379] Risman, P. O. Microwave properties of water in the range +3 to 140° C. *Electromag. Energy Rev.*, 1: 3–5, 1988.

[380] Sokhansanj, S. and Nelson, S. O. Transient dielectric properties of wheat associated with nonequilibrium kernel moisture conditions. *Trans. ASAE*, 31 (4): 1251–1254, 1988.

[381] Stafford, J. V. Remote, non-contact and in-situ measurement of soil moisture content: a review. *J. Agric. Eng. Res.*, 41: 151–172, 1988.

[382] Talanker, V. and Greenwald, M. Microwave measurement of the mass of frozen hydrogen pellets. U.S. Patent 4,899,100; filed Aug. 1, 1988.

[383] Thompson, F. and Clarke, J. R. P. Microwave moisture sensing arrangement. U.S. Patent 4,991,915; filed Aug. 4, 1988.

[384] Tietze, J., Kohler, I. and Sevcik, G. K. Verfahren und Vorrichtung zur kontinuierlichen Messung der Feuchte eines Schuttgutes. BRD Patent 3,805,637 A1, filed Feb. 24, 1988.

[385] Tobler, H., Lehmann, R., Kummer, E. and Mueller, R. Process and device for the continuous determination of moisture content of a bulk material. Int. Patent WO90/07110, filed Dec. 14, 1988.

[386] Topp, C. G., Yanuka, M., Zebchuk, W. D. and Zegelin, S. Determination of electrical conductivity using time-domain reflectometry: Soil and water experiments in coaxial lines. *Water Resources Res.*, 24: 945–952, 1988.

[387] Vichev, B. and Todorov, N. Error analysis methods for adjustment and calibration of direct reading microwave moisture meters (in Bulgarian). *Bulg. J. Phys.*, 15 (2): 188–195, 1988.

[388] Von Hippel, A. The dielectric relaxation spectra of water, ice and aqueous solutions, and their interpretaion. 1. Critical survey of the status-quo for water. *IEEE Trans. Electr. Insul.*, EI-23 (5): 801–816, 1988.

[389] Wayland, J. R. and Persson-Reeves, C. H. Oil/water ratio measurement. U.S. Patent 4,891,969; filed July 7, 1988.

[390] You, T. S. and Nelson, S.O. Microwave dielectric properties of rice kernels. *J. Microwave Power*, 23 (3): 150–159, 1988.

[391] Yukl T. Non-perturbing cavity method and apparatus for measuring certain parameters of fluid within a conduit. U.S. Patent 4,912,982; filed Oct. 11, 1988.

1989

[392] Bresson. J. Water control in concrete mixtures (in German). *Betonwerk Fertigteil-Technik*, (3): 82–88, 1989.

[393] Ceska A. and Svoboda, M. Verfahren und Vorrichtung zur Messung des Wasswergehalts in beliebigen Materialien, BRD Patent 3,941,032 A1; filed Dec. 12, 1989.

[394] Cutmore, N., Abernathy, D. and Evans, T. Microwave technique for on-line determination of moisture in coal. *J. Microwave Power*, 24 (1): 79–90, 1989.

[395] Drungil, C. E., Abt, C. K. and Gish, T. J. Soil moisture determination in gravelly soils with time domain reflectometry. *Trans. ASAE*, 32 (1): 177–180, 1989.

[396] Durrett, M. G., Helms, D. A., Hatton, G. J., Dowty, E. L., Marrelli, J. D., Stafford, J. D. and Stavish, D. J. Microwave water cut monitor with temperature controlled test cell. U.S. Patent 4,977,377; filed April 13, 1989.

[397] Gabriel, C. and Grant, E. H. Dielectric sensors for industrial microwave measurement and control. *Mikrowellen HF Mag.*, 15 (8): 643–645, 1989.

[398] Halabe, U. B., Maser, K. and Kausel, E. Propagation characteristics of electromagnetic waves in concrete. Tech. Report AD-A207-387, Dept. Civil Eng., 94, MIT, Cambridge, MA, 1989.

[399] Hatton, G. J., Helms, D. A., Durrett, M. G., Marrelli, J. D. and Stafford, J. D. Co-variance microwave water cut monitoring means and method. U.S. Patent 4,947,128; filed Feb. 23, 1989.

[400] He, W. Microwave tester for determining moisture content of raw silk (in Chinese). *J. Textile Res. (China)*, 10 (7): 326–327, July 1989.

[401] Helms D. A., Hatton G. J., Durrett M. G., Dowty E. L. and Marrelli J. D. Microwave water cut monitor. U.S. Patent 4,947,127; filed Feb. 23, 1989.

[402] de Jongh, P. F. Moisture measurement with microwaves. *Mikrowellen HF Mag.*, 15 (8): 648–649, 1989.

[403] Kaatze, U. Complex permittivity of water as function of frequency and temperature. *J. Chem. Engng. Data*, 34: 371–374, 1989.

[404] Kara, M., Peltoniemi, H., Fischer, M., Nyfors, E. and Vainikainen, P. A new system for quick paper profiling. *Paper and Timber*, (5): 470–475, 1989.

[405] Kent, M. A simple flowthrough cell for microwave dielectric measurements. *J. Phys. E.: Sci. Instrum.*, 22: 269–271, 1989.

[406] Kent, M. Application of 2-variable microwave techniques to composition analysis problems. *Trans. Inst. Meas. Control*, 11: 58–62, 1989.

452   Sec. V—A Reader's Bibliography

[407] Kent, M. Application of the dielectric properties of foods (in Czech). *Prumysl Potravin*, 40 (7): 383–385, 1989.

[408] Klein, A. and Pesy, W. Experiences with microwave moisture meter "Micro-Moist" (in German). *Aufbereitungs-Technik*, 30 (9): 549–557, 1989.

[409] Kraszewski, A. W. and Nelson, S. O. Composite model of the complex permittivity of cereal grain. *J. Agric. Eng. Res. (G. Britain)*, 43(5): 211–219, 1989.

[410] Kraszewski, A. W., Nelson, S. O. and You, T. S. Sensing dielectric properties of arbitrarily shaped biological objects with microwave resonator. *1989 IEEE MTT-S Int. Microwave Symp. Digest*, vol. 1, 187–190, Long Beach, CA, 1989.

[411] Kraszewski, A. W., Nelson, S. O. and You, T.-S. Microwave resonator technique for moisture content determination in single corn kernels. *Proc. 1989 SBMO Int. Microwave Symp.*, vol. 1, 85–90, Sao Paulo, Brazil, 1989.

[412] Kraszewski, A. W., You, T. S. and Nelson, S. O. Microwave resonator technique for moisture content determination in single soybean seeds. *IEEE Trans. Instrum. Meas.*, IM-38 (1): 79–84, 1989.

[413] Le Toan, T., Laur, H. and Mougin, E. Multitemporal and dual-polarization observations of agricultural vegetation covers by X-band SAR images. *IEEE Trans. Geosci. Remote Sensing*, GE-27: 709–718, Nov. 1989.

[414] McGinn, V. P. and Goldberg, I. B. Meter using a microwave bridge detector for measuring fluid mixtures. U.S. Patent 4,888,547; filed Jan. 23, 1989.

[415] Nelson, S. O. and Kraszewski, A. W. Grain moisture content determination by microwave measurement. *1989 ASAE Int. Winter Meeting*, paper #89-3543, New Orleans, LA, 1989.

[416] Nelson, S. O. and You, T.-S. Dielectric properties of corn and wheat kernels and soybeans at microwave frequencies. *Trans. ASAE*, 32 (1): 242–249, 1989.

[417] Nyfors, E. and Vainikainen, P. *Industrial Microwave Sensors*. Artech House Inc., Norwood, MA, 1989.

[418] Scheurer, A. Determination of moisture content and its distribution in concrete mixture (in German). *Betonwerk Fertigteil-Technik*, (3): 90–96; (4): 114–118; (6): 62–65, 1989.

[419] Scot, J., Bazin, R. and Leveque, J. L. Device for measuring the water content of a substrate, in particular, the skin. U.S. Patent 5,001,436; filed Mar. 27, 1989.

[420] Scott, B. N. and Shortes, S. R. System and method for monitoring substances and reactions. U.S. Patent 5,025,222; filed Nov. 27, 1989.

[421] Scott, B. N. and Yang, Y. S. Microwave apparatus and method for measuring fluid mixtures. U.S. Patent 4,996,490; filed July 7, 1989.

[422] Volgyi, F. Versatile microwave moisture sensor. *Proc. 1989 SBMO Int. Microwave Symp.*, (2): 457–462, Sao Paulo, Brazil, 1989.

[423] Whalley, W. R. The use of time domain reflectance to estimate soil water content. AFRC Institute of Engineering Research, DN1539, 1989.

[424] Wochnowski, W. Verfahren und Vorrichtung zum Messen der Feuchte eines Gutes. BRD Patent 3,905,658 A1; filed Feb. 24, 1989.

[425] Yan, X. and Zhang, Z. Theoretical and experimental investigation on slow-wave type microwave sensor (in Chinese). *J. App. Sci.*, 7 (1): 25–29, 1989.

[426] Zegelin, S. J., White, I. and Jenkins, D. R. Improved field probes for soil water content and electrical conductivity measurements using time domain reflectance. *Water Resources Res.*, 25: 2367–2376, 1989.

[427] Zhang, J., Guan, B. and Lin, S. High sensitivity sensor for microwave moisture content measurements (in Chinese). *Acta Electron. Sinica*, 17 (6): 55–60, 1989.

1990

[428] Baker, J. M. and Allmaras, R. R. System for automating and multiplexing soil moisture measurement by time-domain refelectometry. *Soil Sci. Soc. Am. J.*, 54: 1–6, 1990.

[429] Bereskin, A. B. Microwave test fixture for determining the dielectric properties of a material. U.S. Patent 5,083,088; filed July 24, 1990.

[430] Chrusciel, E., Kopec, M. and Turek, B. Microwave method for moisture content measurement in coal (in Polish). *Przeglad Gorniczy*, 46 (3): 11–12, 1990.

[431] Fisher, M., Vainikainen, P. and Nyfors, E. Dual-mode stripline resonator array for fast error corrected moisture mapping of paper web. *Proc. Int. MTT-S Symp. Digest*, 1133–1136, Dallas, TX, 1990.

[432] Grantz, D. A., Parry, M. H. and Meinzer, F. C. Agroclimatology and modeling: Using time-domain reflectometry to measure soil water in Hawaiian sugarcane. *Agron. J.*, 82: 144–146, 1990.

[433] Heinze, D. Moisture content monitoring in technical processes (in German). *Feingeratetechnik*, 39 (3): 99–101, 1990.

[434] Hurley, R. B., Kaufman, I. and Roy, R. B. Non-contacting microstrip monitor for liquid film thickness. *Rev. Sci. Instrum.*, 61 (9): 2462–2465, 1990.

[435] Jakkula, P. Verfahren und Vorrichtung zur Bestimmung des Wassergehalts von Materialien, BRD Patent, 4,000,925; filed Jan. 15, 1990.

[436] Johri, G. K. and Roberts, J. A. Study of the dielectric response of water using a resonant microwave cavity as a probe. *J. Phys. Chem.*, 94: 7386–7391, 1990.

[437] Kalinski, J. Microwave instrumentation for sample and on-line moisture content monitoring of dielectric media. *Proc. MITEKO 90*, Pardubice, Czechoslovakia, 1990.

[438] Kalinski, J. On-line coal dust moisture content monitoring by means of microwave method and instrumentation. *Proc. 20th Europ. Microwave Conf.*, paper P6.16, 1673–1678, Budapest, Hungary, 1990.

[439] Kent, M. Hand-held instrument for fat/water determination in whole fish. *Food Control*, (1): 47–53, 1990.

[440] Kent, M. Measurement of dielectric properties of herring flesh using transmission time domain spectroscopy. *Int. J. Food Sci. Technol.*, 25: 26–38, 1990.

[441] Kraszewski, A. W. and Nelson, S. O. Microwave resonator techniques for sorting dielectric objects. *Material Research Society Symp. Proc.*, 189: 69–74, San Francisco, CA, 1990.

[442] Kraszewski, A. W. and Nelson, S. O. Microwave technique for single kernel, seed, nut, or fruit moisture content determination. U.S. Patent 5,039,947; filed June 1, 1990.

[443] Kraszewski, A. W. and Nelson, S. O. Study on grain permittivity measurements in free space. *J. Microwave Power*, 25 (4): 202–210, 1990.

[444] Kraszewski, A. W., Nelson, S. O. and You, T. S. Use of a microwave cavity for sensing dielectric properties of arbitrarily shaped biological objects. *IEEE Trans. Microwave Theory Techn.*, MTT-38 (7): 858–863, July 1990.

[445] Lawrence, K. C., Nelson, S. O. and Kraszewski, A. W. Temperature dependence of the dielectric properties of wheat. *Trans. ASAE*, 33(2): 535–540, 1990.

[446] Lawrence, K. C., Nelson, S. O. and Kraszewski, A. W. Temperature-dependent model for the dielectric constant of soft red winter wheat. *1990 ASAE Int. Summer Meeting*, paper 90-6062; Columbus, OH, 1990.

[447] Mlodzka-Stybel, A. Practical verification of the microwave two-parameter method of moisture monitoring in grain in harvest time. *Proc. 20th Europ. Microwave Conf.*, paper P6.17, 1679–1682, Budapest, Hungary, 1990.

[448] Nelson, S. O. and Kraszewski, A. W. Grain moisture content determination by microwave measurements. *Trans. ASAE*, 33 (4): 1303–1307, 1990.

[449] Nelson, S. O., Lawrence, K. C., Kandala, C. V. K., Himmelsbach, D. S., Windham, W. R. and Kraszewski, A. W. Comparison of DC conductance, RF impedance, microwave, and NMR methods for single-kernel moisture measurement in corn. *Trans. ASAE*, 33 (3): 893–898, 1990.

[450] Nelson, S. O., Lawrence, K. C. and Kraszewski, A. W. Sensing moisture content of pecans by RF impedance and microwave resonator measurements. *1990 ASAE Int. Winter Meeting*, paper 90-3554, Chicago, IL, 1990.

[451] Pissis, P. The dielectric relaxation of water in plant tissue. *J. Exper. Botany*, 41 (227): 677–684, 1990.

[452] Rao, P. V. N., Raju, C. S. and Rao, K. S. Microwave remote sensing of soil moisture: elimination of texture effect. *IEEE Trans. Geosci. Remote Sensing*, GE-28 (1): 148–151, 1990.

[453] Roth, K., Schulin, R., Fluhler, H. and Attinger, W. Calibration of time domain reflectometry for water content measurement using a composite dielectric approach. *Water Resources Res.*, 26 (10): 2267–2273, 1990.

[454] Shinyashiki, N., Asaka, N., Mashimo, S., Yagihara, S. and Sasaki, N. Microwave dielectric study on hydration of moist collagen. *Biopolymers*, 29: 1185–1191, 1990.

[455] Tews, M., Sikora, J. and Herrmann, R. Process and device for determining the moisture content of the material using microwaves, BRD Patent 4,004,119; filed Feb. 10, 1990.

[456] Thansandote, A. and Ponukkha, J. Shielded-open-coaxial-line and short-monopole reflection techniques for measuring moisture content of grain and peanuts. *J. Microwave Power*, 25 (4): 195–201, 1990.

[457] Whalley, W. R. The rapid estimation of soil water content using microwave frequencies. AFRC Institute of Engineering Research, DN 1564, 1990.

[458] Yang, Y. S., Scott, B. N. and Cregger, B. B. The design, development and field testing of a water-cut meter based on microwave technique. *Proc. SPE Annual Techn. Conf.*, paper 20697, 775–782; New Orleans, LA, 1990.

1991

[459] Burr, A. Fluid monitoring apparatus. Europ. Patent EP 0499,424 A2; filed Feb. 13, 1991.

[460] Cutmore, N., Evans, T. and McEwan, A. On-conveyer determination of moisture in coal. *J. Microwave Power*, 26 (4): 237–242, 1991.

[461] Ferrazzoli, P., Guerriero, S., Paloscia, S., Pampaloni, P. and Solimini, D. Model analysis of backscatter and emission from vegetated terrains. *J. Electromag. Waves Appl.*, 5 (2): 175–193, 1991.

[462] Göller, A. Sensor zur Bestimmung der Feuchte von Schuttgutern. BRD Patent DE 4106225 A1; filed Feb. 23, 1991.

[463] Göller, A. An improved correction procedure for industrial microwave moisture measurement in grainy bulks. *Proc. 21st Europ. Microwave Conf.*, (1): 441–446, Stuttgart, Germany, 1991.

[464] Helms, D. A., Durrett, M. G. and Harton, G. J. Dual frequency microwave water cut monitoring means and method. Europ. Patent EP 0501,051 A1; filed Feb. 27, 1991.

[465] Herkelrath, W. N., Hamburg, S. P. and Murphy, F. Automatic, real-time monitoring of soil moisture in a remote field area with time domain reflectometry. *Water Resources Res.*, 27 (5): 857–864, 1991.

[466] Holmes, M.G., McCallum, K. and Diament, A. D. Nondestructive measurement of seed moisture content using dielectric properties. *Seed Sci. Technol.*, 19, 413–422, 1991.

[467] Johri, G. K., Johri, M. and Roberts, J. A. Dielectric response of selected ionic solutions using a loaded microwave cavity as a probe. *J. Microwave Power*, 26: 82–89, 1991.

[468] Khalid, K. B. Microwave reflection type moisture meter for lossy liquids. *Proc. Int. Conf. on Instrum. Meas. and Control*, Applied Tech. PTE Ltd., 83–87, Singapore, 1991.

[469] King, R. J. and King, K. V. Microwave moisture measurement of grains. *IEEE 91 Instrum./Meas. Techn. Conf. Record*, 506–512, Atlanta, GA, 1991.

[470] Kraszewski, A. W. Microwave aquametry—needs and perspectives. *IEEE Trans. Microwave Theory Techn.*, MTT-39 (5): 828–835, 1991.

[471] Kraszewski, A. W. and Nelson, S. O. Sorting biological objects with microwave resonant cavities. *IEEE 91/Instrum. Meas. Techn. Conf. Record*, 40–43, Atlanta, GA, 1991.

[472] Kraszewski, A. W. and Nelson, S. O. Density-independent moisture determination in wheat by microwave measurement. *Trans. ASAE*, 34 (4): 1776–1783, July/Aug. 1991.

[473] Kraszewski, A. W. and Nelson, S. O. Microwave resonators as a dielectric object sorting device—advantages and limitations. *Proc. 21st Europ. Microwave Conf.*, (1): 435–440, Stuttgart, Germany, 1991.

[474] Kraszewski, A. W., Nelson, S. O. and You, T.-S. Moisture content determination in single corn kernels by microwave resonator techniques. *J. Agric. Eng. Res. (G. Britain)*, 48 (2): 77–87, 1991.

[475] Kupfer, K. Moisture content measurement in components of concrete using microwave technique (in German). *Betonwerk Fertigteil-Technik*, 57 (7): 86–89, 1991.

[476] Kupfer, K. and Morgeneier, K. D. General problems related to density-corrected moisture content measurements by microwave transmission method (in German). *Mikrowellen HF Mag.*, 17 (1): 34–36, 1988.

[477] Lasri, T., Dujardin, B. and Leroy, Y. Microwave sensor for moisture measurements in solid materials. *Proc. IEE (London)*, pt. H, 138 (5): 481–483, 1991.

[478] Latorre, V. R. and Glenn, H. D. Microwave measurements of the water content in bentonite. *Proc. 1991 Int. Conf. on High-Level Radioactive Waste Management*, Las Vegas, NV, 1991.

[479] Lawrence, K. C., Nelson, S. O. and Kraszewski, A. W. Temperature-dependent model for dielectric constant of soft red winter wheat. *Trans. ASAE*, 34 (5): 2091–2093, Sept./Oct. 1991.

[480] Marrelli, J. D. and Stavish, D. J. Variable mode microwave water cut monitor and method. Europ. Patent EP 0496,144 A1; filed Jan. 25, 1991.

[481] Mashimo, S., Umehara, T. and Redlin, H. Structures of water and primary alcohol studied by microwave dielectric analyses. *J. Chem. Phys.*, 95 (9): 6257–6260, Nov. 1991.

[482] Nelson, S. O. and Kraszewski, A. W. Cereal grain moisture content measurements by microwave techniques. *SBMO Conf. Proc.*, 32–37, Rio de Janeiro, Brazil, 1991.

[483] Nelson, S. O., Kraszewski, A. W., Kandala, C. V. K. and Lawrence, K. C. High-frequency and microwave single kernel moisture sensors. 1991 Int. Winter Meeting of the ASAE, paper 913526, Chicago, IL, 1991.

[484] Puranik, S., Kumbharkhane, A. and Mehrota, S. Dielectric properties of honey-water mixtures between 10 MHz and 10 GHz using time domain technique. *J. Microwave Power*, 26 (4): 196–201, 1991.

[485] Scott, B. N. and Yang, Y. S. Method for measuring water-oil mixtures with relatively high gas content. U.S. Patent 5,157,339; filed April 16, 1991.

[486] Vermeulen, C. and Hancke, G. P. The on-line measurement of the water content of coal on a conveyor belt. *IEEE 91 Instrum./Meas. Techn. Conf. Record*, 117–119, Atlanta, GA, 1991.

[487] Whalley, W. R. Development and evaluation of a microwave soil moisture sensor for incorporating in a narrow cultivator tine. *J. Agric. Eng. Res.*, 50 (1): 25–33, 1991.

[488] Whalley, W. R. and Bull, C. R. An assessement of microwave reflectance as a technique for estimating the volumetric water content in soil. *J. Agric. Eng. Res.*, 50 (4): 315–326, 1991.

[489] Whalley, W. R., Leeds-Harrison, P. B. and Bowman, G. E. Estimation of soil moisture status using near infra-red reflection. *Hydrolo. Process.*, 5: 321–327, 1991.

[490] Yukl, T. Fluid mixture ratio monitoring method and apparatus. U.S. Patent 5,083,089; filed Feb. 20, 1991.

1992

[491] Brandelik, A. and Kraft, G. Measurement of soil moisture and its bound water. *Wissensch. Z.Hochsch. Archit. Bauwesen, Weimar*, 38 (6): 257–260, 1992.

[492] Brisco, B., Pultz, T. J., Brown, R. J., Topp, G. C., Hares, M. A. and Zebchuk, W. D. Soil moisture measurement using portable dielectric probes and time-domain reflectometry. *Water Resources Res.*, 28 (5): 1339–1346, 1992.

[493] Dalton, F. N. Development of time-domain reflectometry for measuring soil water content and bulk soil electrical conductivity. *Soil Sci. Soc. Am. Spec. Publ. Series* (30): 143–167, 1992.

[494] Ferrazzoli, P., Paloscia, S., Pampaloni, P., Schiavon, G., Solimini, D. and Coppo, P. Sensitivity of microwave measurements to vegetation biomass and soil moisture content: a case study. *IEEE Trans. Geosci. Remote Sensing*, GE-30 (4): 750–756, 1992.

[495] Göller, A. Radiating element for microwave transmission measurement in granular bulk material (in Geman). *Wissensch. Z. Hochsch. Archit. Bauwesen, Weimar*, 38 (6): 261–264, 1992.

[496] Hermann, R. and Sikora, J. Experience with using the MW2300 microwave moisture meter operating in resonant mode (in German). *Wissensch. Z. Hochsch. Archit. Bauwesen, Weimar*, 38 (6): 243–249, 1992.

[497] Hokett, S. J., Chapman, J. B. and Cloud, S. D. Time domain reflectometry response to lateral soil water content heterogeneities. *Soil Sci. Soc. Am. J.*, 56: 313–316, 1992.

[498] John, B. Soil moisture detection with airborne passive and active microwave sensors. *Int. J. Remote Sensing*, 13 (3): 481–491, 1992.

[499] Johnson, R. H., Green, J. L., Robinson, M. P., Preece, A. W. and Clarke, R. N. A resonant open-ended coaxial line sensor for measuring complex permittivity. *Proc. IEE*, Pt. A, 139: 179–182, 1992.

[500] Kahle, M. and Illich, B. Structure examination and moisture determination in historic masonry using radar (in German). *Wissensch. Z. Hochsch. Archit. Bauwesen, Weimar*, 38 (6): 237–242, 1988.

[501] Kent, M., Lees, A. and Christie, R. H. Seasonal variation in calibration of a microwave fat/water meter for fish flesh. *Int. J. Food Sci. Technol.*, 27, 137–143, 1992.

[502] Khalid, K. B. Microwave dielectric properties of hevea rubber latex. *Proc. 1992 Asia Pacific Microwave Conf.*, 611–616; Adelaide, Australia, 1992.

[503] Khalid, K. B. and Abbas, Z. B. A microstrip sensor for determination of harvesting time for oil palm fruits (*Tenera: Elaeis guineensis*). *J. Microwave Power*, 27(1): 3–10, 1992.

[504] Khalid, K. B. and Daud, W. M. Dielectric properties of natural rubber latex at frequencies from 200 MHz to 2500 MHz. *J. Nat. Rubber Res.*, 7 (4): 281–289, 1992.

[505] King, R. J. Microwave sensors for process control. Parts I, II, *Sensors*, 9 (9): 68–74; 9 (10): 25–29, 1992.

[506] King, R. J. and King, K. V. Microwave needle dielectric sensor. U.S. Patent 5,227,730; filed Sept. 14, 1992.

[507] King, R. J., King, K. V. and Woo, K. Microwave moisture measurement of grains. *IEEE Trans. Inst. Meas.*, vol. IM-41 (1): 111–115, 1992.

[508] Knight, J. H. The sensitivity of time domain reflectometry measurements to lateral variations in soil water content. *Water Resources Res.*, 28: 2345–2352, 1992.

[509] Kraszewski, A. W. Microwave resonator technique for sorting dielectric objects—advantages and limitations. *J. Wave-Matter Interactions*, 7(1): 39–55, Jan. 1992.

[510] Kraszewski, A. W. and Nelson, S. O. Moisture content determination in single peanut kernels with microwave resonator. ASAE Summer Meeting, paper 923013, Charlotte, NC, 1992.

[511] Kraszewski, A. W. and Nelson, S. O. Microwave resonant cavities for sensing moisture content and mass of single seeds and kernels. *Proc. 1992 Asia Pacific Microwave Conf.*, 555–558; Adelaide, Australia, 1992.

[512] Kraszewski, A. W. and Nelson, S. O. Measuring the moisture content in small dielectric objects of arbitrary shape. *IEE Conf.*, Publ. 363, 362–365, 1992.

[513] Kraszewski, A. W. and Nelson, S. O. Wheat moisture content and bulk density determination by microwave parameters measurement. *Can. Agric. Eng.*, 34 (4): 327–335, 1992.

[514] Kraszewski, A. W. and Nelson, S. O. Microwave sensors for simultaneous measurement of moisture content and mass in single peanut kernels. ASAE Int. Winter Meeting, paper 926505, Nashville, TN, 1992.

[515] Kupfer, K. Moisture content determination of solid and liquid materials using microwave techniques—Bibliography 1961–1991 (in German). *Wissensch. Z. Hochsch. Archit. Bauwesen, Weimar*, 38 (6): 273–284, 1992.

[516] Kupfer, K. and Klein, A. Experiments on the suitability of microwave measuring techniques for moisture measurement in calcium silicate brick production, *Aufbereitungs-Technik*, 33 (4): 213–221, 1992.

[517] Lhiaubert, C., Cottard, G., Ciccotelli, J., Portala, J. F. and Bolomey, J. Ch. On-line control in wood and paper industries by means of rapid microwave linear sensors. *Proc. 22nd Europ. Microwave Conf.*, 1037–1041, Helsinki, Finland, 1992.

[518] Li, L. L., Ismail, N. H., Taylor, L. S. and Davis, C. C. Flanged coaxial microwave probes for measuring thin moisture layers. *IEEE Trans. Biomed. Eng.*, 39 (1): 49–57, Jan. 1992.

[519] Mashimo, S., Miura, N. and Umehara, T. The structure of water determined by microwave dielectric study on water mixtures with glucose, polysaccharides and L-ascorbic acid. *J. Chem. Phys.*, 97 (9): 6759–6765, Nov. 1992.

[520] Mashimo, S., Miura, N., Umehara, T., Yagihara, S. and Higasi, K. The structure of water and methanol in *p*-dioxane as determined by microwave dielectric spectroscopy. *J. Chem. Phys.*, 96 (9): 6358–6361, May 1992.

[521] Pepin, S, Plamondon, A.P. and Stein, J. Peat water measurement using time domain reflectometry. *Can. J. For. Res.*, 22: 534–540, 1992.

[522] Rajkai, K. and Ryden, B. E. Measuring areal soil moisture distribution with the TDR method. *Geoderma*, 52: 73–85, 1992.

[523] Robinson, A. W. and Bialkowski, M. E. An investigation into microwave moisture measurements. *Proc. 1992 Asia Pacific Microwave Conf.*, 571–574; Adelaide, Australia, 1992.

[524] Roth, C. H., Malicki, M. A. and Plagge, R. Empirical evaluation of the relationship between soil dielectric constant and volumetric water content as a basis for calibrating soil moisture measurements. *J. Soil Sci.*, 43: 1–13, 1992.

[525] Sabburg, J., Ball, J. A. R. and Ness, J. B. Broadband permittivity measurements of wet soils. *Proc. 1992 Asia Pacific Microwave Conf.*, 607–610; Adelaide, Australia, 1992.

[526] Toropainen, A. P. A method for measuring properties of grainy materials using depolarized Rayleigh scattering. *Proc. 22nd Europ. Microwave Conf.*, 1041–1045, Helsinki, Finland, 1992.

[527] Trabelsi, S., Ghomi, M., Peuch, J. C. and Baudarand, H. New approach for determining the complex dielectric constant of vegetetion leaves. *Proc. 22nd Europ. Microwave Conf.*, 1234–1239, Helsinki, Finland, 1992.

[528] Vermeulen, C. and Hancke, G. P. Continuous measurement of moisture in non-conducting materials. *IEEE IMTC/92 Conf. Record*, 419-421; New York, 1992.

[529] Vertiy, A. A., Gavrilov, S. P. and Tarapov, S. I. The moisture measurements in industrial materials and articles by using millimeter waves. *Proc. 22nd Europ. Microwave Conf.*, 1052–1055, Helsinki, Finland, 1992.

[530] Volkwein, A. Refraction and diffraction effects in transmission measurement by means of microwaves (in German). *Wissensch. Z. Hochsch. Archit. Bauwesen, Weimar*, 38 (6): 251–256, 1992.

[531] Zegelin, S. J., White, I. and Russell, G. F. A critique of the time domain reflectometry technique for determining soil water content. In *Soil Sci. Soc. Am. Spec. Publ. Series*, (30): 187–208, 1992.

APPLICATION OF MICROWAVE POWER FOR DRYING
WET SAMPLES FOR MOISTURE CONTENT DETERMINATION (ALL YEARS)

[H1] Algee, B. B., Callaghan, J. C. and Creelman, A. E. Rapid determination of moisture content in soil samples using high power microwaves. *IEEE Trans. Geosci. Electr.*, GE-7 (1): 41–43, 1969.

[H2] Ryley, M. D. Use of microwave oven for the rapid determination of moisture content of soils. Road Research Lab. Report No. LR 280, Stationery Office, London, UK, 1969.

[H3] Chevalier, M. Application of microwaves to analysis in the dairy industry (in French). *Indust. Aliment. Agricole*, 87: 243–247, 1970.

[H4] Roshore, E. C. Use of microwave oven to determine water content of fresh concrete. Final Rept. No. WES-MP-C-73-7, U.S. Army Engineer Waterways Experiment Station, Vicksburg, MS, 1973.

[H5] Miller, R. J., Smith, R. B. and Bigger, J. W. Soil water content: Microwave oven method. *Proc. Soil Sci. Soc. Am.*, 38: 535–537, 1974.

[H6] Gorakhpuwalla, H. D. Development of a system for determining moisture content of grains using microwave energy. *IMPI Proc. 10th Microwave Power Symp.*, 194–196, Waterloo, Ont., Canada, 1975.

[H7] Gorakhpuwalla, H. D., McGinty, R. J. and Watson, C. A. Microwave dissipation loss in high moisture grain. *J. Agric. Eng. Res.*, 20 (3): 225–233, 1975.

[H8] Gorakhpuwalla, H. D., McGinty, R. J. and Watson, C. A. Determining moisture content of grain, corn and sorghum using microwave energy for drying. *J. Agric. Eng. Res.*, 20 (3): 319–325, 1975.

[H9] Marchart, H. and Hoffer, H. Rapid estimation of total solids content of milk using a microwave oven (in German). *Osterreichische Milchwirtschaft*, 30: 129, 1975.

[H10] Pettinati, J. D. Microwave oven method for rapid determination of moisture in meat. *J. Assoc. Off. Anal. Chem.*, 58 (6): 1188–1193, 1975.

[H11] Stephenson, M. G. and Gaines, T. P. Microwave drying as a rapid means of sample preparation. *Tobacco Sci.*, 19: 51–52, 1975.

[H12] Lee, J. W. S. and Latham, S. D. Rapid moisture determination by a commercial type microwave oven technique of canned pet food, *J. Food Sci.*, 41 (6): 1487, 1976.

[H13] Routledge, D. B. and Sabey, B. R. Use of a microwave oven for moisture determination in a soil science laboratory. *J. Agronomic Educ.*, 5: 25–27, 1976.

[H14] Steele, D. J. Microwave heating applied to moisture determination. *Lab. Practice*, 25: 515–521, 1976.

[H15] Pieper, H., Stuart, J. A. and Renwick, W. R. Microwave technique for rapid determination of moisture in cheese. *J. Assoc. Anal. Chem.*, 60 (6): 1392–1396, 1977.

[H16] Takahashi, Y., Nagai, A., Kawawa, S., Kubozono, H. and Tanaka, M. Solid or water quantity measurement apparatus using microwaves. U.S. Patent 4,106,329; filed April 26, 1977.

[H17] Diprose, M. F., Lyon, A. J. E., Hackam, R. and Benson, F. A. Partial soil sterilisation and soil and leaf moisture content measurement by microwave radiation. *Proc. Brit. Crop Protection Conf.*, 491–498, Brighton, UK, 1978.

[H18] Hankin, L. and Sawhney, B. J. Soil moisture determination using microwave radiation. *Soil Sci.*, 136 (5): 313–315, 1978.

[H19] Leonhardt, G. F., Gomes, A. M. F., Borzani, W. and Torloni, M. Microwave drying of microorganisms. II: The use of microwave oven for the determination of moisture content of pressed yeast. *J. Microwave Power*, 13: 235–237, 1978.

[H20] Peterson, R. T. and Leftwich, D. Determination of water content of plastic concrete using microwave oven. Final Rept. North Dakota State, Highway Div., 1978.

[H21] Raisanen, W. R. System for measuring moisture content. U.S. Patent 4,165,633; filed Feb. 9, 1978.

[H22] Takahashi, Y., Nagai, A., Suga, N. and Chiba, J. Measurement of total milk solids by microwave heating. *J. Microwave Power*, 13 (2): 167–171, 1978.

[H23] Thien, S. J., Whitney, D. A. and Karlen, D. L. Effect of microwave drying on soil chemical and mineralogical analysis. *Comm. Soil Sci. Plant Anal.*, 9 (3): 231–241, 1978.

[H24] Diprose, M. F., Hackam, R. and Benson, F. A. The measurement of soil and leaf moisture content by 2450 MHz radiation. *Proc. 14th Microwave Power Symp.*, 137–140, IMPI, Monaco, 1979.

[H25] Kallenberger, W. E. and Lollar, R. M. Rapid determination of moisture in cured hides by microwave oven. *J. Am. Leather Chem. Assoc.*, 74 (12): 454–468, 1979.

[H26] Suga, N., Takahashi, Y., Nagai, A., Kubozono, H. and Tanaka, M. Measurement of moisture content by microwave heating. *Anritsu Techn. Bull. (Japan)*, (37): 131–140, 1979.

[H27] Thomas, C. E., Bourlas, M. C., Laszlo, T. S. and Magin, D. F. Automatic microwave moisture meter. *Proc. 14th Microwave Power Symp.*, 150–152, IMPI, Monaco, 1979.

[H28] Click, L. S. and Baker, C. J. Moisture determination of agricultural products using a microwave oven. ASAE, paper 80-3050, St. Joseph, MO, 1980.

[H29] Farmer, G. S. and Brusewitz, G. H. Use of home microwave oven for rapid determination of moisture in wet alfalfa. *Trans. ASAE*, 23 (1): 170–172, 1980.

[H30] Hayward, L. W. and Kropf, D. H. Sample position effects on moisture analysis by microwave oven method. *J. Food Prod.*, 43 (8): 656–657, 1980.

[H31] Koh, T.-S. Microwave drying of biological tissues for trace element determination. *Anal. Chem.*, 52 (12): 1978–1979, 1980.

[H32] Perrin, D. R., Wilhelm, L. R. and Mullins, C. A. Evaluation of microwave energy for rapidly determining moisture in snap beans. ASAE, paper 80-3527, St. Joseph, MO, 1980.

[H33] Smith, M. W. and Gaines, T. P. Microwave drying of nursery leaf samples on experimental analysis. *Hort. Sci.*, 36: 179–187, 1980.

[H34] Thomasow, J. and Paschke, M. Solid determination in dairy products using the Appollo Mark XII microwave instrument (in German). *Deutsche Molkerei-Zeitung*, 101, Oct. 1980.

[H35] Ueda, M. Microwave oven procedure for moisture determination of pomace. *Am. J. Enol. Viticulture*, 31 (2): 202, 1980.

[H36] Abara, A. E. and Hill, M. A. Determination of moisture content of potatoes by microwave energy. *J. Microwave Power*, 16 (3/4): 249–252, 1981.

[H37] Verma, L. R., Noomhorm, A. and Thomas, M. D. Use of microwave oven in moisture determination. ASAE, paper 81-3519, St. Joseph, MO, 1981.

[H38] Kumar, A. Microwave drying of wet polyester fibres. *Int. J. Electron.*, 52; 491–495, 1982.

[H39] Noomhorm, A. and Verma, L. R. A comparison of microwave, air oven and moisture meters with the standard method for rough rice moisture determination. *Trans. ASAE*, 25 (5): 1464–1470, 1982.

[H40] Metaxas, A. C. and Meredith, R. J. *Industrial Microwave Heating*. IEE Power Series (4), Peter Peregrinus, 1983.

[H41] Roussy, G., Zouladian, A., Charreyre, M. and Thiebaut, J.-M. Dehydration of zeolites in microwave oven (in French). *J. Chim. Phys.*, 80: 719–727, 1983.

[H42] Verma, L. R. and Noomhorm, A. Moisture determination by microwave drying. *Trans. ASAE*, 26 (3): 935–939, 1983.

[H43] Davis, A. B. and Lai, C. S. Microwave utilization in the rapid determination of flour moisture. *Cereal Chem.*, 61: 1–4, 1984.

[H44] Vasilakos, N. P. and Magalhaes, F. Microwave drying of polymers. *J. Microwave Power*, 19: 135–144, 1984.

[H45] Backer, L. F. and Walz, A. W. Microwave oven determination of moisture content of sunflower. *Trans. ASAE*, 28 (6): 2063–2065, 1985.

[H46] Casada, M. E. and Walton, L. R. Tobacco moisture content determination by microwave heating. *Trans. ASAE*, 28 (1): 307–309, 1985.

[H47] Decareau, R. V. *Microwaves in the Food Processing Industry*. Academic Press, New York, 1985.

[H48] Mudgett, R. E. Microwave properties and heating characteristics of foods. *Food Technol.*, 40 (6): 84–93, 1986.

[H49] Jackson, D. S. and Rooney, L. W. Rapid determination of moisture in masa with a domestic microwave oven. *Cereal Chem.*, 64 (3): 196–198, 1987.

[H50] Wang, S. L. Microwave oven drying method for total solids determination in tomatoes collaborative survey. *J. Assoc. Off. Anal. Chem.*, 70 (4): 758–759, 1987.

[H51] Canet, W. Determination of the moisture content of some fruits and vegetables by microwave heating. *J. Microwave Power*, 23 (4): 231–235, 1988.

[H52] Hanna, M. A. and Sharma, N. Microwave oven drying for soybean moisture determination. *Trans. ASAE*, 31 (6): 1851–1854, 1988.

[H53] Yoshida, H., Noguchi, T., Ichikawa, M., Nakane, T., Kubo, K. and Kinoshita, K. Moisture content measuring system. U.S. Patent 4,964,734; filed May 10, 1988.

[H54] Ben Souda, K., Akyel, C. and Bilgen, E. Freeze dehydration of milk using microwave energy. *J. Microwave Power*, 24 (4): 195–202, 1989.

[H55] Brusewitz, G. H. and Greenlee, J. G. Microwave oven solids determination of high oil foods. *Appl. Eng. Agric.*, 5 (3): 415–418, 1989.

[H56] Sharma, N. and Hanna, M. A. A microwave oven procedure for soybean moisture content determination. *Cereal Chem.*, 66 (6): 483–485, 1989.

[H57] Tsang, M. M. C. and Furutani, S.C. Rapid moisture content determination of macadamia nuts by microwave drying. *Hortscience*, 24 (4): 694–695, 1989.

[H58] Vaitekunas, D., Raghavan, G. S. V. and van de Voort, F. R. Drying characteristics of soil in a microwave environment. *Can. Agric. Eng.*, 31: 117–123, 1989.

[H59] Akyel, C., Bosisio, R. G., Bose, T. K. and Merabet, M. A comparative study of HF and microwave drying of milk. *IEEE Trans. Electrical Insulation*, EI-25: 493–502, 1990.

[H60] Khalid, K. B. Determination of DRC in hevea rubber latex by microwave drying. *Pertanika (Malaysia)*, 14 (1): 65–67, 1991.

[H61] Zhang, X. and Brusewitz, G. H. Grain moisture measurement by microwave heating. *Trans. ASAE*, 34 (1): 246–250, 1991.

About the Editor

Andrzej Kraszewski was born in Poznan, Poland, on April 22, 1933. He received the M.Sc. degree in electrical engineering from the Technical University of Warsaw, Poland, in 1958 and the D.Sc. degree in technical sciences from the Polish Academy of Sciences (PAN), Warsaw, in 1973. In 1953 he joined the Telecommunications Institute (PIT) in Warsaw, Poland, where he did research and development work on microwave systems and components. In 1963 he joined UNIPAN Scientific Instruments, a subsidiary of the Polish Academy of Sciences, as Head of the Microwave Laboratory. In 1972 he became the Manager of the Microwave Department of WILMER Instruments and Measurements, a subsidiary of the Polish Academy of Sciences in Warsaw, where he codeveloped microwave instruments for moisture content measurement and control. Beginning in November 1980 he was a Visiting Professor at the University of Ottawa, Ontario, Canada, where he did research on RF and microwave dosimetry. In January 1987 he joined the Richard B. Russell Agricultural Research Center (U.S. Department of Agriculture) in Athens, Georgia, where he is involved in research on plant structure and composition using electromagnetic fields. He is a Fellow of the IEEE, member of the International Microwave Power Institute, the Materials Research Society, the International Society for Measurement and Control (ISA), New York Academy of Sciences, Sigma Xi, and the Polish Electricians Association (SEP). He is the author of several books on microwave theory and techniques, has published over 200 papers and holds 19 patents. He has received several professional awards, among them the State Prize in Science in Poland in 1980.

Index

A

absorption
  microwave absorption frequencies, 302, 303
  relaxation time curve for, **15**
acetone, dielectric measurement of, **289**
activation energy, vs. peak temperature, **71**
activation parameters, temperature dependence of, 61–**62**
active integrated antenna (AIA)
  use with integrated moisture sensor, 217, 220
  See also antenna
admittance. *See* load admittance; impedance
agricultural products
  dielectric ring resonator applications for, 205
  oil palm fruit moisture sensor, 239–247
  peanut moisture determination, 183, 184, 187–**192, 188–192, 193,** 197–198, 199
  resonant cavity applications to, 185, 196
  sensor designs for, 167, 215, 223
  See also food; grain
AIA. *See* active integrated antenna

air
  as component of snow, 116
  as host medium for materials, 19
  in mixtures, 124–126, **125**
  permittivity of, **335**
  reflected signal behavior of, **86**
  relation with water in soil, 102
alcohols, affecting frequency, 20
ANA. *See* automatic network analyzer reflectometry
antenna
  active integrated antenna (AIA), 217, 220
  attenuation between, 224, 225, 226
  of electromagnetic propagation tool, 142, **144**
  four-element microstrip, 235
  integrated microwave sensor as, **216,** 231
  material-under-test as, 304
  parameters and values of, **228**
  pyramidal horn type, **228,** 232–**233,** 357–**358,** 368
  relation to reflector, **359**
  relative axial power densities of, 232–**233**
  transmission loss between, **234–235**
  See also resonant cavity; sensor

467